GASTRULATION

Movements, Patterns, and Molecules

BODEGA MARINE LABORATORY
MARINE SCIENCE SERIES

Series Editor: James S. Clegg

Bodega Marine Laboratory
University of California, Davis
Bodega Bay, California

GASTRULATION: Movements, Patterns, and Molecules
Edited by Ray Keller, Wallis H. Clark, Jr., and Frederick Griffin

INVERTEBRATE HISTORECOGNITION
Edited by Richard K. Grosberg, Dennis Hedgecock, and Keith Nelson

MECHANISMS OF EGG ACTIVATION
Edited by Richard Nuccitelli, Gary N. Cherr, and Wallis H. Clark, Jr.

GASTRULATION

Movements, Patterns, and Molecules

Edited by

Ray Keller

University of California, Berkeley
Berkeley, California

Wallis H. Clark, Jr. and
Frederick Griffin

Bodega Marine Laboratory
University of California, Davis
Bodega Bay, California

Springer Science+Business Media, LLC

Library of Congress Cataloging in Publication Data

Bodega Marine Science Colloquium on Gastrulation: Movements, Patterns, and Molecules (3rd: 1990: Bodega Bay, Calif.)

Gastrulation: movements, patterns, and molecules / edited by Ray Keller, Wallis H. Clark, Jr., and Frederick Griffin.

p. cm. — (Bodega Marine Laboratory marine science series)

"Proceedings of the Third Bodega Marine Science Colloquium on Gastrulation: Movements, Patterns, and Molecules, held September 6–8, 1990, in Bodega Bay, California.

Includes bibliographical references and index.

1. Gastrulation—Congresses. I. Keller, Ray. II. Clark, Wallis H. III. Griffin, Frederick. IV. Title. V. Series.

QL955.B64 1990 91-38428

591.3′3—dc20 CIP

Cover: Transverse sections through embryos (stained with anti-*twist* antibodies) depicting four stages of mesoderm invagination and internalization (dorsal is up). **Top left:** Embryo at the completion of cellularization. **Top right:** Early stage of ventral furrow formation. The apical (outer) surfaces of the *twist*-expressing cells have become smooth, and the nuclei of the central 8–10 cells have begun to migrate away from the apical cell surface. **Bottom left:** The apical sides of the central cells have constricted, and the surface of the ventral epithelium is beginning to ident. **Bottom right:** The central cells have invaginated completely, and the peripheral cells are following them into the furrow.

Proceedings of the Third Bodega Marine Science Colloquium on Gastrulation: Movements, Patterns, and Molecules, held September 6–8, 1990, in Bodega Bay, California

ISBN 978-1-4684-6029-2 ISBN 978-1-4684-6027-8 (eBook)
DOI 10.1007/978-1-4684-6027-8

Originally published by Plenum Press, New York in 1991
Softcover reprint of the hardcover 1st edition 1991

CONTRIBUTORS

Brigitte Angres
Max Planck Institute for
Developmental Biology
Spemannstrasse 35
D-7400 Tübingen
Federal Republic of Germany

Jean-Claude Boucaut
Laboratoire de Biologie Experimentale
URA-CNRS-1135
Universitie Pierre et Marie Curie
9, Quai Saint-Bernard 75005
Paris, France

Jonathan Cooke
Laboratory of Embryogenesis
National Institute for
Medical Research
The Ridgeway, Mill Hill
London, NW7 1AA, United Kingdom

Michael Costa
Molecular Biology Department
Princeton University
Princeton, New Jersey 08544

Thierry Darribère
Universite Pierre et Marie Curie
Laboratorie de Biologie Experimentale
URA-CNRS-1135
9, Quai Saint-Bernard 75005
Paris, France

Michel Delarue
Universite Pierre et Marie Curie
Laboratorie de Biologie Experimentale
URA-CNRS-1135
9, Quai Saint-Bernard 75005
Paris, France

Douglas W. DeSimone
Dept. of Anatomy and Cell Biology
University of Virginia
Box 439, Health Sciences Ctr.
Charlottesville, Virginia 22908

Tabitha Doniach
Department of Molecular and
Cell Biology
University of California, Berkeley
Berkeley, California 94720

Charles A. Ettensohn
Department of Biological Sciences
Carnegie Mellon University
4400 Fifth Avenue
Pittsburgh, Pennsylvania 15213

Rachael D. Fink
Department of Biological Sciences
Mount Holyoke College
South Hadley, Massachusetts 01075

Nikolaos C. George
Department of Molecular and
Cell Biology
University of California, Berkeley
Berkeley, California 94720

John Gerhart
Department of Molecular and
Cell Biology
University of California, Berkeley
Berkeley, California 94720

Koichiro Hashimoto
Division of Developmental Biology
Meiji Institute of Health Science
540 Naruda, Odawara 250, Japan

Jonathan J. Henry
Department of Cell and
Structural Biology
University of Illinois
Urbana, Illinois 61801

James E. Howard
Laboratory of Embryogenesis
National Institute for
Medical Research
The Ridgeway, Mill Hill
London NW7 1AA, United Kingdom

William R. Jeffery
Bodega Marine Laboratory
University of California, Davis
P.O. Box 247
Bodega Bay, California 94923

Kurt E. Johnson
Department of Anatomy
The George Washington University
Medical Center
2300 I Street N.W.
Washington, D.C. 20037

Ray Keller
Department of Molecular and
Cell Biology
University of California, Berkeley
Berkeley, California 94720

Oded Khaner
Department of Molecular and
Cell Biology
University of California, Berkeley
Berkeley, California 94720

Mary Constance Lane
Department of Molecular and
Cell Biology
University of California, Berkeley
Berkeley, California 94720

Maria Leptin
Max Planck Institute for
Developmental Biology
Spemannstrasse 35
D-7400 Tübingen
Federal Republic of Germany

David R. McClay
Department of Zoology
Duke University
Durham, North Carolina 27706

Martina Nagel
Max Planck Institute for
Developmental Biology
Spemannstrasse 35
D-7400 Tübingen
Federal Republic of Germany

Norio Nakatsuji
Division of Developmental Biology
Meiji Institute of Health Science
540 Naruda
Odawara, 250 Japan

George Oster
Departments of Molecular and Cell
Biology and Entomology
201 Wellman Hall
University of California, Berkeley
Berkeley, California 94720

Rudolf Raff
Institute for Molecular and
Cellular Biology, and
Department of Biology
Indiana University
Bloomington, Indiana 47405

David G. Ransom
Dept. of Anatomy and Cell Biology
University of Virginia
Bos 439, Health Sciences Ctr.
Charlottesville, Virginia 22908

Jean-Francois Riou
Laboratoire de Biologie Experimentale
URA-CNRS 1135
Universitie Pierre et Marie Curie
9, Quai Saint-Bernard 75005
Paris, France

Gary C. Schoenwolf
Department of Anatomy
School of Medicine
University of Utah
Salt Lake City, Utah 84132

Andreas Selchow
Max Planck Institute for
Developmental Biology
Spemannstrasse 35
D-7400 Tübingen
Federal Republic of Germany

De Li Shi
Laboratoire de Biologie Experimentale
URA-CNRS-1135
Universitie Pierre et Marie Curie
9, Quai Saint-Bernard 75005
Paris, France

John Shih
Department of Molecular and
Cell Biology
University of California, Berkeley
Berkeley, California 94720

Jim C. Smith
Laboratory of Embryogenesis
National Institute for
Medical Research
The Ridgeway, Mill Hill
London NW7 1AA, United Kingdom

Michael Solursh
Department of Biology
University of Iowa
Iowa City, Iowa 52242

Claudio D. Stern
Department of Human Anatomy
University of Oxford
South Parks Road
Oxford OX1 3QX, United Kingdom

Ronald Stewart
Department of Molecular and
Cell Biology
University of California, Berkeley
Berkeley, California 94720

Cornelia Stoltz
Max Planck Institute for
Developmental Biology
Spemannstrasse 35
D-7400 Tübingen
Federal Republic of Germany

Dari Sweeton
Molecular Biology Department
Princeton University
Princeton, New Jersey 08544

Karen Symes
Department of Cell and
Molecular Biology
385 LSA
University of California, Berkeley
Berkeley, California 94720

J.P. Trinkaus
Department of Biology
Yale University
New Haven, Connecticut 06511
and
Marine Biological Laboratory
Woods Hole, Massachusetts 02543

Madeleine Trinkaus
Department of Biology
Yale University
New Haven, Connecticut 06511
and
Marine Biological Laboratory
Woods Hole, Massachusetts 02543

Michael Weliky
Group in Neurobiology
University of California, Berkeley
Berkeley, California 94720

Eric Wieschaus
Department of Biology
Guyot Hall
Princeton University
Princeton, New Jersey 08544-1003

Paul Wilson
Department of Biochemistry and Molecular Biology
Harvard University
Cambridge, Massachusetts 02138

Fred Wilt
Department of Molecular and Cell Biology
371 LSA
University of California, Berkeley
Berkeley, California 94720

Rudolf Winklbauer
Max Planck Institute for Developmental Biology
Spemannstrasse 35
D-7400 Tübingen
Federal Republic of Germany

Gregory A. Wray
Friday Harbor Laboratories
University of Washington
Friday Harbor, Washington 68250

PREFACE

Gastrulation is a fundamental process of early embryonic development. It involves virtually every aspect of cell and developmental biology and results in the formation of fundamental structural elements around which a developing animal's body plan is organized. As such it is not only an important process, but also one that is complicated and not easily dissected into its component parts. To understand the mechanisms of gastrulation one must acknowledge that gastrulation is fundamentally a biomechanical process (that is, a problem of cells generating forces in a three dimensional array, patterned in space and time such that appropriate tissue movements are executed). Three intertwined questions emerge: what cell activities generate forces, how are these cell activities patterned in space and time, and how are the resulting forces harnessed in three dimensional domains? To address these issues it is important to define and characterize regional cell behaviors and to learn how they are patterned in the egg and/or by subsequent cell and tissue interactions. At the biochemical level, what are the cellular and extracellular molecules that control cell behavior? Finally, how are specific patterns of cellular activity integrated to produce tissue behavior? The task of answering the above questions, an immense task in itself, is compounded by the fact that the morphogenetic movements of gastrulation and their underlying mechanisms vary between different organisms. The third Bodega Marine Science Colloquium attempted to examine gastrulation in a number of vertebrate and invertebrate groups and to specifically address the issue of how these early developmental processes appear in organisms at different phylogenetic levels. In addition to phylogenetic considerations we attempted to choose participants who use diverse scientific approaches. Our attempt at the above integration is presented in this volume, and though not comprehensive, we believe it is a succesful and timely treatment of the subject. Each of the authors focuses on his/her particular questions and research directions; however, common themes and approaches are evident throughout the book and we urge the reader to consider each chapter with this in mind.

The third Bodega Marine Science Colloquium was held at the Bodega Marine Laboratory in September of 1990. The three day conference was attended by over 60 people and featured 20 invited speakers, prominent investigators in the field of gastrulation, whose contributions constitute this volume. Both the success of the conference and the compilation of contributed papers into this volume are the result of efforts on the part of many who deserve recognition. The success of the conference itself is due in large part to the efforts of Trisha Pedroia and Vicki Milam who coordinated and oversaw the events and sessions. Eleanor Uhlinger, Bodega Marine Laboratory Librarian, helped proof the manuscript. Diane Cosgrove is responsible for the final package; she formed the camera ready manuscript that is this book. We also thank the remainder of the staff of the Bodega Marine Laboratory for their help and hospitality.

WE GRATEFULLY ACKNOWLEDGE
THE FOLLOWING SOURCES OF FINANCIAL FUNDING

* California Sea Grant College Program *

* Bodega Marine Laboratory *

* Society for Developmental Biology *

CONTENTS

CELL MOVEMENTS IN THE EPIBLAST DURING GASTRULATION AND NEURULATION IN AVIAN EMBRYOS

Gary C. Schoenwolf

Department of Anatomy
School of Medicine
University of Utah
Salt Lake City, Utah 84132

Gastrulation, Edited by R. Keller *et al.*
Plenum Press, New York, 1991

INTRODUCTION

In this essay, I will focus on both gastrulation and neurulation in avian embryos. The two processes are driven by similar cell behaviors: cell shape changes, cell division, and cell rearrangements. Since we have made considerable progress in understanding how these cell behaviors function in avian neurulation, perhaps more so than in gastrulation, I will draw heavily on this material.

The processes of gastrulation and neurulation have been the subjects for hundreds of papers (avian gastrulation: reviewed by Nicolet 1971; Schoenwolf 1983; Vakaet 1984, 1985; Bellairs 1986; Stern and Canning 1988; Harrisson 1989; neurulation: reviewed by Schroeder 1970; Burnside 1973; Karfunkel 1974; Jacobson and Ebendal 1978; Jacobson 1980, 1981; O'Shea 1981; Schoenwolf 1982, 1983, 1991; Gordon 1985; Jacobson *et al.* 1986; Martins-Green 1988; Schoenwolf and Smith 1990a,b). Despite all that has been written, it is probably safe to say that what we know about avian gastrulation and neurulation is considerably less than what still remains to be learned. Rather than discussing these two processes thoroughly, my approach in this essay will be to present summaries of our understanding of avian gastrulation and neurulation, pointing out their interrelationships and emphasizing their coordination. I will then focus on cell movements occurring in the epiblast, describing in detail those cell rearrangements that underlie shaping of the neural plate during neurulation. To better understand the control of such movements, issues concerning the prospective fate and potency of various populations of epiblast cells will be discussed briefly. Finally, I will list some of the partially or totally unanswered questions about avian gastrulation, providing speculation.

SUMMARY OF AVIAN GASTRULATION

Gastrulation in avian embryos involves a number of interesting yet poorly understood events, beginning as early as when the egg is still in the hen's ovary and terminating only after several other processes involved in formation of the axial rudiments (*i.e.*, neural tube and notochord) and associated rudiments (*e.g.*, somites, heart, gut, body wall, and coelom) are largely completed. Gastrulation can be subdivided into the following four stages: (1) formation of the cardinal axes; (2) formation of the area pellucida (area centralis) and area opaca (area marginalis); (3) formation of the bilaminar blastoderm, consisting of a superficial layer (upper layer or *epiblast*) and a deep layer (lower layer or endophyll or *hypoblast*); and (4) formation of the trilaminar blastoderm, consisting of three germ layers—ectoderm, mesoderm, and endoderm.

A few key references will be provided to document the statements in the following summary. Further documentation can be found in the reviews listed in the Introduction (also, see Stern's chapter in this volume).

Formation of the Axes

Some of the cardinal axes form prior to gastrulation and some form during this process. The egg forms an animal-vegetal axis while it is still in the ovary, and therefore prior to gastrulation. The blastoderm contains a dorsal-ventral axis, a cranial-caudal axis, and a medial-lateral axis. The dorsal-ventral axis forms prior to gastrulation, whereas the cranial-caudal and medial-lateral axes are established by the early events of gastrulation. The formation of each of these axes will now be discussed briefly.

Formation of the animal-vegetal axis of the egg occurs while the ovum (actually, the primary oocyte) is developing in a follicle in the hen's ovary. The ovary contains a complex circulatory system which transports yolk to the growing follicle where it accumulates rapidly (Nalbandov and James 1949) and in layers, with yolk of different types occupying various layers (Callebaut 1974, 1983, 1985; reviewed by Bellairs 1964). Because of this layering, and the fact that the cytoplasmic disc (*i.e.*, the blastodisc) floats on top of the yolk mass, an animal-vegetal axis is established. At the time of ovulation, the ovum (actually, the secondary oocyte) is about 35 mm in diameter and consists of a large mass of yolk with a small (about 3.5 mm in diameter) blastodisc crowded off to one side. The yolk and blastodisc together constitute a single, albeit it atypically large, cell enclosed in a plasmalemma. An acellular membrane, the inner vitelline membrane, surrounds the ovum at the time of ovulation.

After ovulation, the ovum enters the oviduct. While it is traversing the first region of the oviduct (*i.e.*, the infundibulum and magnum), it is coated with the outer vitelline membrane and thick and thin albumen. Subsequently, within the remainder of the oviduct (*i.e.*, the isthmus and shell gland), the inner and outer shell membranes and shell are added, following which the egg is laid. As these coats are being deposited, the blastodisc initiates cleavage (assuming fertilization occurred in the first region of the oviduct), becoming a multicellular, stratified blastoderm (also called the germ or germinal disc) consisting of central and marginal blastomeres (cleavage in the chick is meroblastic; *i.e.*, only the blastodisc, not the yolk mass, cleaves; Bellairs *et al.* 1978). The marginal blastomeres are contiguous with the underlying yolk, but the central blastomeres become separated from the yolk by the subgerminal cavity, a space filled with subblastodermic fluid. How the subgerminal cavity forms is not completely understood, but the accumulation of subblastodermic fluid is probably involved. Experiments suggest that this fluid is produced by the blastoderm, which takes up fluid from the albumen and secretes fluid toward the yolk (New 1956).

Formation of the dorsal-ventral (or apical-basal) axis of the blastoderm occurs after the appearance of the subgerminal cavity. In freshly-laid eggs, the albumen is basic (pH 9.5), whereas the subblastodermic fluid is acidic (pH 6.5) (Stern and MacKenzie 1983). In addition, sodium and water are transported from the future apical side of the blastoderm to its future basal side, creating a transepithelial potential of 25 mV (positive at the ventral side) (Jaffe and Stern 1979; Stern and MacKenzie 1983; Stern *et al.* 1985). The dorsal-ventral axis can be reversed by applying a transepithelial potential of opposite polarity or by reversing the pH gradient, suggesting that ionic currents and pH differences determine this axis (reviewed by Stern and Canning 1988).

The blastoderm at the time of its formation is a radially symmetric structure. This radial symmetry is converted to bilateral symmetry as the ovum passes down the oviduct. During this process, the ovum rotates at the rate of 10-15 revolutions per hour. In a series of clever experiments, the effects of rotation were compared to the effects of gravity on bilateralization (Kochav and Eyal-Giladi 1971). It was shown that bilateral symmetry is determined by gravity; rotation plays an indirect role by forcing the blastoderm into an oblique position, with gravity fixing its *highest* point as the future *caudal* end of the blastoderm. Hence, both the cranial-caudal and medial-lateral axes of the blastoderm are determined while the ovum is still in the oviduct.

It should be clear from the above discussion, that by the time the developmental biologist receives the egg from the hen (or nowadays, usually from the supplier), considerable development has occurred already. The ovum is packaged prior to laying as a cleidoic egg, capable of developing on its own, provided that sufficient warmth is given (*i.e.*, it is incubated) to allow further development to occur. The blastoderm floating atop the yolk appears nondescript to the unaided eye, yet it consists already of approximately 60,000 cells (blastomeres) (Spratt 1963) and all its cardinal axes are determined. Surprisingly, besides these complexities, still other complex developmental events occur before the egg is laid. These additional complexities will be discussed below, but before doing so, I will first provide some information on how embryos are staged. This will allow us to categorize embryos according to their level of increasing complexity.

The period ranging from prior to laying through the first few hours after laying has been subdivided into 14 stages (designated as stages I-XIV) by Eyal-Giladi and Kochav (1976). Cleavage (already mentioned) is the principal event occurring during stages I-VI. Stages VII-X and stages X-XIV involve, respectively, the formation of the area pellucida and area opaca, and the formation of the bilaminar blastoderm. The egg is laid usually at stage X; stages X-XIV are progressively more advanced stages, collectively equivalent to what Hamburger and Hamilton (1951) designated as stage 1. Hamburger and Hamilton (1951) stages 1-46 characterize development over the period ranging from when the egg is laid until hatching occurs.

Formation of the Area Pellucida and Area Opaca

Formation of the area pellucida and area opaca begins with the appearance of the subgerminal cavity. Recall that with formation of this cavity, two groups of blastomeres can be identified: the central and marginal blastomeres. The marginal blastomeres are in contact with the yolk; because of this, the marginal region appears opaque and is, therefore, called the area opaca. The central blastomeres undergo changes in their arrangement to establish the area pellucida. This process has been studied by Eyal-Giladi and co-workers (Eyal-Giladi and Kochav 1976; Kochav *et al.* 1980). They describe the formation of the area pellucida as owing to the loss of yolk-ladened deep (lower) cells during stages VII-X from the central portion of the blastoderm (*i.e.*, as owing to polarized cell shedding). Consequently, the thickness of the stratified central portion is reduced from 4-6 cell layers to essentially 1 cell layer. Moreover, during this process the entire blastoderm (including both its central and marginal portions) expands radially over the yolk. As a result of these two processes (thinning and spreading), and because of the presence of the subgerminal cavity, the central area is more translucent than the marginal area (*i.e.*, the area opaca) and is called the area pellucida.

Formation of the Bilaminar Blastoderm

When the egg is laid at stage X (*i.e.*, after formation of the area pellucida and area opaca has occurred), the central portion of the blastoderm consists of an essentially single layer of epithelial cells. Formation of the hypoblast, transforms the unilayered blastoderm into the bilaminar blastoderm. The hypoblast forms during stages X-IV in two manners. The so-called primary (or primitive) hypoblast forms by a process of polyingression (also called "polyinvagination"), namely, the inward movement of individual cells and small islands consisting of groups of cells from the epithelial sheet composing the area pellucida (Weinberger and Brick 1982a,b; Weinberger *et al.* 1984; Penner and Brick 1984; Canning and Stern 1988; Stern 1990; Stern and Canning 1990). As polyingression occurs and the primary hypoblast forms, the upper layer of the blastoderm is called the *epiblast.* Formation of the so-called secondary hypoblast is initiated near the caudal midline portion of the area pellucida from a region called the posterior marginal zone (Spratt and Haas 1960a,b; Stern 1990). The marginal zone is an area located at the juncture between the area pellucida and area opaca. The incipient secondary hypoblast begins its formation beneath or from (its exact origin is controversial) a flap-like basal thickening of epiblast in the posterior marginal zone, called Koller's sickle, so named because it is sicklelike in surface view. Secondary hypoblast cells spread cranially from the posterior marginal zone and join with the primary hypoblast cells to form a complete layer one-cell thick known simply as *the* hypoblast.

Formation of the Trilaminar Blastoderm

The bilaminar blastoderm becomes a trilaminar blastoderm with formation of the primitive streak and the subsequent ingression of cells through the streak, into the interior of the blastoderm. As the secondary hypoblast is forming, a structure called the embryonic shield appears at the caudal end of the area pellucida (at stage 1, Hamburger and Hamilton 1951). The embryonic shield is the first area of the area pellucida to become bilaminar and consists of a thickened epiblast underlain by secondary hypoblast (Spratt 1942). The epiblast is thicker in this area apparently because the precursor cells of the primitive streak are accumulating here. The accumulation of cells results from what has been described as "Polonaise movements" (see Vakaet 1984; Figure 4), in which epiblast cells on each side of the midline (as well as epiblast cells that apparently leave the epiblast and lie just deep to it) stream caudally and begin to pile up. As this process continues, a linear primitive streak forms in the midline and then extends cranially, a process known as "progression" of the primitive streak. Prior to formation and progression of the primitive streak, the area pellucida is circular in outline when viewed from its dorsal or ventral surface. As the primitive streak progresses, its cranial end (*i.e.*, Hensen's node or the primitive knot) moves rostrally and its caudal end (the posterior node) moves caudally. The resulting elongation changes the outline of the area pellucida from circular to pear shaped, with the narrowed portion of the pear pointed caudally. Formation of the primitive streak is a rapid process–in viewing timelapse films or tapes (Vakaet 1970; Stern and Canning 1988; Schoenwolf, unpublished observations) one is given the impression that something suddenly forms out of nothing.

During early stages of primitive streak progression (stages 2, 3a, 3b; modification of the stages of Hamburger and Hamilton 1951, and Vakaet 1962, 1970, as described by Schoenwolf 1988) the innermost germ layer, the endoderm (also called endoblast, tertiary hypoblast, definitive endoderm, or gut endoderm) forms. Prospective endodermal cells contained within the cranial part of the primitive streak ingress into the deep layer, displacing the hypoblast cells cranially to an extraembryonic area called the germ cell crescent (also called Swift's or endophyllic crescent). Ingression of prospective endoderm is followed during later stages (stages 3c, 3d, 4-10) by the ingression of prospective mesodermal cells to form the middle layer, the mesoderm (or mesoblast). Various subdivisions of prospective mesodermal cells have been mapped during ingression by several investigators using a variety of techniques (reviewed by Spratt 1955; Rosenquist 1966; Nicolet 1971; Vakaet 1984, 1985). Ingression of prospective mesodermal cells begins during late stages of primitive streak progression. During this period, the caudal half of the primitive streak contributes extraembryonic mesoderm. During subsequent regression of the primitive streak, all mesoderm of the embryo proper (*i.e.*, prechordal plate mesoderm, notochord, cardiac mesoderm, somites, nephrotomes, and lateral plate mesoderm) ingresses through the cranial half of the primitive streak. The cells remaining within the epiblast after ingression, collectively constitute the ectoderm. Thus, after regression is completed, the embryo consists of a flat trilaminar blastoderm, whose ectodermal, mesodermal, and endodermal layers are formed and are ready to be sculptured into various organ rudiments during subsequent morphogenesis. One of these sculpturing processes, neurulation, results in formation of the neural tube, the rudiment of the central nervous system.

SUMMARY OF AVIAN NEURULATION

Development of the axial structures, including the neural tube, and the paraxial structures occurs in two phases in avian embryos (Holmdahl 1925a,b). *Primary body development* involves the formation of the three germ layers (just described) and the sculpturing of these layers into axial and paraxial organ rudiments. Thus, for example, we speak of the ectodermal neural tube and ectodermal neural crest, the mesodermal notochord and mesodermal somites, and the endodermal gut, all of which originate during primary body development. *Secondary body development* is a later phase in which the process of germ layer formation is omitted; thus, it is basically the antithesis of gastrulation. Instead, the tail bud, a mesenchymal structure formed from remnants of the regressed primitive streak (*e.g.*, Schoenwolf 1979a), *directly* forms certain axial and paraxial organ rudiments. These include the most caudal (so-called secondary) neural tube (and associated neural crest) and the somites, but not the notochord; the latter develops in the cranial region undergoing primary body development and extends caudally into the region of secondary body development (Schoenwolf 1977, 1978a). Below, I will discuss briefly both primary (development from the ectoderm) and secondary (development from the tail bud) neurulation. Primary neurulation occurs in four stages: formation of the neural plate, shaping of the neural plate, bending of the neural plate, and closure of the neural groove. Secondary neurulation occurs in three stages: formation of the tail bud, formation of the medullary cord, and cavitation of the medullary cord.

Primary Neurulation: Formation of the Neural Plate

Formation of the neural plate is one of the least understood aspects of neurulation. Based largely on studies of amphibians, it is clear that the neural plate forms as a result of induction (reviewed by Spemann 1938; Gurdon 1987). The situation in other vertebrates is less clear, but it has been amply demonstrated that Hensen's node, the portion of the primitive streak that contains prospective foregut endodermal cells, prospective prechordal plate mesodermal cells, and prospective notochordal cells

(Nicolet 1970, 1971; Veini and Hara 1975), is capable of inducing a second axis, complete with neural plate, when transplanted subjacent to the ectoderm of the germ cell crescent (reviewed by Gallera 1971; also see Dias and Schoenwolf 1990). As a result of neural induction, the cells of the prospective neural plate increase their height, becoming columnar. Thus, the earliest manifestation of the neural plate is this change in cell shape (*i.e.*, a cell columnarization).

Primary Neurulation: Shaping of the Neural Plate

Shaping of the neural plate is the process during which the neural plate on the average thickens apicobasally, narrows transversely, and lengthens longitudinally. The initial thickening of the neural plate during its induction, owing to the elongation of its cells, continues during shaping. As a result of this elongation, the diameter of the cells of the neural plate decreases (thereby at least partially maintaining cell volume) and such cells acquire a shape that can best be characterized as spindlelike (Schoenwolf and Franks 1984; Schoenwolf 1985). Because cell diameter decreases as cells become taller, both the transverse and longitudinal dimensions of the neural plate would be expected to be reduced by cell elongation. Such reduction in the transverse plane aids in narrowing of the neural plate during its shaping, but it is not sufficient to account for all the narrowing that occurs (Schoenwolf and Powers 1987). Cell elongation is supplemented by at least one other event, the rearrangement of cells. Such rearrangement results in a net movement of cells from the transverse plane of the neural plate to its longitudinal plane. Consequently, this movement assists in narrowing of the neural plate; it also offsets the reduction in the length of the neural plate that would be expected to occur as a result of cell elongation, and it provides additional cells to considerably lengthen the neural plate beyond its original limit (Schoenwolf and Alvarez 1989). Cell rearrangement is accompanied by cell division (all neuroepithelial cells divide at 8-12 h intervals until closure of the neural groove; Martin and Langman 1965; Langman *et al.* 1966; Smith and Schoenwolf 1987, 1988), which provides additional cells for lengthening of the neural plate as well as for localized widening (Schoenwolf and Alvarez 1989).

Primary Neurulation: Bending of the Neural Plate

Bending of the neural plate begins during its shaping and involves two events: neural plate furrowing and neural plate folding. Furrowing of the neural plate occurs within localized regions called hinge points. Three hinge points form during bending (Schoenwolf 1982; Smith and Schoenwolf 1987). The first to form is called the median hinge point (MHP); it consists of a single, midline region of neural plate anchored to underlying tissue (the axial mesoderm). Neuroepithelial cells within the hinge point (*i.e.*, MHP cells) change their shapes as a result of induction by the axial mesoderm (van Straaten *et al.* 1988; Smith and Schoenwolf 1989), namely, MHP cells decrease their height and become wedge shaped. In addition, these cells increase their cell-cycle length, express unique molecular markers, and produce a diffusible chemoattractant (Smith and Schoenwolf 1987, 1988; van Straaten *et al.* 1988; Dodd and Jessell 1988; Jessell *et al.* 1989; Placzek *et al.* 1990). Cells immediately flanking the MHP do not exhibit any of these changes. Rather these cells, now called L cells (for lateral neuroepithelial cells), remain tall and spindle shaped, like the cells of the nascent neural plate (*i.e.*, the neural plate at stages 3 and 4).

The remaining two hinge points consist of a pair called the dorsolateral hinge points (DLHPs). Each DLHP consists of a single dorsolateral region of neural plate (*i.e.*, containing the most lateral neuroepithelial cells) anchored to underlying tissue (the surface ectoderm of the neural fold). As in the MHP, neuroepithelial cells within the DLHP (*i.e.*, DLHP cells) change their shape, namely, they *increase* their height and become wedge shaped. These cells have not been studied as extensively as have MHP cells, so it is unknown whether they are induced to undergo these shape changes and whether they display other properties which make them unique.

Cell wedging within the hinge points causes the neural plate to furrow longitudinally. Formation of the furrows, at least for the MHP, occurs independently of forces generated by more lateral tissues (Schoenwolf 1988). That is, cell wedging and the consequent furrowing are autonomous processes; they are not generated passively by kinking of the apical side of the neural plate as a result of extrinsic forces generated by elevation of the neural folds.

Folding of the neural plate following its furrowing, namely, elevation of the neural folds with rotation of the neural plate on each side around the MHP, and convergence of the neural folds with rotation of the dorsolateral neural plate around each DLHP, is facilitated by cell wedging but is not driven by it; rather, folding is principally the result of forces originating in tissues underlying and lateral to the neural plate (Schoenwolf 1988; Smith and Schoenwolf 1991). Each hinge point is a pivot, allowing the neural plate to rotate around a focus. Cell wedging within each hinge point, as well as the localized anchorage of the neural plate to adjacent tissue, facilitates and directs rotation and contributes to the establishment of the neural tube's characteristic cross-sectional morphology.

Primary Neurulation: Closure of the Neural Groove

The paired neural folds are brought into apposition in the dorsal midline as a result of their elevation and convergence. There, they adhere to one another, presumably because of the character of their surface coats or cell surfaces (Moran and Rice 1975; Lee *et al.* 1976a,b; Rice and Moran 1977; Mak 1978; Silver and Kerns 1978; Sadler 1978; Smits-van Prooije 1986; Takahashi and Howes 1986; Takahashi 1988), allowing fusion to occur. Fusion has not been thoroughly studied, but sufficient work has been done to show that species differences exist. For example, closure of the neural groove in cranial regions of mouse and hamster embryos involves the formation of numerous filopodia, blebs, and ruffles (Waterman 1975, 1976), whereas much less elaborate protrusions are formed in chick embryos (Schoenwolf 1978b, 1979b; Silver and Kerns 1978).

Closure of the neural groove is associated with formation of the neural crest. In chick embryos, neural crest cells at a particular craniocaudal level do not leave the neuroepithelium until the neural groove has closed at that level (reviewed by LeDouarin 1982). By contrast, in cranial regions of mouse embryos, formation of the neural crest is precocious, beginning as the neural folds are elevating and converging (Nichols 1981; Tan and Morriss-Kay 1985; Chan and Tam 1988); however, at spinal cord levels, the neural crest is delayed in its formation, and does not initiate its migration until after the neural tube has formed (Erickson and Weston 1983; Schoenwolf and Nichols 1984).

Secondary Neurulation: Formation of the Tail Bud

Formation of the tail bud is the first stage in secondary neurulation. The tail bud consists of a mesenchymal mass of cells covered by surface ectoderm dorsally and hindgut endoderm ventrally. It is formed during stages 11-14, ventral and caudal to the closing caudal neuropore, by consolidation of persisting remnants of the cranial part of the primitive streak; this consolidation occurs owing to the action of the caudal (tail) body fold (Schoenwolf 1979a). In avian embryos, the transition between primary and secondary neurulation occurs at the future lumbosacral level (Criley 1969; Schoenwolf 1977); here, an overlap zone forms so that primary neurulation (*i.e.*, closure of the caudal neuropore) occurs dorsally and secondary neurulation occurs ventrally (Criley 1969; Dryden 1980).

Secondary Neurulation: Formation of the Medullary Cord

Shortly after formation of the tail bud, the most dorsal cells of this structure undergo a mesenchymal-epithelial transformation (Criley 1969; Schoenwolf and DeLongo 1980), which results in the formation of the medullary cord, the rudiment of the secondary neural tube. During this process, mesenchymal change their shape from stellate to columnar and become connected to adjacent columnarizing cells *via* gap junctions (Schoenwolf and DeLongo 1980; Schoenwolf and Kelley 1980). This establishes the medullary cord as a solid epithelial structure, which soon becomes demarcated from adjacent mesenchymal cells (which form somites; Schoenwolf 1977; Sanders *et al.* 1986) by the formation of a basal lamina (Schoenwolf and DeLongo 1980; Griffith and Wiley 1990).

Secondary Neurulation: Cavitation of the Medullary Cord

Formation of the secondary neural tube from the medullary cord involves cavitation. During cavitation, gap junctions are formed between the forming apices of columnar medullary-cord cells, and small, irregularly shaped spaces soon appear, each developing into an isolated lumen (Schoenwolf and DeLongo 1980; Schoenwolf and Kelley 1980). Ultimately, all lumina coalesce to create a single, central lumen, the neurocele. With appearance of the neurocele, formation of the secondary neural tube is completed. This structure consists of a tube whose walls are composed of pseudostratified columnar epithelium; it is morphologically identical to the more cranial primary neural tube. Formation of the secondary neural crest then occurs as the most dorsal cells of the secondary neural tube leave this structure and migrate laterally (Schoenwolf *et al.* 1985).

The processes of primary and secondary neurulation are autonomous; that is, development of the secondary neural tube is not dependent upon proper development of the primary neural tube. This was shown by inducing defects of the primary neural tube in the caudal neuropore region of chick embryos developing *in ovo*; secondary neurulation occurred normally even though the primary neural tube was dysraphic (Costanzo *et al.* 1982).

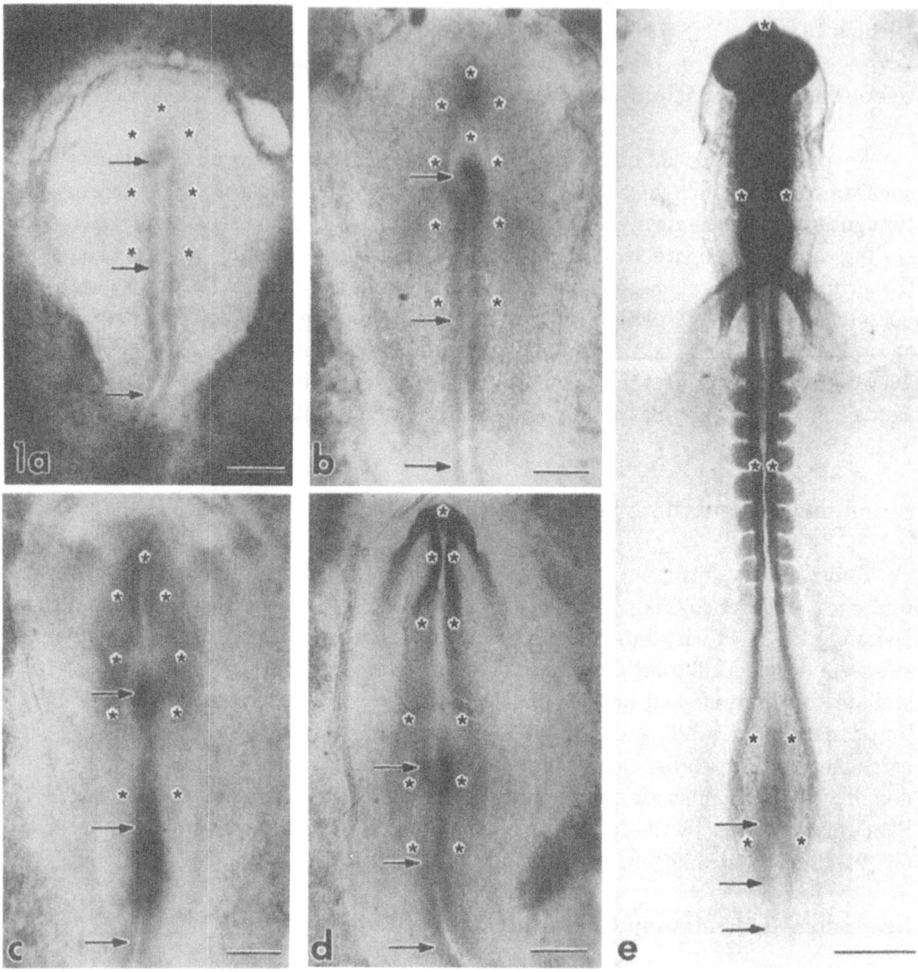

Figure 1. Light micrographs of the dorsal surface of the chick epiblast. Part **a**, stage 4; part **b**, stage 5; part **c**, stage 6; part **d**, stage 7; part **e**, stage 10. Arrows, primitive streak; asterisks, neural plate (and neural tube in parts d and e). Bars = 200 μm.

RELATIONSHIP BETWEEN GASTRULATION AND NEURULATION

Temporal and Spatial Overlap

The processes of gastrulation and neurulation are intimately associated both temporally and spatially. Temporal overlap occurs between these two processes because formation of the neural plate is induced by cells contained within the cranial end of the forming and progressing primitive streak (at about stage 2) (reviewed by Gallera 1971). Furthermore, shaping and bending of the neural plate and closure of the neural groove occur during stages of primitive streak regression (stages 4-11) (Figure 1).

Spatial overlap between areas undergoing gastrulation and neurulation also exists. This was revealed by fate mapping (*e.g.*, Spratt 1952; Schoenwolf and Sheard 1990). The neural plate consists of two zones: a single prenodal region and a paired postnodal region. Each half of the postnodal neural plate flanks the primitive streak and lies in close vicinity to (and perhaps is intermixed with) the prospective mesodermal cells still in the epiblast. Thus, neurulation occurs both prenodally and postnodally in close association with preingression, ingressing, and ingressed prospective endodermal and mesodermal cells (Figure 2).

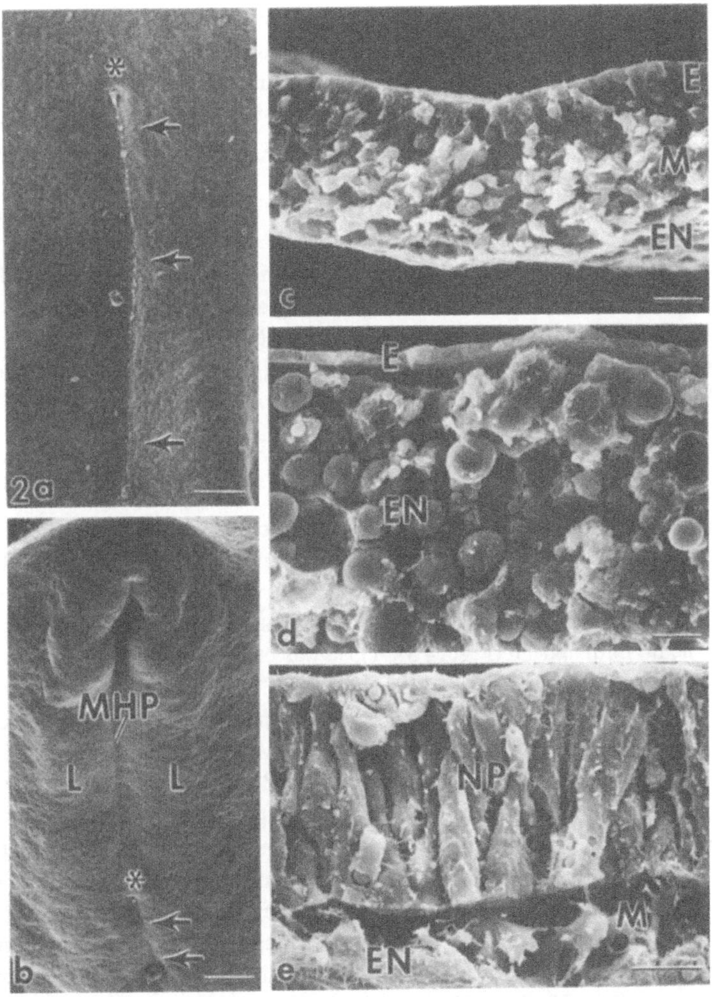

Figure 2. Scanning electron micrographs of either the dorsal surface of the chick epiblast (part **a**, stage 4; part **b**, stage 7) or transverse slices through the primitive streak (part **c**), yolk-filled area opaca (part **d**), and neural plate with underlying ingressed mesoderm and endoderm (part **e**). Asterisks (parts **a**, **b**), Hensen's node; arrows (parts **a**, **b**), primitive streak. **E** (parts **c**, **d**), epiblast; **EN** (parts **c-e**), endoderm; **L** (part **b**), lateral neural plate; **M** (parts **c**, **e**), mesoderm; **MHP** (part **b**), median hinge point region of neural plate; **NP** (part **e**), neural plate. Bars = 100 μm (parts **a**, **b**); 20 μm (parts **c**, **d**); 10 μm (part **e**).

Degree of Autonomy

Although temporal and spatial overlap exists between areas undergoing gastrulation and neurulation, these two processes exhibit a considerable degree of autonomy; that is, although these processes are coordinated with one another, they are partially separable causally. This was revealed by experiments in which blastoderms were transected near the level of Hensen's node at stages 3-8 to partially uncouple neurulating areas (principally prenodally) from gastrulating areas (principally postnodally) (Schoenwolf *et al.* 1989b). Both pre- and postnodal areas of the blastoderm undergo normal morphogenesis after transection, regardless of whether Hensen's node is included with the prenodal or postnodal piece. For example, in the absence of cranial tissues, including Hensen's node, the primitive streak remaining in the caudal isolate still undergoes regression and produces mesoderm, which subdivides into paraxial mesoderm, nephrotome, and lateral plate (but not into notochord because its progenitor cells were removed with Hensen's node). Similarly, in the absence of caudal tissues, including Hensen's node, the prenodal neural plate can undergo shaping (including longitudinal lengthening) and bending. Thus, although the morphogenetic movements of gastrulation and neurulation are temporally and spatially coordinated, each process exhibits a substantial degree of autonomy. Nevertheless, mesoderm derived from the primitive streak during gastrulation (and generally already present beneath and lateral to the neural plate at the time of transection) is thought to play an important role during neurulation (see review by Schoenwolf and Smith 1990b), again showing the interrelatedness of gastrulation and neurulation.

Similarity of Cell Movements

Displacements of different areas of the epiblast have been followed by microinjecting small pools of a fluorescent-histochemical marker into gastrulating and neurulating embryos at stages 3 and 4 (Schoenwolf and Sheard 1989). These experiments have shown that similar patterns of displacement occur during gastrulation and neurulation (Figure 3). A grid was projected onto epiblasts, which were then injected at several defined pre- and postnodal locations. Epiblasts were injected at five mediolateral positions, with injections grouped into the following "columns": a single midline column, which marked prospective MHP cells prenodally and primitive streak, including Hensen's node, postnodally; paired medial columns (each 250-300 μm lateral to the midline), which marked on each side prospective L cells both prenodally and postnodally as well as prospective mesodermal cells postnodally; and paired lateral columns (each 500-600 μm lateral to the midline), which marked on each side prospective surface ectodermal cells both prenodally and postnodally. Directions of displacements were similar prenodally and postnodally. Marks in the midline column were displaced mainly caudally. Marks in the medial columns were displaced mainly medially and caudally. Marks in the lateral column were displaced mainly cranially and medially.

Although the patterns of displacement were similar prenodally and postnodally, the role of such displacements is very different for regions undergoing gastrulation and neurulation. The prenodal midline column forms MHP cells, which move caudally in the wake of the regressing Hensen's node. Thus, this movement participates in shaping of the neural plate. By contrast, cells in the postnodal midline column are prospective

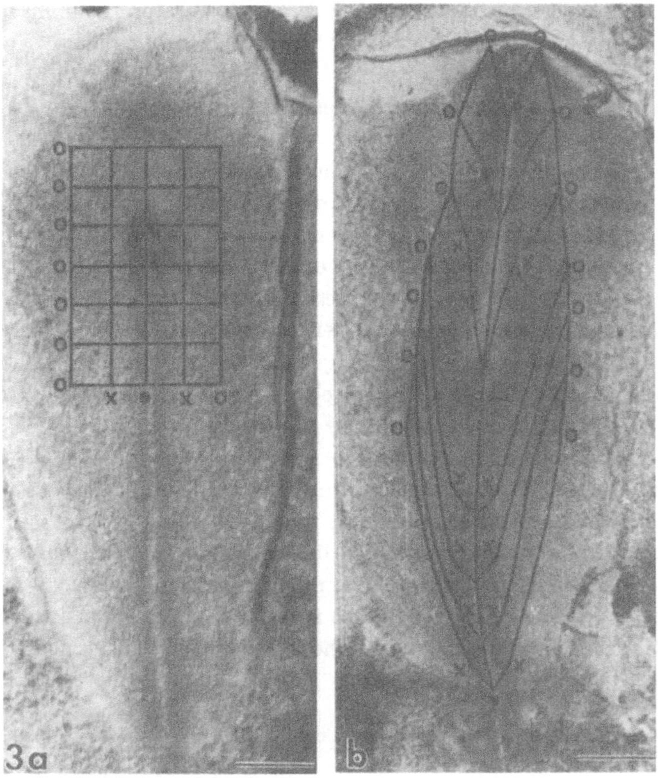

Figure 3. Light micrographs of the dorsal surface of the chick epiblast at stages 4 and 7 (parts **a** and **b**, respectively) summarizing the data of Schoenwolf and Sheard (1989). Note that a similar deformation of the grid occurs prenodally and postnodally (lines interconnecting the injection sites in the medial columns of part **b** have been omitted for the sake of clarity). Open circles indicate the lateral columns, x's indicate the medial columns, and closed circles indicate the midline column. Bars = 200 μm.

mesodermal cells, which ultimately ingress through the primitive streak. Their movement caudally with regression of the primitive streak allows the mesoderm to be laid down as the axis elongates cranial to Hensen's node. The prenodal medial columns as well as much of the postnodal medial columns form L cells, which also move caudally as both shaping of the neural plate and regression of the primitive streak occur. Movement of these cells medially contributes to narrowing of the neural plate, whereas their caudal movement contributes to its lengthening. Prospective mesodermal cells in the medial postnodal column move medially to enter the streak and eventually ingress at a more caudal level owing to the regression of the primitive streak. The prenodal and postnodal lateral columns give rise to surface ectoderm. These cells spread medially, probably providing extrinsic forces for neurulation (see review by Schoenwolf and Smith 1990b). Also, medial spread of the surface ectoderm is required to "fill in the space" created by narrowing of the neural plate and ingression of prospective mesodermal cells. The entire area pellucida becomes narrowed mediolaterally and lengthened craniocaudally during these processes as revealed by timelapse videomicroscopy (Schoenwolf and Sheard 1989). The pre- and postnodal areas that have been followed in the lateral columns move principally cranially and medially as the shape of the area pellucida changes; presumably, more caudal areas of the lateral columns move caudally and medially, but such areas have not yet been sufficiently studied.

The Blastoderm as a Convergent-Extension Machine

Based on the above description of epiblast displacements, and the realization that similar movements occur in the endoderm and mesoderm after ingression, it seems likely that the entire blastoderm functions like a convergent-extension machine during axial morphogenesis. That is, cells, as part of the three germ layers, move coordinately toward the midline (converge) and then craniocaudally (extend) so that the forming embryo narrows transversely and lengthens craniocaudally. The extreme degree of coordination that must occur among the germ layers is especially evident during the induction of an ectopic embryo after transplantation of Hensen's node to the germ cell crescent (Dias and Schoenwolf 1990). In this process, a new center for convergent-extension is established in the extraembryonic area over the course of just a few hours. Perfect, although miniature, embryos form at the ectopic site, each complete with a neural tube, notochord, paired columns of somites, nephrotomes, lateral plates, guts, etc. The body folds form normally in such embryos, and they progressively separate the embryo from the underlying blastoderm. How such cooperation occurs among diverse tissues is unknown, but it constitutes a source of considerable fascination for those of us who like to watch "embryos at work."

CELL MOVEMENTS IN THE EPIBLAST

In the previous section, I discussed some of the general patterns of epiblast displacement that occur during gastrulation and neurulation. Here, I will provide some details about the movements of three types of epiblast cells: prospective MHP cells, prospective L cells, and prospective surface ectodermal cells.

Cell Rearrangement and Division

The movement of small patches of epiblast cells (about 300 cells/patch) has been followed over time by constructing quail/chick transplantation chimeras (Schoenwolf and Alvarez 1989; Schoenwolf and Alvarez, in preparation; Schoenwolf *et al.* 1989a; Alvarez and Schoenwolf 1991). All epiblast cells exhibit two behaviors during their displacement: rearrangement and division. Prospective MHP cells arise from a central prenodal area that overlaps Hensen's node (Figure 4; Rosenquist 1966; Nicolet 1970 1971; Schoenwolf *et al.* 1989a; Schoenwolf and Sheard 1990). Prospective MHP cells rearrange with one another by intercalating, such that the transverse dimension of the patch decreases while its longitudinal dimension increases. Prospective MHP cells undergo on the average about 2 rounds of cell rearrangement from the flat neural plate stage (stages 3, 4) to the closed neural groove stage (stages 10, 11), with each round of rearrangement being defined as a halving of the width of the patch and a doubling of its length. Additionally, prospective MHP cells undergo 2-3 rounds of division during this time, with each round resulting in a doubling of cell number. Much of this division is directed to the longitudinal plane of the neural plate, thereby contributing to neural plate lengthening. The population of prospective MHP cells as a whole always spans the midline during cell rearrangement, and it ultimately contributes to the floor plate of the central nervous system.

Prospective L cells also undergo rearrangement and division of a similar magnitude, namely, 2 rounds of rearrangement and 2-3 rounds of division. These cells arise from bilateral areas flanking the cranial primitive streak (Figure 4; Schoenwolf *et al.* 1989a; Schoenwolf and Sheard 1990). As the primitive streak regresses, each group of flanking prospective L cells remains as a whole lateral to the forming MHP region, ultimately contributing to the lateral walls of the central nervous system. A few prospective L cells, however, typically leave the group and enter the caudal MHP region; here, they are induced to form MHP cells by the underlying notochord.

Analysis of prospective surface ectodermal cells has not been completed, but some preliminary observations can be made. These cells, like the prospective MHP and L cells of the neural plate, also exhibit rearrangement and division. Prospective surface ectodermal cells undergo a remarkable journey during neurulation. For example, grafts of cells to the most lateral region of the area pellucida often end up in the neural folds. Thus, extensive displacements occur in this layer, which are precisely coordinated with (and to some degree constitute) the concurrent morphogenetic movements of gastrulation and neurulation.

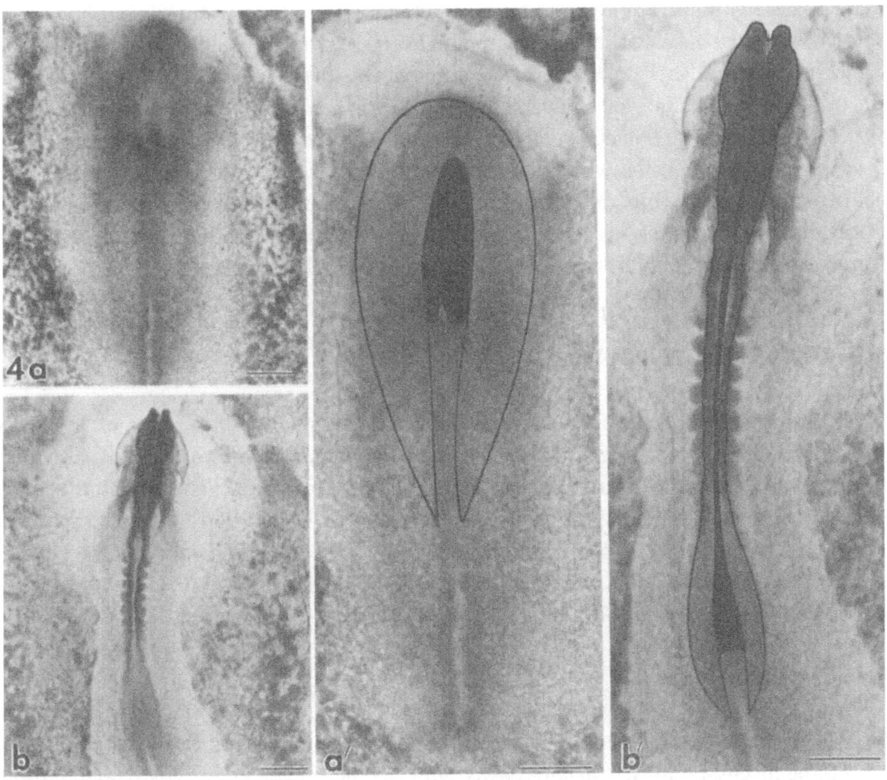

Figure 4. Light micrograph showing the displacements that occur in the MHP and L regions (filled and open areas, respectively) during shaping and bending of the neural plate (based principally on the data of Schoenwolf and Alvarez 1989; Schoenwolf *et al.* 1989a). Part **a**, **a′**, stage 4; part **b**, **b′**, stage 8. Bars = 200 μm.

Restriction of Cell Intermingling

Heterotopic transplantations have been performed to analyze how patterns of cell movement are established in the epiblast during neurulation (Alvarez and Schoenwolf 1991; Schoenwolf and Alvarez, in preparation). These experiments provide evidence that the pattern of cell movement results at least in part from the restriction of intermingling of cells of adjacent populations. For example, when prospective L cells are grafted into prospective MHP-cell territory, some of them spread *laterally,* intermixing with other L cells, rather than remaining on the midline and intercalating with host MHP cells. Similarly, when prospective MHP cells are grafted into prospective L-cell territory, some of them spread *medially*, intermixing with host MHP cells in the forming MHP region, rather than intercalating with host L cells. These experiments suggest not only that there is a restriction of mingling among heterologous populations of cells, but that this restriction is established prior to inductive interactions with the notochord. Recall that such interactions are necessary to establish the morphological differences characterizing MHP and L cells. The experiments just discussed show that rearrangement behavior is independent of such interactions, because these experiments were done prior to formation of the notochord.

Experiments to examine whether a restriction of cell intermingling is also involved in the rearrangement behavior of surface ectodermal cells are under way, but again some preliminary observations can be made. Prospective L cells grafted to prospective surface-ectodermal territory convert (morphologically) to prospective surface ectodermal cells and seemingly (not yet analyzed in detail) intermix with host surface ectodermal cells. Furthermore, prospective surface ectodermal cells grafted to prospective L-cell territory can form L cells, which seemingly intermix with other L cells. Thus, these experiments suggest that differences in the rearrangement behavior of prospective neural plate and surface ectoderm arise during neurulation as a result of inductive interactions, presumably by the underlying mesodermal tissues.

PROSPECTIVE FATE MAPS

Several methods have been used to construct prospective fate maps of the early avian blastoderm. Each method has its own particular strengths and weaknesses (discussed by Schoenwolf and Sheard 1990), which affect the results obtained. My approach here will be to highlight what I believe to be the three most important approaches, those of (1) Spratt (1952, 1955); (2) Rosenquist (review 1966), Nicolet (review 1971), Vakaet (reviews 1984, 1985), and Schoenwolf *et al.* (1989a); and (3) Schoenwolf and Sheard (1990).

Three Types of Maps

The first type of fate map is based on carbon marking and is the least reliable of those to be discussed here. Carbon marks tend to float on the suprablastodermic fluid, rather than being firmly attached to cell surfaces, and unless this technique is combined with time-lapse microscopy (which Spratt did not use), actual cell or tissue displacements cannot be discriminated from spurious movements of the particle. Also, cells that are dividing, undergoing extensive migration, ingressing, or being subjected to deformation by morphogenetic movements might not retain their carbon particles. Consequently, this technique is most reliable for "passive" cells in "quiescent" areas and least reliable for "active" cells in "dynamic" areas.

The basic conclusion that has been drawn from these maps is that the early epiblast is a composite of spatially segregated groups of cells contributing, either by ingressing or by remaining on the surface, to various organ rudiments and their subdivisions. For example, the four major craniocaudal subdivisions of the central nervous system, the forebrain, midbrain, hindbrain, and spinal cord, are shown in these maps as the stripes of a chevron lying cranial to and flanking Hensen's node. Furthermore, prospective mesoderm, which could be mapped only crudely (by combining a number of techniques with carbon marking), is shown as being partially organized into discrete notochord- and somite-forming centers.

The second type of fate map is based on the transplantation of identifiable cells–either cells labeled with tritiated thymidine or quail cells grafted to chick hosts (quail cells can be discriminated from chick cells by the presence of a heterochromatin nucleolar marker after staining according to the Feulgen procedure; LeDouarin 1973). Transplantation is associated with trauma; new morphogenetic events may occur during healing of the graft; and graft age (*i.e.*, its stage of development), position, and orientation might not perfectly match those of the host. Furthermore, tritiated thymidine is diluted with each cell division (some rapidly-dividing cells may be no longer labeled at the stages studied) and can be transferred from dying cells to their neighbors; quail cells placed in a chick-cell environment may behave atypically owing to species differences (incuding the rate of development). Transplantation of identifiable cells, however, would seem to be unaffected by the state of the cell's "activity," at least in a short-term experiment, in contrast to carbon marking.

The fate maps obtained from transplantation reveal a different pattern for the prospective neural plate than that found by Spratt based on carbon marking. Rather than showing segregation into distinct craniocaudal subdivisions, the neural plate is shown to be subdivided mediolaterally, either into MHP and L cells (Schoenwolf *et al.* 1989a) or into "dorsal" and "ventral" neural tube (Rosenquist 1966); each population of these mediolateral subdivisions arises from a discrete area and moves down the length of the axis to contribute to multiple craniocaudal subdivisions of the central nervous system. The maps based on transplantation do provide some confirmation of those based on carbon marking in that they show that the most cranial cells *tend* to form more cranial regions of the neuraxis than do more caudal cells, again suggesting that carbon marking identifies mainly "passive" cells (*e.g.*, cranial cells that form forebrain and undergo little migration probably retain their label, but cranial cells that migrate down the length of the neuraxis probably lose it).

The maps based on transplantation also better define the location of prospective endodermal and mesodermal areas (Figure 5). These areas are shown to be arranged within and alongside the streak in craniocaudal sequence; namely, prospective endoderm, prechordal plate mesoderm, and notochord occupy the level of the cranial part of the primitive streak; prospective paraxial mesoderm, nephrotome, and lateral plate mesoderm (including prospective heart mesoderm) occupy the level of the middle part of the primitive streak; and prospective extraembryonic mesoderm occupies the level of the caudal part of the primitive streak.

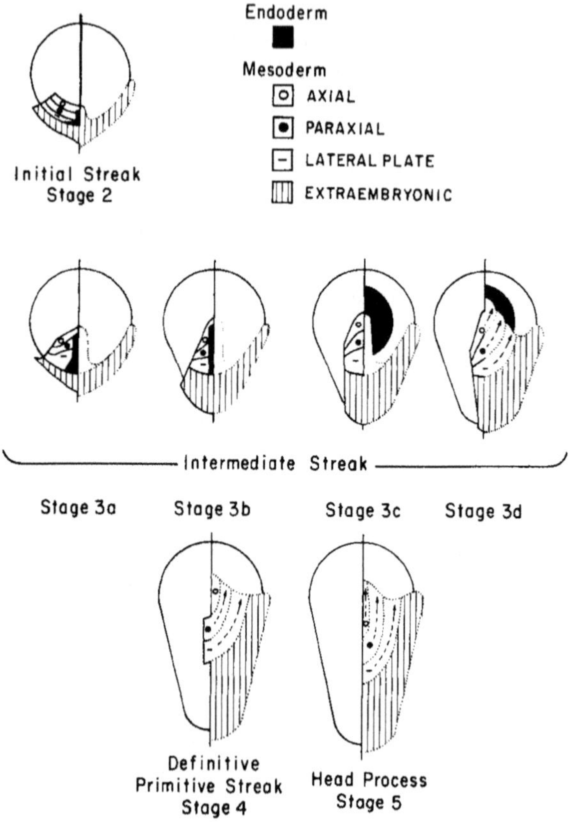

Figure 5. Prospective fate maps showing the endodermal and mesodermal derivatives of the epiblast (modified from Vakaet 1984, 1985).

The third type of fate map is based on localized microinjection of the epiblast with a recognizable cell marker. This technique causes much less trauma than does transplantation, and it allows the marking of small groups of contiguous cells. How cells are labeled is unclear, that is, it is unknown whether they are filled directly by the injection or whether they are filled indirectly by endocytosis over time, but regardless of how cells are labeled, this technique has the potential to provide the best resolution obtained so far (considerably better than that obtained by transplantation of tissues consisting of tens or hundreds of cells; single cell labeling has even been reported: Fraser *et al.* 1990). The high localization of labeling, the main advantage of this technique, is also somewhat of a disadvantage in that, because so few cells are labeled, hundreds of injections are required to map each stage of development completely. A further disadvantage is that it is difficult to control the depth of the injection in a flat blastoderm. Thus, for example, labeled mesoderm could result from the injection of prospective mesodermal cells still contained in the epiblast or to the injection of ingressed mesodermal cells already underlying the epiblast.

Comparison of fate maps obtained from microinjection with those obtained from transplantation reveals a strong similarity. For example, injection of the prenodal site or of Hensen's node yields labeled MHP cells, the same cells obtained from grafts to the midline prenodal area. Similarly, injection of the area flanking the cranial primitive streak yields labeled L cells, the same cells obtained from grafts to the paranodal areas. Mesodermal areas have not been mapped yet by microinjection, owing largely to the technical difficulties described above.

In summary, as a result of fate mapping studies over the last few decades, a clearer picture of the organization of the avian epiblast is emerging. Fate mapping is tedious, slow work, but the rewards of the labor are a better understanding of how the embryo is organized at very early stages prior to the appearance of overt anatomical structure.

Prospective Fate and Prospective Potency

A major question raised by fate mapping studies concerns the state of determination of the mapped cells. There are two mutually exclusive possibilities. First, the fate of these cells might be already determined. In which case they require only a proper environment in which to differentiate according to their prospective fate. Therefore, the prospective potency of these cells (*i.e.*, what they are capable of forming) is now equivalent to their prospective fate (*i.e.*, what they are destined to form). Second, the fate of these cells might not be determined yet. Rather, later events, such as a cell's birth order in a particular lineage (*i.e.*, whether it arises, for example, after one, two, or three divisions) or its local interactions with other cells (*i.e.*, inductions), specify its fate. Therefore, the prospective potency of these cells is still greater than their prospective fate.

An ideal way to address the relationship between prospective fate and potency would be to follow individual cells and to determine when they are capable of forming only a single cell type. For technical reasons, such a paradigm has not been accomplished yet for cells of the chick epiblast. Instead, heterotopic grafts of epiblast areas have been used to determine whether *groups* of cells at various sites as a whole exhibit pluripotentiality. In experiments mentioned above (see section entitled, "Restriction of Cell Intermingling"), prospective MHP cells were transplanted to prospective L-cell territory. Those trapped in the L region formed typical L cells as well as neural crest cells; the latter cells are thought to derive only from the lateral neural plate during normal development. This result indicates that the prospective potency of cells of the early epiblast still exceeds their prospective fate (see Alvarez and Schoenwolf 1991; Schoenwolf and Alvarez, in preparation, for more examples of this). Thus, it is important to keep in mind when studying prospective fate maps, that cells in different locations either may be truly different from one another or they may be identical; in the latter case, they are destined to different fates only because they occupy different spatial or lineal positions within the neural plate.

QUERIES AND SPECULATIONS

In this essay, I have tried to paint a broad picture of our general understanding of gastrulation and neurulation in avian embryos. Here, I wish to raise a number of questions about gastrulation that remain to be answered, at least partially, as well as to freely speculate on the roles of cell rearrangement in this process. My purpose in speculating is not "to set" my ideas "in stone," but rather to provide a stimulus for future studies of the very complex problem of axial morphogenesis.

There are a number of important questions about avian gastrulation. These include the following: (1) How does the subgerminal cavity form? (2) How does cell shedding occur during formation of the area pellucida, and are other processes also involved in the formation of this region? (3) How are cells specified to be hypoblast? (4) Where are prospective mesodermal, endodermal, and ectodermal cells located prior to formation of the primitive streak? (5) Is the primitive streak and/or the prospective endodermal and mesodermal cells induced, and if so, by what tissue? (6) What cell behaviors underlie primitive streak formation and progression? (7) To what extent does the primitive streak act as a blastema (*i.e.*, a mass of cells that *proliferates* new cells during ingression), a blastopore (*i.e.*, an area through which cells *move* into the interior of the blastoderm during ingression), or a queue (*i.e.*, an area where cells line up and *wait* to ingress)? (8) What mechanisms underlie the epithelial-mesenchymal transformation that occurs during ingression of cells through the primitive streak? (9) Is there a limit of ingression (*i.e.*, a distinct boundary, on one side of which cells remain in the epiblast and on the other side of which cells ingress), or are those cells that are destined to ingress, intermixed with those that are destined to remain on the surface? (10) When do cells become committed to a particular fate? (11) To what extent are cell fates specified by lineage and by inductive interactions? and (12) How does primitive streak regression occur? Perusing this list of questions, which is by no means exhaustive, could leave one daunted by the state of our ignorance. Yet, I hope that the unknowns will instead be seen as opportunities and will serve as challenges evoking further study.

Before ending this essay, I wish to briefly speculate on the possible role of cell rearrangement in regards to three of the above questions. Pursuant to the first question (numbered above as 2), let us consider the possibility that cell rearrangement plays a role in the formation of the area pellucida. This possibility seems likely to me for the following reasons. As the area pellucida forms, the entire blastoderm expands radially, increasing its surface area, while its central portion thins from 4-6 cell layers to 1 (see Eyal-Giladi and Kochav 1976, cf., Figures 12-20; Kochav *et al.* 1980, cf., Figures 7-10). A similar process occurs during epiboly (*i.e.*, the spreading of a sheet of cells to enclose deeper cells) in amphibian gastrulae (Keller 1975, 1978, 1980), where it has been demonstrated that cell intercalation occurs radially (*i.e.*, deep cells intercalate with superficial cells) (see Figure 1: Keller 1980). Based on these observations, and on the belief that developmental mechanisms have been conserved during evolution, it would seem prudent to explore the possible role of radial intercalation in formation of the area pellucida during avian gastrulation.

Pursuant to the second (numbered above as 6) and third (numbered above as 12) questions, let us consider the possibility that cell rearrangement is involved in the formation, progression, and regression of the primitive streak. Figure 6 shows one model of how cell rearrangement might be involved. Formation and progression of the streak could occur owing to convergent-extension. Each prospective primitive-streak half could converge with its counterpart on the caudal midline through cell-cell intercalation. As a result of this process the forming streak would elongate (*i.e.*, it would undergo progression), with its cranial end extending rostrally and its caudal end caudally. This extension could result in the change in the outline of the area pellucida that occurs during this period (*i.e.*, a change from circular to pear-shaped). Ingression

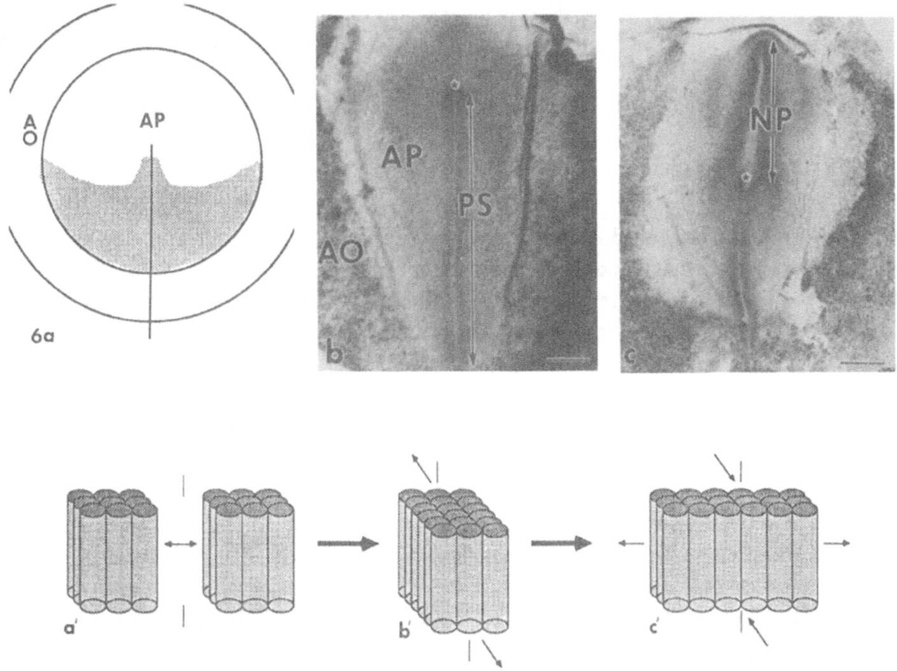

Figure 6. Diagrams and light micrographs showing stages of primitive streak formation (**a**, **a'**), progression (**b**, **b'**), and regression (**c**, **c'**). Parts **a'**, **b'**, and **c'** are hypothetical and indicate possible cell rearrangements within the primitive streak; large arrows indicate that **a'** transforms into **b'**, which in turn transforms in **c'**; small arrows indicate directions of cell rearrangements; vertical lines indicate the midline. Asterisk (part **b**), Hensen's node; shading (part **a**) primitive-streak-forming cells; vertical line (part **a**), midline. **AO** (parts **a**, **b**), area opaca; **AP** (parts **a**, **b**), area pellucida; **NP** (part **c**), neural plate; **PS** (part **b**), primitive streak. Bars = 200 μm (parts **b**, **c**).

of cells through the primitive streak, and regression of the streak, could result from a "reversed" convergent-extension (perhaps called a divergent-retraction). During this process, Hensen's node would move caudally (*i.e.*, it would undergo regression) and the posterior node would move cranially as cell-cell intercalation occurred. This would broaden the streak and if combined with an epithelial-mesenchymal transformation, would allow the prospective mesodermal cells to spread as a sheet into the interior of the blastoderm between the outer ectoderm and inner endoderm. These latter possibilities are supported by the experiments of Ooi *et al.* (1986), which showed that two movements are associated with regression of the primitive streak: craniocaudal movement within the streak during its regression, and mediolateral movements as mesodermal cells migrate away from the streak as it undergoes regression. I must emphasize that this model is largely speculative so I urge caution. Nevertheless, good experimental analysis of such a testable model is certain to provide us with new information about avian gastrulation.

New information is indeed what we seek. We have won many battles against the embryo but the war is far from being over. Let us muster our troops, for there is still much to learn.

ACKNOWLEDGMENTS

The original research described above was funded by grants from principally the NIH. I wish to acknowledge the constructive editorial criticisms of Dr. Ignacio S. Alvarez, the technical assistance of Fahima Rahman, and the secretarial assistance of Jennifer Parsons. Finally, I wish to dedicate this paper to Dr. Ray Keller, my scientific big brother (that is, in the nurturing sense not in the "1984" sense).

REFERENCES

Alvarez, I.S. and G.C. Schoenwolf. 1991. Patterns of neuroepithelial cell rearrangement during avian neurulation are established independently of notochordal inductive interactions. *Dev. Biol.* 143:78-92.

Bellairs, R. 1964. Biological aspects of the yolk of the hen's egg. *Adv. Morphog.* 4:217-272.

Bellairs, R. 1986. The primitive streak. *Anat. Embryol.* 174:1-14.

Bellairs, R., F.W. Lorenz, and T. Dunlap. 1978. Cleavage in the chick embryo. *J. Embryol. Exp. Morphol.* 43:55-69.

Burnside, B. 1973. Microtubules and microfilaments in amphibian neurulation. *Am. Zool.* 13:989-1006.

Callebaut, M. 1974. La formation de l'oocyte d'oiseau. Etude autoradiographique chez la caille japonaise pondeuse à l'aide de la leucine tritiée. *Arch. Biol.* 85:201-233.

Callebaut, M. 1983. The constituent oocytal layers of the avian germ and the origin of the primordial germ cell yolk. *Arch. Anat. Microsc. Morphol. Exp.* 72:199-214.

Callebaut, M. 1985. Link between avian oogenesis and gastrulation: demonstration of a cytoplasmic pre-embryonic fate map by trypan blue induced fluorescence. *IRCS Med. Sci.* 13:711-712.

Canning, D.R. and C.D. Stern. 1988. Changes in the expression of the carbohydrate epitope HKN-1 associated with mesoderm induction in the chick embryo. *Development* 104:643-655.

Chan, W.Y. and P.P.L. Tam. 1988. A morphological and experimental study of the mesencephalic neural crest cells in the mouse embryo using wheat germ agglutinin-gold conjugate as the cell marker. *Development* 102:427-442.

Costanzo, R., R.L. Watterson, and G.C. Schoenwolf. 1982. Evidence that secondary neurulation occurs autonomously in the chick embryo. *J. Exp. Zool.* 219:233-240.

Criley, B.B. 1969. Analysis of the embryonic sources and mechanisms of development of posterior levels of chick neural tubes. *J. Morphol.* 128:465-501.

Dias, M.S. and G.C. Schoenwolf. 1990. Formation of ectopic neurepithelium in chick blastoderms: Age-related capacities for induction and self-differentiation following transplantation of quail Hensen's nodes. *Anat. Rec.* 229:437-448.

Dodd, J. and T.M. Jessell. 1988. Axon guidance and the patterning of neuronal projections in vertebrates. *Science* 242:692-699.

Dryden, R. J. 1980. Spina bifida in chick embryos: Ultrastructure of open neural defects in the transitional region between primary and secondary modes of neural tube formation. p. 75-100. *In: Advances in the Study of Birth Defects*. T.V.N. Persaud (Ed.). MTP Press Ltd., Lancaster.

Erickson, C.A. and J.A. Weston. 1983. An SEM analysis of neural crest migration in the mouse. *J. Embryol. Exp. Morphol.* 74:97-118.

Eyal-Giladi, H. and S. Kochav. 1976. From cleavage to primitive streak formation: A complementary normal table and a new look at the first stages of the development of the chick. I. General morphology. *Dev. Biol.* 49:321-337.

Fraser, S., R. Keynes, and A. Lumsden. 1990. Segmentation in the chick embryo hindbrain is defined by cell lineage restrictions. *Nature* 344:431-435.

Gallera, J. 1971. Primary induction in birds. *Adv. Morphog.* 9:149-180.

Gordon, R. 1985. A review of the theories of vertebrate neurulation and their relationship to the mechanics of neural tube birth defects. *J. Embryol. Exp. Morphol. Suppl.* 89:229-255.

Griffith, C.M. and M.J. Wiley. 1990. Distribution of cell surface glycoconjugates during chick secondary neurulation. *Anat. Rec.* 226:81-90.

Gurdon, J.B. 1987. Embryonic induction—molecular prospects. *Development* 99:285-306.

Hamburger, V. and H.L. Hamilton. 1951. A series of normal stages in the development of the chick embryo. *J. Morphol.* 88:49-92.

Harrisson, F. 1989. The extracellular matrix and cell surface, mediators of cell interactions in chicken gastrulation. *Int. J. Dev. Biol.* 33:417-438.

Holmdahl, D.E. 1925a. Die erste Entwicklung des Körpers bei den Vögeln und Säugetieren, inkl. dem Menschen, besonders mit Rücksicht auf die Bildung des Rückenmarks, des Zöloms und der entodermalen Kloake nebst einem Exkurs über die Entstehung der Spina bifida in der Lumbosakralregion. *Gegenbaurs Morphol. Jahrb.* I. 54:333-384.

Holmdahl, D.E. 1925b. Experimentelle Untersuchungen über die Lage der Grenze zwischen primärer und sekundärer Körperentwicklung beim Huhn. *Anat. Anz.* 59:393-396.

Jacobson, A.G. 1980. Computer modeling of morphogenesis. *Am. Zool.* 20:669-677.

Jacobson, A.G. 1981. Morphogenesis of the neural plate and tube. p. 233-263. *In: Morphogenesis and Pattern Formation*. T.G. Connelly, L.L. Brinkley, and B.M. Carlson (Eds.). Raven Press, New York.

Jacobson, A.G., G.F. Oster, G.M. Odell, and L.Y. Cheng. 1986. Neurulation and the cortical tractor model for epithelial folding. *J. Embryol. Exp. Morphol.* 96:19-49.

Jacobson, C.-O. and T. Ebendal (Eds.). 1978. *Formshaping Movements in Neurogenesis*. Almqvist & Wiksell International, Stockholm.

Jaffe, L.F. and C.D. Stern. 1979. Strong electrical currents leave the primitive streak region of chick embryos. *Science* 206:569-571.

Jessell, T.M., P. Bovolenta, M. Placzek, M. Tessier-Lavigne, and J. Dodd. 1989. Polarity and patterning in the neural tube: the origin and function of the floor plate. p. 257-282. *In: Cellular Basis of Morphogenesis, Ciba Foundation Symposium*. Wiley, Chichester.

Karfunkel, P. 1974. The mechanisms of neural tube formation. *Int. Rev. Cytol.* 38:245-271.

Keller, R.E. 1975. Vital dye mapping of the gastrula and neurula of *Xenopus laevis*. I. Prospective areas and morphogenetic movements of the superficial layer. *Dev. Biol.* 42:222-241.

Keller, R.E. 1978. Time-lapse cinemicrographic analysis of superficial cell behavior during and prior to gastrulation in *Xenopus laevis*. *J. Morphol.* 157:223-247.

Keller, R.E. 1980. The cellular basis of epiboly: An SEM study of deep cell rearrangement during gastrulation in *Xenopus laevis*. *J. Embryol. Exp. Morphol.* 60:201-234.

Kochav, S. and H. Eyal-Giladi. 1971. Bilateral symmetry in chick embryo determination by gravity. *Science* 171:1027-1029.

Kochav, S., M. Ginsburg, and H. Eyal-Giladi. 1980. From cleavage to primitive streak formation: A complementary normal table and a new look at the first stages of the development of the chick. II. Microscopic anatomy and cell population dynamics. *Dev. Biol.* 79:296-308.

Langman, J., R.L. Guerrant, and B.G. Freeman. 1966. Behavior of neuro-epithelial cells during closure of the neural tube. *J. Comp. Neurol.* 127:399-412.

LeDouarin, N.M. 1973. A Feulgen-positive nucleolus. *Exp. Cell Res.* 77:459-468.

LeDouarin, N.M. 1982. *The Neural Crest.* Cambridge University Press, London.

Lee, H.-Y., R.G. Nagele, and G.W. Kalmus. 1976a. Further studies on neural tube defects caused by Concanavalin A in early chick embryos. *Experientia* 32:1050-1052.

Lee, H.-Y., J.B. Sheffield, R.G. Nagele, and G.W. Kalmus. 1976b. The role of extracellular material in chick neurulation. I. Effects of Concanavalin A. *J. Exp. Zool.* 198:261-266.

Mak, L.L. 1978. Ultrastructural studies of amphibian neural fold fusion. *Dev. Biol.* 65:435-446.

Martin, A. and J. Langman. 1965. The development of the spinal cord examined by autoradiography. *J. Embryol. Exp. Morphol.* 14:25-35.

Martins-Green, M. 1988. Origin of the dorsal surface of the neural tube by progressive delamination of epidermal ectoderm and neuroepithelium: Implications for neurulation and neural tube defects. *Development* 103:687-706.

Moran, D. and R.W. Rice. 1975. An ultrastructural examination of the role of cell membrane surface coat material during neurulation. *J. Cell Biol.* 64:172-181.

Nalbandov, A.V. and M.F. James. 1949. The blood-vascular system of the chicken ovary. *Am. J. Anat.* 85:347-378.

New, D.A.T. 1956. The formation of sub-blastodermic fluid in hens' eggs. *J. Embryol. Exp. Morphol.* 4:221-227.

Nichols, D.H. 1981. Neural crest formation in the head of the mouse embryo as observed using a new histological technique. *J. Embryol. Exp. Morphol.* 64:105-120.

Nicolet, G. 1970. Analyse autoradiographique de la localisation des différentes ébauches présomptives dans la ligne primitive de l'embryon de Poulet. *J. Embryol. Exp. Morphol.* 23:79-108.

Nicolet, G. 1971. Avian gastrulation. *Adv. Morphog.* 9:231-262.

Ooi, V.E.C., E.J. Sanders, and R. Bellairs. 1986. The contribution of the primitive streak to the somites in the avian embryo. *J. Embryol. Exp. Morphol.* 92:193-206.

O'Shea, S. 1981. The cytoskeleton in neurulation: Role of cations. p. 35-60. *In: Progress in Anatomy.* R.J. Harrison (Ed.). Cambridge University Press, London.

Penner, P.L. and I. Brick. 1984. Acetylcholinesterase and polyingression in the epiblast of the primitive streak chick embryo. *Wilhelm Roux's Arch. Dev. Biol.* 193:234-241.

Placzek, M., M. Tessier-Lavigne, T. Yamada, T. Jessell, and J. Dodd. 1991. Mesodermal control of neural cell identity: Floor plate induction by the notochord. *Science* 250:985-988.

Rice, R.W. and D.J. Moran. 1977. A scanning electron microscope and X-ray microanalytic study of cell surface material during amphibian neurulation. *J. Exp. Zool.* 201:471-478.

Rosenquist, G.C. 1966. A radioautographic study of labeled grafts in the chick blastoderm. Development from primitive-streak stages to stage 12. *Carnegie Inst. Wash. Contrib. Embryol.* 38:31-110.

Sadler, T.W. 1978. Distribution of surface coat material on fusing neural folds of mouse embryos during neurulation. *Anat. Rec.* 191:345-350.

Sanders, E.J., M.K. Khare, V.C. Ooi, and R. Bellairs. 1986. An experimental and morphological analysis of the tail bud mesenchyme of the chick embryo. *Anat. Embryol.* 174:179-185.

Schoenwolf, G.C. 1977. Tail (end) bud contributions to the posterior region of the chick embryo. *J. Exp. Zool.* 201:227-246.

Schoenwolf, G.C. 1978a. Effects of complete tail bud extirpation on early development of the posterior region of the chick embryo. *Anat. Rec.* 192:289-296.

Schoenwolf, G.C. 1978b. An SEM study of posterior spinal cord development in the chick embryo. *Scanning Electron Microsc.* 1978/II:739-746.

Schoenwolf, G.C. 1979a. Histological and ultrastructural observations of tail bud formation in the chick embryo. *Anat. Rec.* 193:131-148.

Schoenwolf, G.C. 1979b. Observations on closure of the neuropores in the chick embryo. *Am. J. Anat.* 155:445-466.

Schoenwolf, G.C. 1982. On the morphogenesis of the early rudiments of the developing central nervous system. *Scanning Electron Microsc.* 1982/I:289-308.

Schoenwolf, G.C. 1983. The chick epiblast: A model for examining epithelial morphogenesis. *Scanning Electron Microsc.* 1983/III:1371-1385.

Schoenwolf, G.C. 1985. Shaping and bending of the avian neuroepithelium: Morphometric analyses. *Dev. Biol.* 109:127-139.

Schoenwolf, G.C. 1988. Microsurgical analyses of avian neurulation: Separation of medial and lateral tissues. *J. Comp. Neurol.* 276:498-507.

Schoenwolf, G.C. 1991. Neurepithelial cell behavior during avian neurulation. *In: Cell-Cell Interactions in Early Development.* J. Gerhart (Ed.). Alan R. Liss, Inc., New York. In press.

Schoenwolf, G.C. and I.S. Alvarez. 1989. Roles of neuroepithelial cell rearrangement and division in shaping of the avian neural plate. *Development* 106:427-439.

Schoenwolf, G.C., H. Bortier, and L. Vakaet. 1989a. Fate mapping the avian neural plate with quail/chick chimeras: Origin of prospective median wedge cells. *J. Exp. Zool.* 249:271-278.

Schoenwolf, G.C., N.B. Chandler, and J. Smith. 1985. Analysis of the origins and early fates of neural crest cells in caudal regions of avian embryos. *Dev. Biol.* 110:467-479.

Schoenwolf, G.C. and J. DeLongo. 1980. Ultrastructure of secondary neurulation in the chick embryo. *Am. J. Anat.* 158:43-63.

Schoenwolf, G.C., S. Everaert, H. Bortier, and L. Vakaet. 1989b. Neural plate- and neural tube-forming potential of isolated epiblast areas in avian embryos. *Anat. Embryol.* 179:541-549.

Schoenwolf, G.C. and M.V. Franks. 1984. Quantitative analyses of changes in cell shapes during bending of the avian neural plate. *Dev. Biol.* 105:257-272.

Schoenwolf, G.C. and R.O. Kelley. 1980. Characterization of intercellular junctions in the caudal portion of the developing neural tube of the chick embryo. *Am. J. Anat.* 158:29-41.

Schoenwolf, G.C. and D.H. Nichols. 1984. Histological and ultrastructural studies on the origin of caudal neural crest cells in mouse embryos. *J. Comp. Neurol.* 222:496-505.

Schoenwolf, G.C. and M.L. Powers. 1987. Shaping of the chick neuroepithelium during primary and secondary neurulation: Role of cell elongation. *Anat. Rec.* 218:182-195.

Schoenwolf, G.C. and P. Sheard. 1989. Shaping and bending of the avian neural plate as analysed with a fluorescent-histochemical marker. *Development* 105:17-25.

Schoenwolf, G.C. and P. Sheard. 1990. Fate mapping the avian epiblast with focal injections of a fluorescent-histochemical marker: Ectodermal derivatives. *J. Exp. Zool.* 255:323-339.

Schoenwolf, G.C. and J.L. Smith. 1990a. Epithelial cell wedging: A fundamental cell behavior contributing to hinge point formation during epithelial morphogenesis. *In: Control of Morphogenesis by Specific Cell Behaviors*. R.E. Keller and D. Fristrom (Eds.). W.B. Saunders Co., London. 1:325-334.

Schoenwolf, G.C. and J.L. Smith. 1990b. Mechanisms of neurulation: Traditional viewpoint and recent advances. *Development* 109:243-270.

Schroeder, T.E. 1970. Neurulation in *Xenopus laevis*. An analysis and model based upon light and electron microscopy. *J. Embryol. Exp. Morphol.* 23:427-462.

Silver, M.H. and J.M. Kerns. 1978. Ultrastructure of neural fold fusion in chick embryos. *Scanning Electron Microsc.* 1978/II:209-215.

Smith, J.L. and G.C. Schoenwolf. 1987. Cell cycle and neuroepithelial cell shape during bending of the chick neural plate. *Anat. Rec.* 218:196-206.

Smith, J.L. and G.C. Schoenwolf. 1988. Role of cell-cycle in regulating neuroepithelial cell shape during bending of the chick neural plate. *Cell Tissue Res.* 252:491-500.

Smith, J.L. and G.C. Schoenwolf. 1989. Notochordal induction of cell wedging in the chick neural plate and its role in neural tube formation. *J. Exp. Zool.* 250:49-62.

Smith, J.L. and G.C. Schoenwolf. 1991. Further evidence of extrinsic forces in bending of the neural plate. *J. Comp. Neurol.* 307:225-236.

Smits-van Prooije, A.E., R.E. Poelmann, A.F. Gesink, M.J. van Groeningen, and C. Vermeij-Keers. 1986. The cell surface coat in neurulating mouse and rat embryos, studied with lectins. *Anat. Embryol.* 175:111-117.

Spemann, H. 1938. *Embryonic Development and Induction*. Yale University Press, New Haven.

Spratt, N.T., Jr. 1942. Location of organ-specific regions and their relationship to the development of the primitive streak in the early chick blastoderm. *J. Exp. Zool.* 89:69-101.

Spratt, N.T., Jr. 1952. Localization of the prospective neural plate in the early chick blastoderm. *J. Exp. Zool.* 120:109-130.

Spratt, N.T., Jr. 1955. Analysis of the organizer center in the early chick embryo. I. Localization of prospective notochord and somite cells. *J. Exp. Zool.* 128:121-164.

Spratt, N.T., Jr. 1963. Role of substratum, supracellular continuity, and differential growth in morphogenetic cell movements. *Dev. Biol.* 7:51-63.

Spratt, N.T., Jr. and H. Haas. 1960a. Importance of morphogenetic movements in the lower surface of the young chick blastoderm. *J. Exp. Zool.* 144:257-276.

Spratt, N.T., Jr. and H. Haas. 1960b. Morphogenetic movements in the lower surface of the unincubated and early chick blastoderm. *J. Exp. Zool.* 144:139-157.

Stern, C.D. 1990. The marginal zone and its contribution to the hypoblast and primitive streak of the chick embryo. *Development* 109:667-682.

Stern, C. 1991. Mesoderm Formation in the Chick Embryo, Revisited. p. 29-42. *In: Gastrulation: Movements, Patterns, and Molecules*. R. Keller, W.H. Clark, Jr., F. Griffin (Eds.). Plenum Press, New York.

Stern, C.D. and D.R. Canning. 1988. Gastrulation in birds: A model system for the study of animal morphogenesis. *Experientia* 44:651-657.

Stern, C.D. and D.R. Canning. 1990. Origin of cells giving rise to mesoderm and endoderm in chick embryo. *Nature* 343:273-275.

Stern, C.D. and D.O. MacKenzie. 1983. Sodium transport and the control of epiblast polarity in the early chick embryo. *J. Embryol. Exp. Morphol.* 77:73-98.

Stern, C.D., S. Manning, and J.I. Gillespie. 1985. Fluid transport across the epiblast of the early chick embryo. *J. Embryol. Exp. Morphol.* 88:365-384.

Takahashi, H. 1988. Changes in peanut lectin binding sites on the neuroectoderm during neural tube formation in the bantam chick embryo. *Anat. Embryol.* 178:353-358.

Takahashi, H. and R.I. Howes. 1986. Binding pattern of ferritin-labeled lectins (RCA_1 and WGA) during neural tube closure in the bantam embryo. *Anat. Embryol.* 174:283-288.

Tan, S.S. and G. Morriss-Kay. 1985. The development and distribution of cranial neural crest in the rat embryo. *Cell Tissue Res.* 240:403-416.

Vakaet, L. 1962. Some new data concerning the formation of the definitive endoblast in the chick embryo. *J. Embryol. Exp. Morphol.* 10:38-57.

Vakaet, L. 1970. Cinephotomicrographic investigations of gastrulation in the chick blastoderm. *Arch. Biol.* 81:387-426.

Vakaet, L. 1984. Early development of birds. p. 71-88. *In: Chimeras in Developmental Biology*. N. LeDouarin and A. McLaren (Eds.). Academic Press, London.

Vakaet, L. 1985. Morphogenetic movements and fate maps in the avian blastoderm. p. 99-109. *In: Molecular Determinants of Animal Form*. G.M. Edelman (Ed.). Alan R. Liss, New York.

van Straaten, H.W.M., J.W.M. Hekking, E.J.L.M. Wiertz-Hoessels, F. Thors, and J. Drukker. 1988. Effect of the notochord on the differentiation of a floor plate area in the neural tube of the chick embryo. *Anat. Embryol.* 177:317-324.

Veini, M. and K. Hara. 1975. Changes in the differentiation tendencies of the hypoblast-free Hensen's node during "gastrulation" in the chick embryo. *Wilhelm Roux' Arch. Entwicklungsmech. Org.* 177:89-100.

Waterman, R.E. 1975. SEM observations of surface alterations associated with neural tube closure in the mouse and hamster. *Anat. Rec.* 183:95-98.

Waterman, R.E. 1976. Topographical changes along the neural fold associated with neurulation in the hamster and mouse. *Am. J. Anat.* 146:151-172.

Weinberger, C. and I. Brick. 1982a. Primary hypoblast development in the chick. I. Scanning electron microscopy of normal development. *Wilhelm Roux's Arch. Dev. Biol.* 191:119-126.

Weinberger, C. and I. Brick. 1982b. Primary hypoblast development in the chick. II. The role of cell division. *Wilhelm Roux's Arch. Dev. Biol.* 191:127-133.

Weinberger, C., P.L. Penner, and I. Brick. 1984. Polyingression, an important morphogenetic movement in chick gastrulation. *Am. Zool.* 24:545-554.

MESODERM FORMATION IN THE CHICK EMBRYO, REVISITED

Claudio D. Stern

Department of Human Anatomy
University of Oxford
South Parks Road
Oxford OX1 3QX U.K.

INTRODUCTION

Germ Layers and Cell Movement Patterns During Primitive Streak Formation in the Chick Embryo

At the time of egg laying, the chick blastoderm consists of a disc, some 2mm in diameter, comprising an inner, translucent *area pellucida* and an outer, more opaque *area opaca* (Figures 1, 2). The latter region contributes only to extraembryonic structures. The first layer of cells to be present as such is the *epiblast*, which is continuous over both *areae opaca* and *pellucida*. It is a one-cell thick epithelium which soon becomes pseudostratified and columnar, the apices of the cells facing the albumen. From the center of this initial layer arises a second layer of cells, the *primary hypoblast*, consisting of several unconnected islands of 5-20 cells (for more detailed explanation of the terminology see Stern 1990 and Figures 1 and 2). Subsequently, more cells are added to the primary hypoblast, and by 6 hr of incubation it becomes a loose but continuous epithelium, the *secondary hypoblast*. The source of these new hypoblast cells is the deep (endodermal) portion of a crescent-shaped region, the

Gastrulation, Edited by R. Keller *et al.*
Plenum Press, New York, 1991

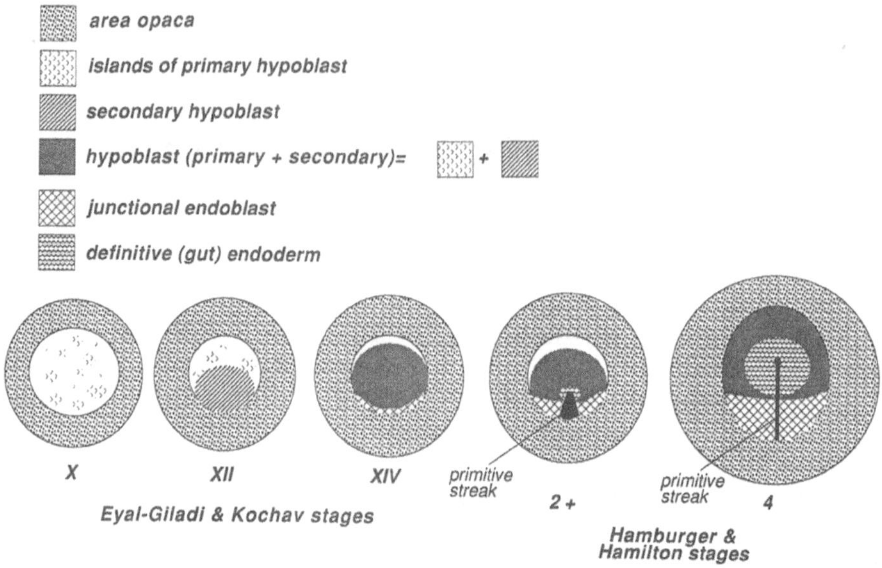

Figure 1. Components of the lower (endodermal) layer of cells of the chick embryo during gastrulation. Only the definitive (gut) endoderm contributes cells to the embryo proper; the remaining components are extraembryonic. From Stern (1990), reproduced with permission from the Company of Biologists Ltd.

marginal zone (Figure 2), which separates *areae pellucida* and *opaca* at the future posterior (caudal) end of the embryo.

The embryo now consists of two layers: the epiblast proper, facing the albumen, from which will arise all of the embryonic tissues, and the hypoblast (primary and secondary), facing the yolk, which will give rise only to extraembryonic tissues (mainly the yolk sac stalk) although it may also contain some primordial germ cells (see Bellairs 1986; Ginsburg and Eyal-Giladi 1987). As the hypoblast continues to spread from posterior to anterior parts of the blastodisc, cells appear between the epiblast and hypoblast. These are the first cells of the *mesoderm* (Vakaet 1984). As more accumulate, they coalesce into the first axial structure of the embryo, the *primitive streak*, which appears at the posterior margin of the *area pellucida* at about 10h of incubation (see Bellairs 1986 for review). Later, more mesodermal cells are recruited into the primitive streak as this structure elongates along the antero-posterior axis of the embryo (Vakaet 1984). The mesoderm eventually migrates out of the streak and gives rise to the lateral plates, intermediate mesoderm, paraxial (somitic) mesoderm and the axial notochord.

Primitive streak cells, soon after the streak forms, begin to insert into the hypoblast, displacing it towards the edges of the *area pellucida* (Figure 1). These primitive streak derived cells form the *definitive endoderm*, which in turn gives rise to the lining of the gut. The elongation of the primitive streak and the expansion of the blastodisc, together with further recruitment of cells derived from the posterior marginal zone (this contribution now forming the *junctional endoblast*, which is also extraembryonic; see Figures 1 and 2), confine the original hypoblast to a crescent shaped region underlying the anterior portion of the *area pellucida*; this region, known

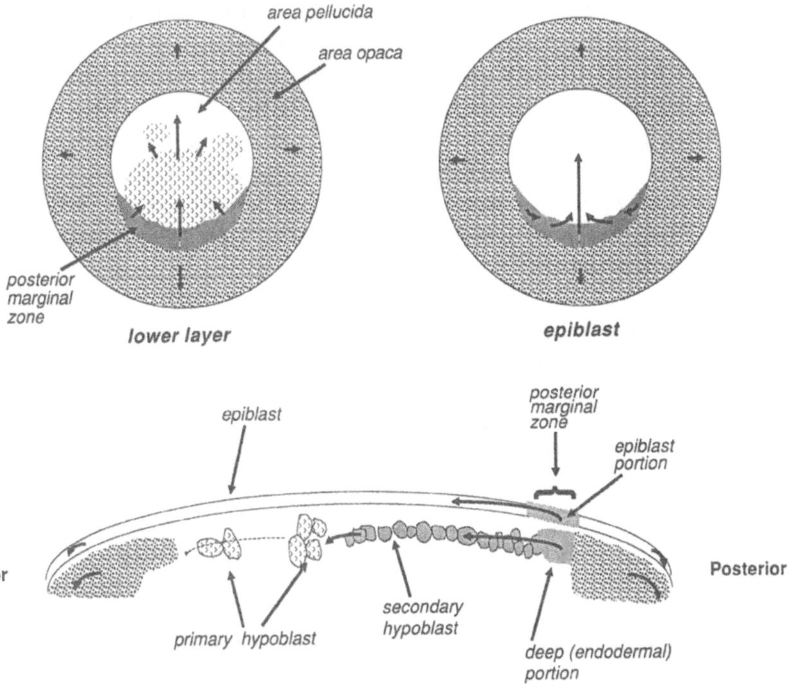

Figure 2. Contributions from the marginal zone of the early chick embryo to various structures during gastrulation. The two diagrams above represent embryos at pre-primitive streak stages, viewed from below (left; lower layer) and above (right; epiblast). The diagram below is a diagrammatic sagittal section. Arrows indicate the direction of cell movements in the upper and lower cell layers during the early stages of primitive streak formation. Modified from Stern (1990), with permission from the Company of Biologists Ltd.

as the *germinal crescent*, contains the primordial germ cells. These cells will later migrate into the gonads (see Bellairs 1986; Ginsburg and Eyal-Giladi 1987).

Regulative potential of the early embryo. Higher vertebrate embryos are capable of extensive regulation. If a pre-streak chick blastoderm is cut into several portions, most of the portions are able to produce a complete, albeit somewhat smaller embryo (see Spratt and Haas 1960; Khaner *et al.* 1985; Nieuwkoop *et al.* 1985; and Bellairs 1986). There are, in theory, two possible interpretations for this phenomenon: in the first, cells in the embryo are pluripotent and are allocated their ultimate fates by cell interactions; when a fragment is separated from the embryo, cells in the fragment interact so that the correct proportions of different cell types are generated. The second interpretation of the phenomenon of regulation requires the embryo to consist of a fairly random mix of cells already specified to their ultimate fates, which later sort to their correct destinations. According to this second interpretation, isolation of a fragment of embryo leads to sorting of the cell types to newly-defined sites, but the cells do not change fates.

One problem with the second interpretation is that it does not account for our ability to construct fate maps. If embryos consist of random mixtures of different cell types, there should not be discrete regions with defined fates unless cells sort soon after becoming specified. Such a 'mosaic' arrangement of two or more cell types would explain the ill-defined boundaries between different regions that characterize the fate maps of many vertebrates. Nevertheless, the first interpretation is the one usually accepted; it is generally assumed that cells become determined as a result of inductive interactions with other cells, so that the correct cell types are generated at least near their ultimate locations.

A note on the staging system. Stages of development after the appearance of the primitive streak are classified according to the system of Hamburger and Hamilton (1951) and use Arabic numerals. The primitive streak appears at stage 2 and reaches its full length at stage 4 (Figure 1). Before the appearance of the primitive streak, the staging system of Eyal-Giladi and Kochav (1976) is used, in Roman numerals. The last stage in this system before appearance of the streak is stage XIV, characterized by a full hypoblast. The hypoblast begins to form as islands at stage IX and the contribution of posterior marginal zone cells to it (secondary hypoblast) begins at about stage XI (Figure 1).

When Does Cell Diversity Become Established? Expression of Cell-Type-Specific Markers During Primitive Streak Formation

The apparent complexity of the processes that lead to germ layer formation and of the subsequent movements that rearrange cells leads us to address the questions: when do different embryonic cells become committed to their fates, and what are the mechanisms responsible?

To determine the time during embryonic development at which cell diversity becomes established, it is advantageous to possess cell-type-specific markers allowing different cell types to be identified as soon as possible after their appearance. Until recently, no such markers were available in the avian embryo; different cell types could only be recognized by their morphology and their position. As a result of a thorough search, several molecular probes for different tissues or regions of the primitive-streak-stage embryo have been reported (Stern and Canning 1988).

In our recent studies, we have concentrated mainly on the expression of immunoreactivity to a monoclonal antibody, HNK-1. This antibody was originally raised by Abo and Balch (1981) against a cell surface determinant on human natural killer cells. Since then, it has been used extensively as a marker for cells of the developing peripheral nervous system of avian embryos, particularly neural crest cells (e.g. Tucker *et al.* 1984, 1988; Rickmann *et al.* 1985; Bronner-Fraser 1986; Stern *et al.* 1986). In chick development, HNK-1 recognizes cells of the hypoblast from the time when they ingress from the overlying epiblast to form the islands of primary hypoblast. It also recognizes the cells of the secondary hypoblast and their precursors in the posterior marginal zone (Canning and Stern 1988; Stern 1990). The cells of the hypoblast remain positive for this antibody even after they have become confined to the anterior germinal crescent.

Just before primitive streak formation (stages XII-XIV), a 'mosaic' of HNK-1-positive and -negative cells is found in the epiblast; positive cells gradually become concentrated at the posterior end of the blastodisc and eventually disappear from the epiblast as the primitive streak forms. After this, the antibody recognizes all of the middle layer cells; these cells continue to express immunoreactivity until the end of primitive streak formation (stage 4), when some of them leave the streak region. These findings led to the suggestion (Canning and Stern 1988; Stern and Canning 1990), which will be discussed in detail below, that the HNK-1-positive cells of the epiblast prior to streak formation are the progenitors of the (also HNK-1-positive) cells of the streak, which will give rise to the mesoderm and endoderm of the embryo.

At later stages of development, the notochord, splanchnopleural mesoderm (especially in cranial regions), neural crest cells, motor axons and associated glial cells express the epitope (see Lim *et al.* 1987; Lunn *et al.* 1987; Canning and Stern 1988; Tucker *et al.* 1988). In addition, other structures, like the posterior half of the otic vesicle (Lim *et al.* 1987), portions of the developing retina and lens and of the aortic endothelium also express immunoreactivity (unpublished observations). It has been suggested (Canning and Stern 1988) that HNK-1 immunoreactivity is coincident with cells involved in inductive interactions during development.

Nature of Macromolecules Carrying the L2/HNK-1 Epitope and their Possible Involvement in Gastrulation

The epitope recognized by the HNK-1 antibody has been studied extensively. It is a complex, sulfated carbohydrate structure (Chou *et al.* 1986) known as L2/HNK-1 and is common to many molecules thought to play a role in cell-cell or cell-substrate adhesion (Kruse *et al.* 1984, 1985; Keilhauer *et al.* 1985; Bollensen and Schachner 1987; Cole and Schachner 1987; Hoffman and Edelman 1987; Pesheva *et al.* 1987; Hoffman *et al.* 1988; Künemund *et al.* 1988). They include: the neural cell adhesion molecule (NCAM), the neural-glial cell adhesion molecule (Ng-CAM) (also known as G4 in the chick and L1 in the mouse), a chondroitin sulfate proteoglycan receptor for cytotactin (CTB-proteoglycan), the 'adhesion molecule on glia' (AMOG) and the laminin/fibronectin receptor, integrin. It is also present on other molecules expressed in association with the nervous system such as the glycoprotein P0, myelin-associated glycoprotein (MAG), myelin basic protein (MBP) and the enzyme acetylcholinesterase (Bon *et al.* 1987). The epitope has also been reported to form part of certain complex glycolipids (e.g. Chou *et al.* 1986; Dennis *et al.* 1988).

The promiscuity of the L2/HNK-1 epitope among molecules with putative roles in adhesion has led several investigators (Keilhauer *et al.* 1985; Bronner-Fraser 1987; Künemund *et al.* 1988) to suggest that it may play a role in modulating the adhesive properties of its host molecules. There is, in fact, some direct evidence supporting this proposal: sulfated oligosaccharides contained within the epitope can disrupt cell-substrate adhesion of certain cultured cells (Künemund *et al.* 1988). However, how this modulation is achieved is not yet understood. The epitope may also be involved in conferring specificity to cell recognition because the bonds between the component sugar groups may display some variation, as they do in blood group specific oligosaccharides (T. Rademacher, personal communication).

Because of this complexity, it becomes important to determine the nature of the molecules bearing the L2/HNK-1 epitope during early development of the chick embryo. Initial investigations (Canning and Stern 1988 and unpublished) used affinity purification of detergent-solubilized proteins followed by SDS-polyacrylamide electrophoresis and immunoblotting with antibodies against glycoproteins known to express the epitope.

Several bands were seen in gels run from eluted material. The most prominent is a triplet of bands at about 70kD, which can be obtained from samples of all germ layers. Other bands are more tissue specific; the epiblast from stage XIV embryos shows a major unique band in excess of 300kD. This polypeptide is recognized by antibodies against J1/tenascin and against chondroitin sulfate (unpublished observations). This suggests that it is related or identical to the cytotactin binding proteoglycan (CTB-proteoglycan), which is known to contain the L2/HNK-1 epitope (Hoffman and Edelman 1987; Hoffman *et al.* 1988). Antibodies directed to J1/tenascin or to chondroitin sulfate only stain the basal lamina underlying the epiblast from stage XIV. Samples of hypoblast show a more minor specific component of about 140kD. This reacts with mono- and polyclonal antibodies directed to G4 (the chick equivalent of L1, or Ng-CAM), which carries the L2/HNK-1 epitope (see Fushiki and Schachner 1986). In frozen sections, antibodies to G4 stain the hypoblast, the basal lamina of the epiblast, and cells in the posterior marginal zone. Samples of primitive streak are characterized by a band of 32kD, which does not change in mobility upon reduction, and which is not recognized by antibodies to any of the known HNK-1-related molecules.

Possible roles of HNK-1-related antigens in gastrulation. In an effort to determine whether the L2/HNK-1 epitope or its carrying molecules play a role in the formation of the primitive streak or the development of the hypoblast in the chick embryo, pre-streak embryos (stage XII-XIII) were grafted with a pellet of hybridoma cells secreting the antibody (unpublished observations). None of the grafted embryos developed an embryonic axis; instead, a thick 'button' of tissue formed on the dorsal side of the epiblast. The cells of this button had mesenchymal appearance; this structure may be analogous to amphibian exogastrulae. None of several control hybridomas secreting antibodies to other molecules had any effect on development. This result is consistent with the view that the L2/HNK-1 epitope, or its carrying molecule(s), might play some role in the normal development of the embryonic axis and in the formation of the mesoderm in the chick embryo.

Fate of HNK-1-Positive Cells

Although we do not yet know the precise functions of the L2/HNK-1 epitope, some recent experiments have used this antibody for following the fates of HNK-1-positive cells. Expression of region-specific markers is not a sufficient criterion to decide whether cells that differ in their expression of antibody immunoreactivity are developmentally different from one another. For this reason, two simple experiments were performed to establish whether or not the HNK-1-positive cells found in the epiblast prior to primitive streak formation are in fact the precursors of the (HNK-1-positive) cells of the primitive streak, which in turn give rise to the mesoderm and embryonic endoderm (Stern and Canning 1990).

Fate of HNK-1-positive cells of the epiblast. In the first experiment, the HNK-1 antibody was directly coupled to colloidal gold particles. Living embryos (stage XIII) from which the hypoblast had been removed were incubated briefly in this reagent, during which time the HNK-1-positive cells become labeled; during subsequent incubation at 37°C these labeled cells endocytose the antibody-gold complex and can subsequently be recognized by the presence of gold particles in their cytoplasm. After further incubation, the fates of the cells that were HNK-1-positive at the time of labeling can be identified. It was found that gold labeled cells are found in the primitive streak and its mesodermal and endodermal derivatives, but not in the epiblast. This result suggests that the HNK-1-positive cells of the epiblast found prior to primitive streak formation are in fact the precursors of the primitive streak derived mesendoderm.

The second experiment of Stern and Canning (1990) was designed to test whether the HNK-1-negative cells of the epiblast have the ability to contribute to the primitive streak in the absence of HNK-1-positive cells. Embryos at stage XIII (before streak formation) without their hypoblasts were incubated in a mixture of the HNK-1 antibody and guinea pig complement to kill the epiblast HNK-1-positive cells. Subsequent incubation of the embryo led to normal expansion of the treated embryo, but no axial structures formed. However, such embryos could be rescued by a graft of quail primitive streak cells. In this case, a normal axis formed, and the mesoderm and endoderm of such chimeric embryos were composed entirely of quail cells. This result suggests that the HNK-1-positive cells of the epiblast at stage XIII are necessary for the formation of a normal primitive streak.

Fate of HNK-1-positive cells of the posterior marginal zone. In another series of experiments (Stern 1990), the fate of the HNK-1-positive cells of the posterior marginal zone was followed using the HNK-1-gold technique in combination with grafting. A donor embryo was incubated in the HNK-1-gold reagent and its posterior marginal zone grafted into an unlabeled recipient at the same stage of development (stage XII). It was found that the gold-labeled cells contribute to the secondary hypoblast, but not to other tissues. When the HNK-1-positive cells of the posterior margin were ablated with a mixture of HNK-1 and complement, the grafted margin did not contribute to the secondary hypoblast. These experiments suggest that the posterior marginal zone contains special, HNK-1-positive cells which are the precursors of the secondary hypoblast, and that the remaining (HNK-1-negative) cells are unable to contribute to this layer.

The Marginal Zone, 'Organizer' of the Early Embryo

In 1933, C.H. Waddington (see also Azar and Eyal-Giladi 1981) rotated the hypoblast of a pre-streak chick embryo to reverse its antero-posterior orientation. He found that the primitive streak now formed from the opposite (anterior) margin of the blastodisc, and proposed that the hypoblast induces the primitive streak.

Spratt and Haas (1960) demonstrated the importance of the posterior marginal zone in the formation of the embryonic axis, based on a study of the regulative ability of isolated pieces of blastoderm. They postulated the existence of a gradient in 'embryo forming potential' along the circumference of the marginal zone ring, with its highest point at the posterior margin. Since then, Eyal-Giladi and her colleagues have expanded on this suggestion (Eyal-Giladi and Spratt 1965; Azar and Eyal-Giladi 1979; Mitrani *et al.* 1983; Khaner *et al.* 1985; Khaner and Eyal-Giladi 1986, 1989; Eyal-Giladi and Khaner 1989); they found that rotation of the marginal zone about its antero-posterior axis resulted in reversal of the orientation of the streak, as Waddington had found after hypoblast rotation. They also suggested that the posterior marginal zone acts both to induce the primitive streak and to prevent the formation of secondary axes (Eyal-Giladi and Khaner 1989).

However, the epiblast of the early chick embryo can give rise to mesoderm cells even in the absence of both hypoblast and marginal zone (Mitrani and Shimoni 1990; Stern 1990). It therefore seems unlikely that the interaction between marginal zone or hypoblast and epiblast represents an *induction* of mesoderm by these tissues as defined by Gurdon (1987). It is clear, nevertheless, that the marginal zone is required both for the formation of a primitive streak with normal morphology which undergoes normal elongation and to prevent the formation of secondary axes (Eyal-Giladi and Khaner 1989; Khaner and Eyal-Giladi 1989), but not for the differentiation of mesodermal cells.

Induction of the Primitive Streak: A Revised View

Based in part on the above findings, I propose that primitive streak formation occurs in three stages, and that this axial structure contains cells derived from three different sources (see Figure 3). First, HNK-1-positive cells, which appear as a randomly distributed population in the epiblast, ingress individually or in small groups into the interior of the blastoderm, starting at stages XII-XIII. In the second stage (stages XIV-3), these cells accumulate at the posterior margin of the blastoderm, where they interact with the overlying epiblast. I suggest that this interaction is required in order to allow the cells of the epiblast of this region to undergo a process analogous to the *convergent extension* of the amphibian marginal zone cells (see contribution by Keller *et al.*, this volume) which leads to elongation of the primitive streak. When the cells of the epiblast portion of the streak, derived from the posterior margin, have reached maximum elongation along the future antero-posterior axis (stage 3^+), the third and final stage begins: a groove develops in the center of this structure, and this is accompanied by the dissolution of the basement membrane underlying this region of epiblast. This allows further cells (many of which may be HNK-1-negative) to be recruited into the primitive streak, where they will mix with the earlier HNK-1-positive cells and will contribute, together, to the embryonic mesoderm and definitive endoderm.

The above model does not explain three features: first, it fails to provide a mechanism for the convergence of the early HNK-1-positive epiblast cells towards the posterior marginal zone. Second, it does not account for the effects of hypoblast rotation on the orientation of the primitive streak. Finally, it does not explain why or how some cells are allocated as HNK-1-positive, early streak precursors while others are not. The following discussion addresses these three issues.

How do HNK-1-positive cells find the posterior marginal zone? Evidence for chemotaxis. Given that the epiblast contains a seemingly random mixture of HNK-1-positive and -negative cells, in which the former are the precursors of the primitive streak derived mesoderm and endoderm, how do these primitive streak cells find their destination in the embryo? One possibility is that one of the roles of the posterior marginal zone

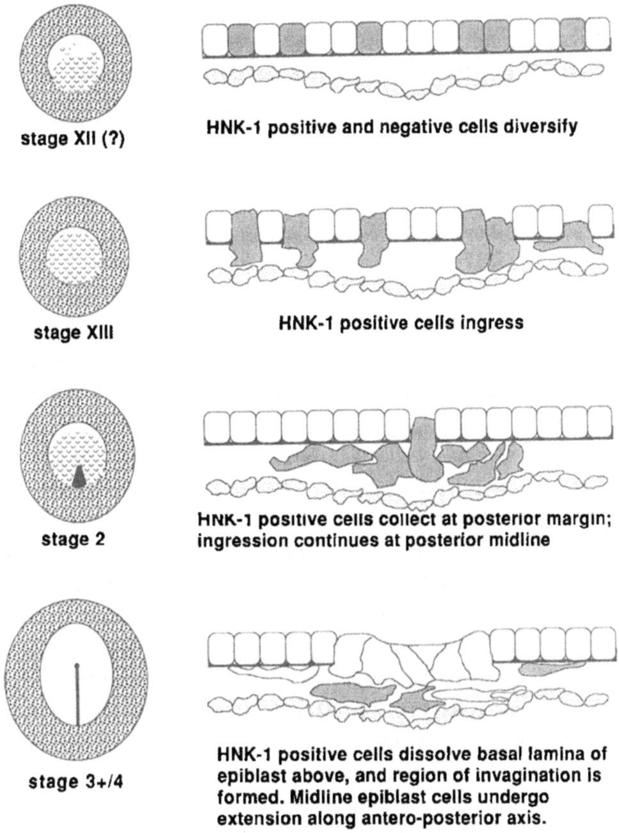

Figure 3. A revised view of chick gastrulation.

is to direct the migration of the HNK-1-positive cells to its proximity, where the primitive streak starts to form. To test this hypothesis, a collagen gel assay was used to find out if chemotaxis is involved (Jephcott and Stern, in preparation).

A small explant of central epiblast cells was placed in the center of a collagen gel matrix, and confronted on one side with an explant of posterior marginal tissue and on the other with a piece of anterior marginal tissue. It was found that cells of the central epiblast migrated towards the posterior marginal explant and to some extent away from the anterior explant. These results suggest that the posterior marginal zone emits a chemoattractant, whilst the anterior margin may emit a chemorepellant.

It is important to determine that it is the HNK-1-positive cells and not others that respond to these chemotactic signals. To address this, we pre-labeled the central epiblast explant using HNK-1-gold complex as described above, and examined the cultures using a confocal scanning laser microscope. It was found that HNK-1-positive cells displayed chemoattraction to the posterior margin, whilst HNK-1-negative cells spread evenly away from the explant.

In an attempt to establish the molecular nature of the chemotactic signals, we incorporated heparin immobilized on agarose beads into the collagen gel. This inhibited both the attraction of the posterior margin and the repulsion from the anterior/lateral margins. This result suggests that both signals are heparin-binding and provides an explanation for recent results of Mitrani and his colleagues (Mitrani and Shimoni 1990; Mitrani *et al.* 1990), in which it was suggested that heparin-binding molecules (*e.g.* basic-FGF, activin-A) play a role in the 'inductive' effects of the posterior marginal zone.

Role of the hypoblast in antero-posterior polarity of the embryo. The hypoblast may play a role in the process discussed above, and contribute to the guidance of HNK-1-positive pre-mesendodermal cells to the posterior marginal zone. It is possible, for example, that the chemoattractant and/or the chemorepellant bind to its extracellular matrix, as several heparin-binding factors are known to do. Rotation of the hypoblast, which will include the gradient of attractant/repellant, will therefore result in migration of HNK-1-positive cells in reversed direction. Thus, rotation of the hypoblast (gradient) and rotation of the marginal zone (source of these chemicals) will have the same effects.

The origins of cell diversity. The experiments that have been conducted to date on the early chick embryo cannot address the question of how the HNK-1-positive cells of the epiblast become different from their HNK-1-negative counterparts, because once they can be recognized they have already diversified. However, we might speculate on the type of mechanism that could lead to a constant proportion of cells, within a large population, to become different from the rest irrespective of their position in the embryo. In principle, there are two main ways in which this could be achieved: by local cell interactions, or by some intrinsic, pre-existing diversity among the cells.

Local cell interactions could lead to diversification, if, for example, random cells within the population became different at some particular time in development, and then these cells inhibited their neighbors from undergoing a similar change. By limiting the range of such an inhibition, the proportions of the two cell types can be controlled. This mechanism will also prevent cells of the divergent cell type from forming large clusters and would distribute them evenly in the embryo.

For the second mechanism we have to assume some parameter that differs between different cells of the population at the time of diversification. An obvious candidate is the cell cycle. Suppose, for example, that at some time in development, a signal is produced (perhaps by the marginal zone, or by the hypoblast, or even by all cells) which is an inductor of mesoderm, and which diffuses easily all over the embryo. However, competence to respond at any one time is restricted to a sub-population of cells, for example those that are in one particular phase of the cell cycle.

Within a large population of cells (the early blastoderm contains about 20,000 cells), both of the mechanisms proposed above will regulate perfectly for the size of the embryo, even within very large limits. I suggest that most vertebrate embryos will be found to contain restrictions to the competence of cells for inductive signals, and that it is these restrictions that are primarily responsible for embryonic size regulation during mesoderm induction.

ACKNOWLEDGEMENTS

My research on this subject is funded by the Wellcome Trust.

REFERENCES

Abo, T. and C.M. Balch. 1981. A differentiation antigen of human NK and K cells identified by a monoclonal antibody (HNK-1). *J. Immunol.* 127:1024-1029.

Azar, Y. and H. Eyal-Giladi. 1979. Marginal zone cells: The primitive streak-inducing component of the primary hypoblast in the chick. *J. Embryol. Exp. Morphol.* 52:79-88.

Azar, Y. and H. Eyal-Giladi. 1981. Interaction of epiblast and hypoblast in the formation of the primitive streak and the embryonic axis in the chick, as revealed by hypoblast rotation experiments. *J. Embryol. Exp. Morphol.* 61:133-144.

Bellairs, R. 1986. The primitive streak. *Anat. Embryol.* 174:1-14.

Bollensen, E. and M. Schachner. 1987. The peripheral myelin glycoprotein P_0 expresses the L2/HNK-1 and L3 carbohydrate structures shared by neural cell adhesion molecules. *Neurosci. Lett.* 82:77-82.

Bon, S., K. Meflah, F. Musset, J. Grassi, and J. Massoulie. 1987. An IgM monoclonal antibody, recognising a subset of acetylcholinesterase molecules from electric organs of *Electrophorus* and *Torpedo*, belongs to the HNK-1 anti-carbohydrate family. *J. Neurochem.* 49:1720-1731.

Bronner-Fraser, M. 1986. Analysis of the early stages of trunk neural crest cell migration in avian embryos using monoclonal antibody HNK-1. *Dev. Biol.* 115:44-55.

Bronner-Fraser, M. 1987. Perturbation of cranial neural crest migration by the HNK-1 antibody. *Dev. Biol.* 123:321-331.

Canning, D.R. and C.D. Stern. 1988. Changes in the expression of the carbohydrate epitope HNK-1 associated with mesoderm induction in the chick embryo. *Development* 104:643-656.

Chou, D.K., A.A. Ilyas, J.E. Evans, C. Costello, R.H. Quarles, and F.B. Jungalwala. 1986. Structure of sulfated glucuronyl glycolipids in the nervous system reacting with HNK-1 antibody and some IgM paraproteins in neuropathy. *J. Biol. Chem.* 261:11717-11725.

Cole, G.J. and M. Schachner. 1987. Localization of the L2 monoclonal antibody binding site on chicken neural cell adhesion molecule (NCAM) and evidence for its role in NCAM-mediated adhesion. *Neurosci. Lett.* 78:227-232.

Dennis, R.D., H. Antonicek, H. Weigandt, and M. Schachner. 1988. Detection of the L2/HNK-1 carbohydrate epitope on glycoproteins and acidic glycolipids of the insect *Calliphora vicina*. *J. Neurochem.* 51:1490-1496.

Eyal-Giladi, H. and O. Khaner. 1989. The chick's marginal zone and primitive streak formation. II. Quantification of the marginal zone's potencies: Temporal and spatial aspects. *Dev. Biol.* 49:321-337.

Eyal-Giladi, H. and S. Kochav. 1976. From cleavage to primitive streak formation: A complementary normal table and a new look at the first stages of the development of the chick. *Dev. Biol.* 49:321-337.

Eyal-Giladi, H. and N.T. Spratt. 1965. The embryo forming potencies of the young chick blastoderm. *J. Embryol. Exp. Morphol.* 13:267-273.

Fushiki, S. and M. Schachner. 1986. Immunocytological localisation of cell adhesion molecules L1 and N-CAM and the shared carbohydrate epitope L2 during development of the mouse neocortex. *Dev. Brain Res.* 24:153-167.

Ginsburg, M. and H. Eyal-Giladi. 1987. Primordial germ cells of the young chick blastoderm originate from the central zone of the area pellucida irrespective of the embryo-forming process. *Development* 101:209-220.

Gurdon, J.B. 1987. Embryonic induction – molecular prospects. *Development* 99:285-306.

Hamburger, V. and H.L. Hamilton. 1951. A series of normal stages in the development of the chick. *J. Morphol.* 88:49-92.

Hoffman, S., K.L. Crossin, and G.M. Edelman. 1988. Molecular forms, binding functions and developmental expression patterns of cytotactin and cytotactin binding proteoglycan, an interactive pair of extracellular matrix molecules. *J. Cell Biol.* 106:519-532.

Hoffman, S. and G. Edelman. 1987. A proteoglycan with HNK-1 antigenic determinants is a neuron associated ligand for cytotactin. *Proc. Natl. Acad. Sci. USA* 84:2523-2527.

Keilhauer, G., A. Faissner, and M. Schachner. 1985. Differential inhibition of neurone-neurone, neurone-astrocyte and astrocyte-astrocyte adhesion by L1, L2 and N-CAM antibodies. *Nature* 316:728-730.

Keller, R.E., J. Shih, and P. Wilson. 1991. Cell Motility, Control and Function of Convergence and Extension During Gastrulation in *Xenopus*. p. 101-120. *In: Gastrulation: Movements, Patterns, and Molecules*. R. Keller, W.H. Clark, Jr., F. Griffin (Eds.). Plenum Press, New York.

Khaner, O. and H. Eyal-Giladi. 1986. The embryo-forming potency of the posterior marginal zone in stages X through XII of the chick. *Dev. Biol.* 115:275-281.

Khaner, O. and H. Eyal-Giladi. 1989. The chick's marginal zone and primitive streak formation. I. Coordinative effect of induction and inhibition. *Dev. Biol.* 134:206-214.

Khaner, O., E. Mitrani, and H. Eyal-Giladi. 1985. Developmental potencies of area opaca and marginal zone areas of early chick blastoderms. *J. Embryol. Exp. Morphol.* 89:235-241.

Kruse, J., G. Keilhauer, A. Faissner, R. Timpl, and M. Schachner. 1985. The J1 glycoprotein—a novel nervous system cell adhesion molecule of the L2/HNK-1 family. *Nature* 316:146-148.

Kruse, J., R. Mailhammer, H. Wernecke, A. Faissner, I. Sommer, C. Goridis, and M. Schachner. 1984. Neural cell adhesion molecules and myelin-associated glycoprotein share a carbohydrate moiety recognised by monoclonal antibodies L2 and HNK-1. *Nature* 311:153-155.

Künemund, V., F.B. Jungalwala, G. Fischer, D.K.H. Chou, G. Keilhauer, and M. Schachner. 1988. The L2/HNK-1 carbohydrate of neural cell adhesion molecules is involved in cell interactions. *J. Cell Biol.* 106:213-223.

Lim, T.M., E.R. Lunn, R.J. Keynes, and C.D. Stern. 1987. The differing effects of occipital and trunk somites on neural development in the chick embryo. *Development* 100:525-534.

Lunn, E.R., J. Scourfield, R.J. Keynes, and C.D. Stern. 1987. The neural tube origin of ventral root sheath cells in the chick embryo. *Development* 101:247-254.

Mitrani, E., Y. Shimoni, and H. Eyal-Giladi. 1983. Nature of the hypoblastic influence on the chick embryo epiblast. *J. Embryol. Exp. Morphol.* 75:21-30.

Mitrani, E. and Y. Shimoni. 1990. Induction by soluble factors of organized axial structures in chick epiblasts. *Science* 247:1092-1094.

Mitrani, E., Y. Gruenbaum, H. Shohat, and T. Ziv. 1990. Fibroblast growth factor during mesoderm induction in the early chick embryo. *Development* 109:387-393.

Nieuwkoop, P.D., A.G. Johnen, and B. Albers. 1985. *The Epigenetic Nature of Early Chordate Development*. Cambridge University Press, Cambridge.

Pesheva, P., A.F. Horwitz, and M. Schachner. 1987. Integrin, the cell surface receptor for fibronectin and laminin, expresses the L2/HNK-1 and L3 carbohydrate structures shared by adhesion molecules. *Neurosci. Lett.* 83:303-306.

Rickmann, M., J.W. Fawcett, and R.J. Keynes. 1985. The migration of neural crest cells and the outgrowth of motor axons through the rostral half of the chick somite. *J. Embryol. Exp. Morphol.* 90:437-455.

Spratt, N.T. and H. Haas. 1960. Integrative mechanisms in development of early chick blastoderm. I. Regulative potentiality of separated parts. *J. Exp. Zool.* 145:97-137.

Stern, C.D. 1990. The marginal zone and its contribution to the hypoblast and primitive streak of the chick embryo. *Development* 109:667-682.

Stern, C.D. and D.R. Canning. 1988. Gastrulation in birds: A model system for the study of animal morphogenesis. *Experientia* 44:61-67.

Stern, C.D. and D.R. Canning. 1990. Origin of cells giving rise to mesoderm and endoderm in chick embryo. *Nature* 343:273-275.

Stern, C.D., S.M. Sisodiya, and R.J. Keynes. 1986. Interactions between neurites and somite cells: Inhibition and stimulation of axon outgrowth in the chick embryo. *J. Embryol. Exp. Morphol.* 91:209-226.

Tucker, G.C., H. Aoyama, M. Lipinski, T. Tursz, and J.P. Thiery. 1984. Identical reactivity of the monoclonal antibodies HNK-1 and NC-1: Conservation in vertebrates on cells derived from the neural primordium and on some leucocytes. *Cell Differ.* 14:223-230.

Tucker, G.C., M. Delarue, S. Zada, J.C. Boucaut, and J.P. Thiery. 1988. Expression of the HNK-1/NC-1 epitope in early vertebrate neurogenesis. *Cell Tissue Res.* 251:457-465.

Vakaet, L. 1984. The initiation of gastrular ingression in the chick blastoderm. *Am. Zool.* 24:555-562.

Waddington, C.H. 1933. Induction by the endoderm in birds. *Wilhelm Roux' Arch. Entwicklungsmech. Org.* 128:502-521.

CULTURE OF EMBRYONIC CELLS FOR ANALYSIS OF AMPHIBIAN AND MAMMALIAN EARLY EMBRYOGENESIS

NORIO NAKATSUJI AND KOICHIRO HASHIMOTO

Division of Developmental Biology
Meiji Institute of Health Science
540 Naruda, Odawara 250, Japan

* Present address: National Institute of Genetics
1111 Yata
Mishima 411, Japan

ABSTRACT

Culture of embryonic cells isolated from early embryos allows detailed analysis of the cell motility, interactions between cells, and between cells and the extracellular matrix (ECM). There is always the risk that cell behavior in culture might be artifacts not related to the behavior inside embryos. In many situations, one must use cells immediately after isolation from embryos, and try to prepare culture conditions similar to those inside the embryo.

Migration of the presumptive mesodermal cells during amphibian gastrulation was studied by such methods. It was first necessary to find culture conditions to allow mesodermal cells to behave in a manner similar to that found inside embryos. By the use of culture medium with a pH and calcium ion concentration found in embryos, as well as coating of the substratum with extracellular matrix (ECM) components, we were able to perform detailed analysis of cell behavior. Our analysis revealed an important role for the ECM fibril network containing fibronectin as a substratum which guided mesodermal cell migration by contact guidance.

A similar strategy was used for the analysis of mammalian gastrulation. Mesodermal cells isolated from the primitive-streak-stage mouse embryos attached to and migrated on the ECM produced by endothelial cells. This culture system revealed a deficiency in the mesodermal cells isolated from Brachyury (*T*) mutant embryos. Cell attachment to ECM was further analyzed by using antibodies against fibronectin or laminin and synthetic peptides, or by coating of the culture substratum with various ECM components. These studies indicated that both fibronectin and laminin play roles in adhesion and migration of the mesodermal cells.

For longer range analysis, however, it would be advantageous if isolated embryonic cells could be re-introduced into embryos after manipulations in culture. Genetical manipulation of the isolated cells would produce insight into the molecular basis of embryogenesis, but it requires a long term culture of the embryonic cells without losing the ability to re-integrate into the embryo. Mouse embryonic stem (ES) cells have such characteristics. We introduced a marker gene (*Lac Z*) into ES cells and analyzed early embryogenesis by producing chimaeric mouse embryos.

INTRODUCTION

During gastrulation in vertebrate embryos, coordinated cell migrations occur inside the multicellular mass of the embryo. At the same time, determination of cell fates is carried out by various cell-cell interactions and production of soluble factors. In order to analyze such processes at the cellular and molecular level, an obvious approach involves the isolation of embryonic cells and subsequent investigations of their behavior under defined culture conditions.

Since the isolation and culture of embryonic cells could produce cell behavioral artifacts, observations in culture should be reconciled to the events occurring inside normal embryos.

We have used this approach to analyze migration of mesodermal cells during amphibian and mammalian gastrulation. Since the studies on amphibian mesodermal cells have been published previously (for reviews, see Nakatsuji 1984; Johnson *et al.* 1990), we will only review the essential results. We will then describe our observations of migrating mesodermal cells in mouse embryos. Finally, we will describe some results using mouse embryonic stem (ES) cells, which represent an interesting extension of the methodology; involving not only the isolation and culture of embryonic cells but also the re-introduction of cultured cells into embryos.

MESODERMAL CELLS ISOLATED FROM AMPHIBIAN GASTRULAE

Culture Conditions for Mesodermal Cells

Presumptive mesodermal cells in amphibian gastrulae migrate from the blastopore region toward the animal pole by using the inner surface of the ectodermal layer as their substratum of translocation. Amphibian mesodermal cells extend lamellipodia and filopodia during migration (Nakatsuji *et al.* 1982) that are attached to ECM fibrils on the inner surface of the ectodermal layer (Nakatsuji *et al.* 1982; Nakatsuji and Johnson 1983a). These ECM fibrils contain fibronectin (Boucaut and Darribère 1983a,b; Nakatsuji *et al.* 1985; Darribère *et al.* 1985) and laminin (Nakatsuji *et al.* 1985; Darribère *et al.* 1986).

When dissociated mesodermal cells were cultured on glass or tissue-culture plastic surfaces, they did not translocate. We found that by lowering calcium ion concentration and raising the pH of the medium, we were able to get mesodermal cells to attach to and move on substrata coated with collagen and fetal bovine serum (Nakatsuji and Johnson 1982). Later studies showed that fibronectin is the major component responsible for the coating effects, because coating with purified fibronectin provided similar cell adhesion and movement (Nakatsuji 1986). Under appropriate culture conditions, the presumptive mesodermal cells translocated with filopodia and lamellipodia at a similar rate to that observed during normal gastrulation (Nakatsuji 1975).

Orientation of the Cell Movement

Transfer of the extracellular matrix (ECM) fibril network from the inner surface of the ectodermal layer of embryos to culture dish surfaces ("conditioning") provided adequate substrata for adhesion and movement of the mesodermal cells (Nakatsuji and Johnson 1983b; see also Boucaut *et al.*; this volume). Trajectories of cell locomotion were analyzed by time-lapse microcinematography and a computer morphometry program. We found that mesodermal cell translocation was oriented along the animal pole-blastopore (AP-BP) axis of the ectodermal layer that had conditioned the substratum. Moreover, directionality of the movement toward the animal pole was revealed by a statistical analysis in some experiments (Nakatsuji and Johnson 1983b). This observation was supported by related experiments (Shi *et al.* 1989), although in these experiments, directionality was displayed by a cell population rather than individual cells. Examination of the conditioned surface by scanning electron microscopy (SEM) after filming cell trajectories showed that the orientation of cell movement was closely related to the orientation of the ECM fibrils on the conditioned substratum (Nakatsuji and Johnson 1983b). More importantly, the ECM fibril network in gastrulae is aligned along the AP-BP axis, suggesting that there is contact guidance of mesodermal cell migration in the embryo (Nakatsuji *et al.* 1982; Nakatsuji and Johnson 1983b).

The ectodermal layer undergoes extension along the same axis during the epibolic movement of gastrulation. Such extension should create alignment along the AP-BP axis even in the previously random fibril network. If this hypothesis is true, conditioning of the culture substratum with an ectodermal layer stretched by mechanical tension would produce oriented cell movements along the axis of tension. If the distortion caused by stretching was large enough, it would erase the alignment along the AP-BP axis and create a new orientation. Such conditioning experiments (Nakatsuji and Johnson 1984) indeed demonstrated the new orientation of the cell movement along the axis of tension.

MESODERMAL CELLS ISOLATED FROM PRIMITIVE-STREAK-STAGE MOUSE EMBRYOS

Gastrulation in Mouse Embryos

During gastrulation in mammalian embryos, mesodermal cells ingress from the epiblast (embryonic ectodermal) layer into the space between the epiblast and visceral endodermal layers in the primitive streak region (Figure 1). The ingressed cells assume a polygonal shape, generate filopodia and lamellipodia and migrate from the primitive streak toward the anterior and lateral portions of the gastrula. These ingressed cells are destined to form not only mesodermal tissues (*e.g.* somites) but also may contribute to the definitive endoderm (Lawson *et al.* 1986; Kadokawa *et al.* 1987).

Cell migration in primitive-streak-stage mouse embryos was studied by time-lapse microcinematography of whole-embryos in culture. Mouse embryos at 7.0 dpc (days post coitum) were dissected from the uterus, cultured in a glass plate chamber, and photographed through Nomarski differential interference contrast optics (Nakatsuji *et al.* 1986). We found that individual mesodermal cells protruded filopodia and other cell processes and migrated actively in a zigzag fashion away from the primitive streak. The average rate of movement was 46 μm/h.

Morphological studies using SEM of fixed and fractured embryos (Figure 2) revealed that filopodia and other cell processes of migrating cells were attached to the basal lamina and ECM fibrils on the basal surface of the epiblast layer (Hashimoto and Nakatsuji 1989). Association between the migrating cells and the visceral endodermal layer seemed to be weak, because fracturing of the fixed and dried samples usually detached migrating cells from the endodermal layer (Figure 2).

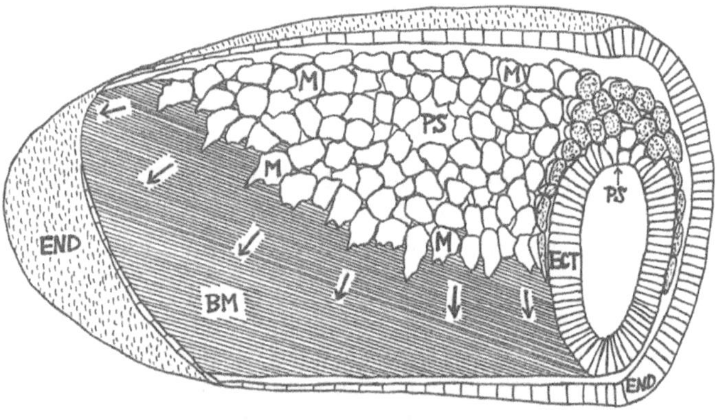

Figure 1. Diagram of gastrulation in the primitive-streak-stage mouse embryos. Mesodermal cells (**M**) migrate from the primitive streak (**PS**) toward anterior (distal) and lateral portions of the embryos through the space between the embryonic ectodermal (**ECT**) (epiblast) and visceral endodermal (**END**) layers. Migrating mesodermal cells are mostly attached to the basal lamina (**BM**) on the basal surface of the epiblast layer.

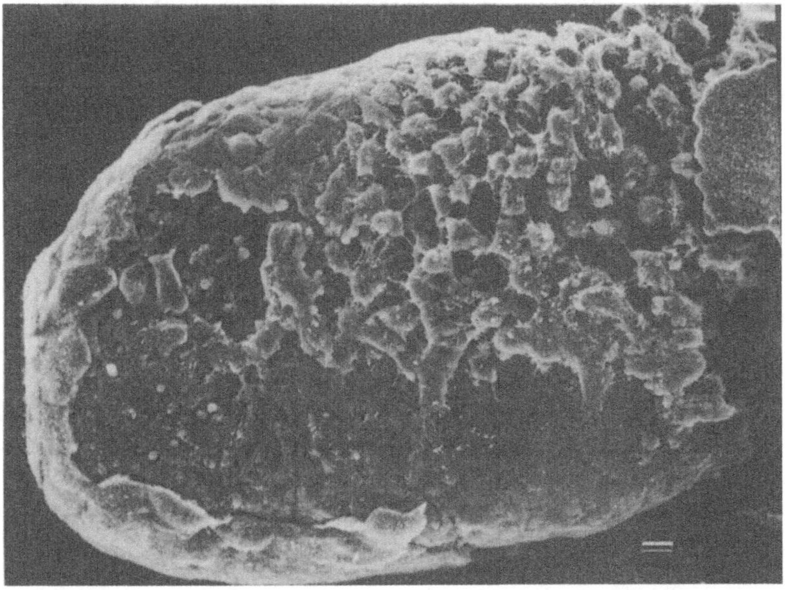

Figure 2. A SEM photograph of a 7.5 dpc mouse embryo, which was orientated to correspond to the diagram in Figure 1. A part of the visceral endodermal layer was removed to reveal the migrating mesodermal cells. Scale bar, 10 μm.

Culture Conditions for Mesodermal Cells

The mesodermal cell layer can be isolated from primitive-streak-stage mouse embryos by microsurgery and enzyme treatment (Snow 1978). We prepared micro-explants of small fragments of primitive-streak regions, and attempted to develop culture conditions in which these cells assumed shapes and migration rates similar to those *in situ* (Hashimoto *et al.* 1987). On glass or tissue-culture plastic, small groups of cells flattened and formed a coherent cell sheet. Individual cells did not migrate out of the explant. The mean rate of movement (28-30 μm/h) was smaller than that inside embryos (ca. 50 μm/h). Next we used commercially available ECM-coated dishes (International Bio-Technologies Ltd, Jerusalem, Israel) which were covered by bovine corneal endothelial cells produced basal lamina containing type III and IV collagen, fibronectin and laminin. On these ECM-coated dishes, mesodermal cells migrated out of the micro-explants as individual cells which were less flattened than on untreated glass or plastic. In addition, the mean rate of movement was similar to that inside the embryo. Thus, the ECM-coated substratum seemed to provide conditions similar to those inside embryos (Hashimoto *et al.* 1987).

We used this culture system to search for differences in cell behavior between the mesodermal cells isolated from the Brachyury (*T*) mutant embryos and normal embryos. In Brachyury (*T*) homozygotes (*T/T*), epiblast cells fail to move through the primitive streak region (Yanagisawa *et al.* 1981). When mesodermal cells from *T/T* or normal embryos were cultured on glass or tissue-culture plastic surface, the rates of movement were similar (both less than 50 μm/h) and there were no other recognizable differences. Culture on ECM, however, did reveal a difference between the mutant and normal cells (Hashimoto *et al.* 1987). Movement of mutant cells was slower than the normal cells (25-40 *vs.* 40-50 μm/h). The gene for *T* mutation has been cloned (Herrmann *et al.* 1990), and thus there will be more progress concerning the function of the gene and its primary effect.

Effects of Antibodies Against Fibronectin or Laminin

To investigate which components of the ECM are important for cell adhesion and migration, the behavior of cells on ECM-coated substrates was examined in the presence of ECM antibodies. Cells were isolated and cultured as previously described (Hashimoto *et al.* 1987) using Dulbecco's modified minimum essential medium (DMEM) supplemented with glucose (4.5 g/1) and 20% fetal bovine serum. The time-lapse videomicrography and analysis of cell trajectories were performed as described by Hashimoto *et al.* (1987). Addition of F(ab')2 fragments of an antibody against fibronectin caused drastic cell detachment and rounding-up of the mesodermal cells (Figure 3). Addition of anti-laminin antibody F(ab')2 fragments caused a degree of cell detachment, but the effect was much less dramatic than with the anti-fibronectin antibody. Measurement of the mean rate of cell translocation also confirmed a strong effect of the anti-fibronectin antibody and a weaker effect of the anti-laminin antibody (Table 1). Thus, fibronectin seems to be a major component responsible for cell migration on the ECM.

Methods: Small fragments of the mesodermal cell layer isolated from 7.5 dpc embryos were explanted on ECM-coated plastic tissue culture plates. After 15-18 hrs of culture, antibodies were added to the culture medium. Then, cell behavior was recorded with a time-lapse video system using 20× phase contrast objective lens for a period of 5 hrs. Analysis of cell trajectories and calculation of the mean rate was carried

Table 1. Effects of Antibodies on Cell Movement

F(ab')2 of antibodies	Concentration mg/ml	Rate of translocation mean ± S.D. μm/h (no.)	*t*-test
anti-fibronectin	0.5	13.1 ± 5.1 (24)	b1
	2.5	9.3 ± 6.2 (26)	d1
anti-laminin	0.5	38.7 ± 14.4 (23)	b2
	1.75	32.2 ± 18.0 (24)	d2
preimmune	0.5	52.8 ± 21.9 (20)	a
	2.5	48.6 ± 16.2 (21)	c
none		54.8 ± 18.0 (43)	

t-test; bl *vs.* a, $P<0.001$; b2 *vs.* a, $P < 0.025$; d1 *vs.* c, $P<0.001$; d2 *vs.* c, $P<0.005$.

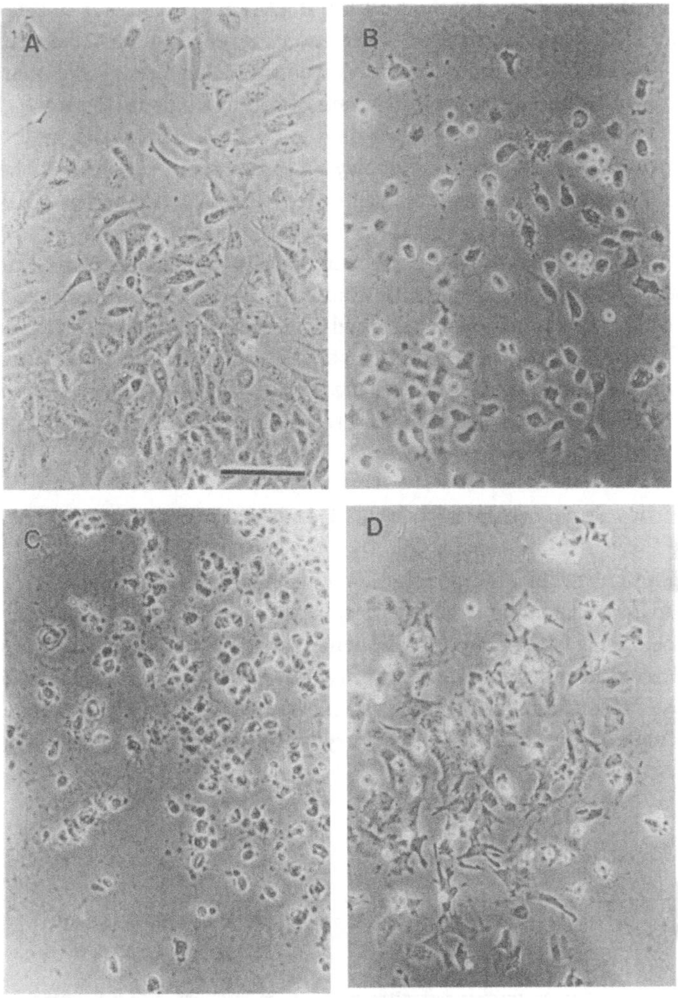

Figure 3. Phase contrast micrographs showing the effects of F(ab')2 fragments of antibodies on the mesodermal cells isolated from 7.5 dpc mouse embryos. **A.** Cells at 6 hrs after the addition of preimmune fragments (0.5 mg/ml). **B.** 4 hrs after the addition of anti-fibronectin fragments (0.5 mg/ml). **C.** 1 h after the addition of anti-fibronectin fragments (2.5 mg/ml). **D.** 1 h after the addition of anti-laminin fragments (1.75 mg/ml). Scale bar, 100 μm (A-D, same magnification).

out as described previously (Hashimoto *et al.* 1987). Rabbit anti-laminin antibody (Nakatsuji *et al.* 1985) was a gift of Dr. M. Hayashi. F(ab')2 fragments of rabbit anti-fibronectin and preimmune antibodies were purchased from Cappel (Cochranville, PA, USA).

Effects of Synthetic Peptides

We examined the effects of synthetic peptides related to the cell adhesion sites of fibronectin (for review, see Ruoslahti and Pierschbacher 1987; Dufour *et al.* 1988) in order to analyze which domain of fibronectin is important in adhesion of the mesodermal cells. Synthetic peptides containing the RGD sequence caused a reduction in flattening of the cells and a small decrease in rate of cell movement (Figure 4A,B Table 2). Peptides containing REDV sequence caused a much more dramatic detachment of cells and decrease in the rate of cell movement (Figure 4C,D, Table 2). Peptides unrelated to these domains had no effect. Thus, although the central cell-binding domain (RGD site) maybe playing some role in adhesion of the mesoderm cells to ECM, the type III connecting segment (IIICS) containing the REDV sequence is a more likely candidate for the cell adhesion site used by these cells. The IIICS region is also known to be the cell adhesion site of B16 melanoma cells (Humphries *et al.* 1986; 1987) and peripheral nervous system neurons (Humphries *et al.* 1988).

Methods: REDV peptides were obtained from Peninsula Laboratories (Belmont, CA, USA). RGDSPASSKP peptides were a gift of Dr. M. Hayashi. Other peptides were obtained from Bachem (Bubendorf, Switzerland). They were added to the culture medium in the same way as described in case of the antibodies. Succeeding analysis was also carried out using the same methods.

Coating of the Substratum with ECM Components

We coated substrata with purified ECM components to analyze the role of each individual ECM component in the adhesion and motility of mesodermal cells. Our culture system always contained fetal bovine serum (20%) in the medium, and thus, always contained fibronectin. On laminin coated petri dish surfaces mesodermal cells assumed less flattened shapes (Figure 5), and increased translocation rates to around 50 μm/h (Table 3). Coating with collagen type IV yielded cell adhesions but not high

Table 2. Effects of Synthetic Peptides

Peptides	Concentration mg/ml	Rate of translocation mean ± S.D. μm/h (no.)	*t*-test
REDV	1.0	27.6 ± 14.8 (23)	
	2.5	13.7 ± 5.5 (78)	b
RGDS	2.5	29.3 ± 12.1 (49)	b
GRGD	2.5	32.8 ± 13.5 (74)	b
RGDSPASSKP	1.0	40.1 ± 13.5 (22)	
	2.5	36.1 ± 20.4 (20)	c
CQDSETRTFY	2.5	46.8 ± 10.4 (25)	a
none		54.8 ± 18.0 (43)	

t-test; *b vs.* a, $P<0.001$; c *vs.* a, $P<0.05$.

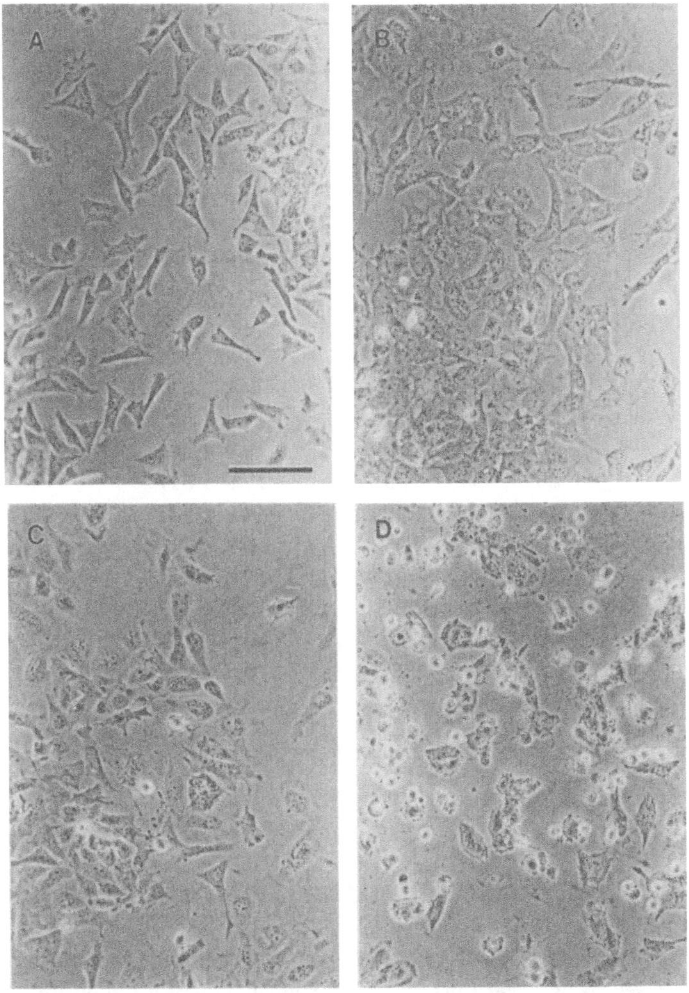

Figure 4. Phase contrast micrographs showing the effects of synthetic peptides GRGD (**A, B**) or REDV (**C, D**) added (2.5 mg/ml) to the culture medium of the mesodermal cells from 7.5 dpc mouse embryos. **A** and **C** show cells immediately after the addition of peptides; **B** and **D**, the same fields at 6 hrs after the addition. Scale bar, 100 μm (**A-D**, same magnification).

rates of cell movement. Combining laminin and collagen type IV did not enhance motility (Table 3).

Methods: Plastic petri dishes (diameter, 35 mm) were coated by applying 100 μl of an ECM component solution in DMEM onto a circular area approximately 10 mm in diameter. Coated dishes were incubated at 37°C for 12 hrs in a humidified CO2-incubator. They were rinsed twice with DMEM before use for culture of the mesodermal cells. Time-lapse video recording during a period from 15 to 24 hrs after start of the explant culture was analyzed as described before (Hashimoto *et al.* 1987).

Thus although fibronectin is the major ECM component responsible for mesodermal cell adhesion, it appears that fibronectin and laminin both function in cell migration.

Table 3. Effects of Substratum Coating with ECM Components

Coating	Amount μg/dish	Rate of translocation mean	±	S.D. μm/h (no.)	*t*-test
none		26.6	±	9.1 (47)	
ECM		51.7	±	19.3 (167)	a
fibronectin	10	38.3	±	21.6 (72)	c
laminin	5	63.2	±	17.2 (21)	
	10	50.2	±	19.4 (60)	b
	20	50.7	±	17.9 (31)	
	40	69.4	±	44.9 (23)	
collagen type IV	5	25.6	±	8.7 (21)	
	10	22.2	±	12.3 (21)	c
	20	21.0	±	8.3 (51)	
	40	25.4	±	10.0 (24)	
laminin	10 + 10	20.8	±	9.0 (21)	c
+ collage type IV	10 + 20	22.6	±	9.3 (23)	

t-test; b *vs.* a, not significant; c *vs.* a, $P<0.001$.

RE-INTRODUCTION OF CULTURED EMBRYONIC CELLS INTO EMBRYOS

The ability to re-introduce cultured, manipulated embryonic cells back into embryos promises to be a very useful tool for developmental biologists. A relatively simple example of this approach has been to label isolated embryonic cells with vital fluorescent dye so that their fate can be analyzed after re-introduction into embryos (Heasman *et al.* 1984). More extensive manipulation of embryonic cells could include genetic manipulation, such as gene transfection.

Such genetic manipulations, however, require relatively long culture periods for proliferation and selection of the transformed cells. These cells must then be able to re-incorporate into normal embryos. Therefore, embryonic cell lines that maintain their basic characteristics after prolonged culture must be utilized. Mouse ES cells meet these requirements (Evans and Kaufman 1981; Martin 1981; Robertson 1987).

We have established ES cell lines from the inner cell mass (ICM) of C57BL/6 strain mouse blastocysts. ES cells were transfected with a *E. coli* β-galactosidase gene (*Lac Z*) construct. One cell line (MS1-EL4) that expressed strong β-galactosidase activity ubiquitously at least until mid-gestation stages (9-10 dpc), was chosen (Suemori *et al.* 1990). Chimaeric embryos, made by injecting these MS1-EL4 cells into blastocysts have given various insights into histogenesis during early embryogenesis (Suemori *et al.* 1990; Kadokawa *et al.* 1990).

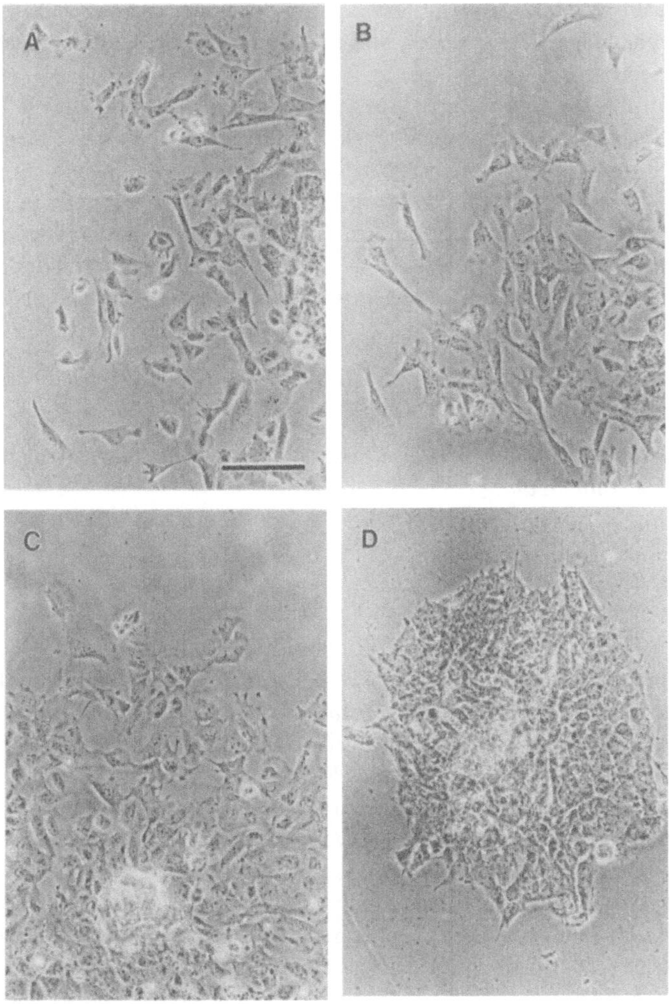

Figure 5. Phase contrast micrographs showing the effects of substratum coating with ECM (**A**) or 10 μg of laminin (**B**), fibronectin (**C**) or collagen type IV (**D**) on the mesodermal cells isolated from 7.5 dpc mouse embryos. Photographs were taken at 15 hrs after the start of explant culture. Scale bar, 100 μm (**A-D**, same magnification).

One example is the analysis of the fate of the MS1-EL4 cells in chimaeric blastocysts. ES cells contributed not only the ICM-derived tissues but also the trophectoderm-derived tissues (Suemori *et al.* 1990). Thus, ES cells seem to regain the potency to differentiate into all cell lineages of the preimplantation mouse embryos, although the ICM cells themselves lose potency to differentiate into the trophectoderm lineage during the blastocyst stage.

Another example of analysis is that the injected MS1-EL4 cells were present mostly as single cells intermingled with non-stained host cells in the embryonic ectoderm layer of the egg cylinder stage embryos (Kadokawa *et al.* 1990). At later stages (8-9 dpc), the stained cells formed small cohesive patches. These results indicate that there is extensive cell mixing in the embryonic ectoderm layer until the late primitive-streak-stage (late gastrula stage). These cells stop intermingling at later stages.

In the future, we plan to introduce other genes into MS1-EL4 cells or other ES cells expressing a cell marker gene. By making chimaeric embryos and mice, the effects of introduced genes can be analyzed in various tissues as local effects around the stained cells. Ectopic expression of genes important in embryogenesis probably cause overall embryonic death if they are introduced into the whole embryo by injection into fertilized eggs. The strategy using labelled ES cells and chimaera production may well give an alternative method to analyze the functions of such genes during embryogenesis.

ACKNOWLEDGEMENTS

Authors would like to thank Prof. Kurt E. Johnson for reading the manuscript and giving valuable comments.

REFERENCES

Boucaut, J.-C., T. Darribère, D. Shi, J.-F. Riou, K.E. Johnson, and M. Delarue. 1991. Amphibian Gastrulation: The Molecular Bases of Mesodermal Cell Migration in Urodel Embryos. p. 169-184. *In: Gastrulation: Movements, Patterns, and Molecules*. R. Keller, W.H. Clark, Jr., F. Griffin (Eds.). Plenum Press, New York.

Boucaut, J.-C. and T. Darribère. 1983a. Fibronectin in early amphibian embryos: Migrating mesodermal cells are in contact with a fibronectin-rich fibrillar matrix established prior to gastrulation. *Cell Tissue Res.* 234:135-145.

Boucaut, J.-C. and T. Darribère. 1983b. Presence of fibronectin during early embryogenesis in the amphibian *Pleurodeles waltl. Cell Differ.* 12:77-83.

Darribère, T., H. Boulekbache, D.L. Shi, and J.-C. Boucaut. 1985. Immunoelectron microscopic study of fibronectin in gastrulating amphibian embryos. *Cell Tissue Res.* 239:75-80.

Darribère, T., J.-F. Riou, D.L. Shi, M. Delarue, and J.-C. Boucaut. 1986. Synthesis and distribution of laminin related polypeptides in early amphibian embryos. *Cell Tissue Res.* 246:45-51.

Dufour, S., J.-L. Duband, A.R. Kornblihtt, and J.P. Thiery. 1988. The role of fibronectins in embryonic cell migration. *Trends Genet.* 4:198-203.

Hashimoto, K., H. Fujimoto, and N. Nakatsuji. 1987. An ECM substratum allows mouse mesodermal cells isolated from the primitive streak to exhibit motility similar to that inside embryo, and it reveals a deficiency in the *T/T* mutant cells. *Development* 100:587-598.

Hashimoto, K. and N. Nakatsuji. 1989. Formation of the primitive streak and mesoderm cells in mouse embryos–detailed scanning electron microscopical study. *Dev. Growth & Differ.* 31:209-218.

Heasman, J., C.C. Wylie, P. Hausen, and J.C. Smith. 1984. Fates and states of determination of single vegetal pole blastomeres of *Xenopus laevis. Cell* 37:185-194.

Herrmann, B.G., S. Labeit, A. Poustka, T.R. King, and H. Lehrach. 1990. Cloning of the *T* gene required in mesoderm formation in the mouse. *Nature* 343:617-622.

Humphries, M.J., S.K. Akiyama, A. Komoriya, K. Olden, and K.M. Yamada. 1986. Identification of an alternatively spliced site in human plasma fibronectin that mediates cell type-specific adhesion. *J. Cell Biol.* 103:2637-2647.

Humphries, M.J., S.K. Akiyama, A. Komoriya, K. Olden, and K.M. Yamada. 1988. Neurite extension of chicken peripheral nervous system neurons on fibronectin: Relative importance of specific adhesion sites in the central cell-binding domain and the alternatively spliced type III connecting segment. *J. Cell Biol.* 106:1289-1297.

Humphries, M.J., A. Komoriya, S.K. Akiyama, K. Olden, and K.M. Yamada. 1987. Identification of two distinct regions of the type III connecting segment of human plasma fibronectin that promote cell type-specific adhesion. *J. Biol. Chem.* 262: 6886-6892.

Johnson, K.E., N. Nakatsuji, and J.-C. Boucaut. 1990. Extracellular matrix control of cell migration during amphibian gastrulation. p. 349-374. *In: Cytoplasmic Organization Systems: Primers in Developmental Biology*. G.M. Malacinski (Ed.). McGraw-Hill, New York.

Kadokawa, Y., Y. Kato, and G. Eguchi. 1987. Cell lineage analysis of the primitive and visceral endoderm of mouse embryos cultured *in vitro*. *Cell Differ.* 21:69-76.

Kadokawa, Y., H. Suemori, and N. Nakatsuji. 1990. Cell lineage analyses of epithelia and blood vessels in chimeric mouse embryos by use of an embryonic stem cell line expressing the β-galactosidase gene. *Cell Differ. Dev.* 29:187-194.

Lawson, K.A., J.J. Meneses, and R.A. Pedersen. 1986. Cell fate and cell lineage in the endoderm of the presomite mouse embryo, studied with an intracellular tracer. *Dev. Biol.* 115:325-339.

Martin, G.R. 1981. Isolation of a pluripotent cell line from early mouse embryos cultured in medium conditioned by *Teratocarcinoma* stem cells. *Proc. Natl. Acad. Sci. USA* 78:7634-7638.

Nakatsuji, N. 1975. Studies on the gastrulation of amphibian embryos: Cell movement during gastrulation in *Xenopus laevis* embryos. *Wilhelm Roux's Arch. Dev. Biol.* 178:1-14.

Nakatsuji, N. 1984. Cell locomotion and contact guidance in amphibian gastrulation. *Am. Zool.* 24:615-627.

Nakatsuji, N. 1986. Presumptive mesodermal cells from *Xenopus laevis* gastrulae attach to and migrate on substrata coated with fibronectin or laminin. *J. Cell Sci.* 86:109-118.

Nakatsuji, N., A. Gould, and K.E. Johnson. 1982. Movement and guidance of migrating mesodermal cells in *Ambystoma maculatum* gastrulae. *J. Cell Sci.* 56:207-222.

Nakatsuji, N., K. Hashimoto, and M. Hayashi. 1985. Laminin fibrils in newt gastrulae visualized by immunofluorescent staining. *Dev. Growth & Differ.* 27:639-643.

Nakatsuji, N. and K.E. Johnson. 1982. Cell locomotion *in vitro* by *Xenopus laevis* gastrula mesodermal cells. *Cell Motil.* 2:149-161.

Nakatsuji, N. and K.E. Johnson. 1983a. Comparative study of extracellular fibrils on the ectodermal layer in gastrulae of five amphibian species. *J. Cell Sci.* 59:61-70.

Nakatsuji, N. and K.E. Johnson. 1983b. Conditioning of a culture substratum by the ectodermal layer promotes attachment and oriented locomotion by amphibian gastrula mesodermal cells. *J. Cell Sci.* 59:43-60.

Nakatsuji, N. and K.E. Johnson. 1984. Experimental manipulation of a contact guidance system in amphibian gastrulation by mechanical tension. *Nature* 307:453-455.

Nakatsuji, N., M.A. Smolira, and C.C. Wylie. 1985. Fibronectin visualized by scanning electron microscopy immunocytochemistry on the substratum for cell migration in *Xenopus laevis* gastrulae. *Dev. Biol.* 107:264-268.

Nakatsuji, N., M.H.L. Snow, and C.C. Wylie. 1986. Cinemicrographic study of the cell movement in the primitive-streak-stage mouse embryo. *J. Embryol. Exp. Morphol.* 96:99-109.

Robertson, E.J. 1987. Embryo-derived stem cell lines. p. 71-112. *In: Teratocarcinomas and Embryonic Stem Cells: A Practical Approach.* E.J. Robertson (Ed.). IRL Press, Oxford.

Ruoslahti, E. and M.D. Pierschbacher. 1987. New perspectives in cell adhesion: RGD and integrins. *Science* 238:491-497.

Shi, D.-L., T. Darribère, K.E. Johnson, and J.-C. Boucaut. 1989. Initiation of mesodermal cell migration and spreading relative to gastrulation in the urodele amphibian *Pleurodeles waltl. Development* 105:351-363.

Snow, M.H.L. 1978. Techniques for separating early embryonic tissues. p. 167-178. *In: Methods in Mammalian Reproduction.* J.C. Daniel, Jr. (Ed.). Academic Press, New York.

Suemori, H., Y. Kadokawa, K. Goto, I. Araki, H. Kondoh, and N. Nakatsuji. 1990. A mouse embryonic stem cell line showing pluripotency of differentiation in early embryos and ubiquitous β-galactosidase expression. *Cell Differ. Dev.* 29:181-186.

Yanagisawa, K.O., H. Fujimoto, and H. Urushibara. 1981. Effects of the Brachyury (*T*) mutation on morphogenetic movement in the mouse embryo. *Dev. Biol.* 87:242-248.

ORGANIZING THE *XENOPUS* ORGANIZER

John Gerhart, Tabitha Doniach, and Ronald Stewart

Department of Molecular and Cell Biology
University of California,
Berkeley CA 94720

INTRODUCTION

In chordates in general and in *Xenopus* in particular, the animal's body axis is organized largely as a dorsal collinear array of notochord, nerve cord, and somites, divided antero-posteriorly into head, trunk, and tail. Ventral aspects of the body axis are less obvious than dorsal, but include an antero-posterior ordering of branchial and gut regions, and heart and blood islands. This organization arises at the gastrula stage of development, through the action of a small population of cells collectively called the "gastrula organizer" which we wish to discuss in terms of its formation and function.

So great is the increase in the embryo's anatomical complexity during gastrulation that it is probably inaccurate to say the body axis exists before this event. Just before gastrulation the embryo (the late blastula stage of 20,000 cells; Gerhart and Keller 1986) contains six cellularized regions arranged according to the egg's initial animal-vegetal and bilateral organization (Figure 1): the animal cap which is divided into (1) a narrow dorsal part (DAC) and (2) a wide lateroventral (LVAC) part; the marginal zone which is divided into (3) a narrow dorsal part (DMZ) which is the site of the gastrula organizer and (4) a wide lateroventral part (LVMZ); and a vegetal base which is divided into (5) a narrow dorsal part (DV) and (6) a wide lateroventral part

Gastrulation, Edited by R. Keller *et al.*
Plenum Press, New York, 1991

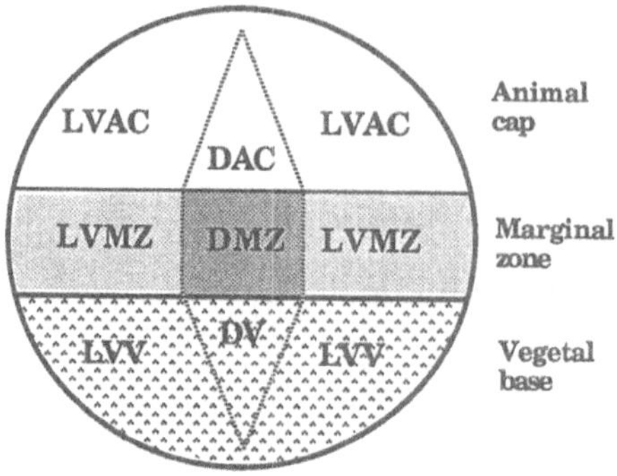

Figure 1. Schematic view of the surface organization of the *Xenopus* late blastula. The animal pole is at the top and vegetal pole at the bottom. The white area is animal cap. It is divided into a narrow dorsal region (DAC) and a wide lateroventral region (LVAC) which is continuous around the back of the figure. The grey region is the marginal zone. It is divided into a narrow dorsal region (DMZ) and a wide lateroventral region (LVMZ) extending around the back. The pocked region is the vegetal base divided into a narrow dorsal region (DV) and a wide lateroventral region (LVV) extending around the back. The DAC, DMZ, and DV lie in the "dorsal sector" of cytoplasm activated in the egg during cortical rotation in the first cell cycle. A vertical meridian bisecting these regions coincides with the dorsal midline of the embryo. The DV region contains at least part of the early/mid blastula organizer. The DMZ contains the late blastula and gastrula organizers; notice the neighboring regions surrounding the organizer.

(LVV). Two of these regions (DAC and DV; see later) are dispensable at the late blastula stage, and so four regions can suffice for normal development of the fully patterned body axis: the organizer and three regions surrounding it.

The processes generating the body axis are to be found in the behaviors and interactions of these regions. Even though the egg is extensively remodeled in gastrulation, the displacements of parts are reproducible from embryo to embryo, and so it is possible to make good fate maps at pre-gastrula stages (Dale and Slack 1987a; Moody 1987). In *Xenopus* the marginal zone cells take the most active role in shaping the embryo, especially those of the DMZ and organizer (Keller and Danilchik 1988; Keller *et al.* this volume). Internal movements of DMZ cells begin an hour or more before the blastopore is externally visible, in what is defined as still the late blastula stage. DMZ cells are initially unique as a population in their capacity for repacking by convergent extension. This behavior collects cells toward the dorsal midline and lengthens the midline in the antero-posterior dimension (see Keller *et al.*, this volume). Bottle cells of the blastopore first contract and invaginate at the vegetal border of the DMZ, and this "dorsal lip" identifies the position of the DMZ in the marginal zone and defines the start of gastrulation. LVMZ cells begin their gastrulation movements 30-60 min after the DMZ has begun, and their capacity for convergent extension is weak or negligible, except in regions close to the DMZ.

Different developmental stages have different organizers distinguishable by their location and mode of action, though all share the ability to initiate the formation of a secondary body axis if transplanted from a donor embryo into the ventral midline of a host embryo of the same age. The host embryo contributes most of the cells of the secondary axis; the fate of these cells is altered by inductive signals from the organizer. Such activity was first discovered by Spemann and Mangold for the gastrula organizer (or "Spemann's organizer"; Spemann 1938), and since that time it has become recognized that gastrula organizer cells have three functions: 1) initiating and executing convergent extension, 2) signaling neighboring cells of the LVMZ to form somites and kidneys (a process known as "dorsalization" of the mesoderm; Yamada 1950; Smith and Slack 1983), and 3) signaling neighboring cells of the animal cap to form the neural plate (the well known process of "neural induction"). Organizer cells exercise these functions when organizing a primary or secondary axis. Descendants of gastrula organizer cells eventually differentiate as the notochord and perhaps also as head mesoderm. They are surrounded by the somites and neural tube, both composed of the differentiated progeny of cells initially induced by the organizer.

Our questions include: 1) how organized is the early gastrula organizer, 2) does it obtain this organization by self-organization or by interactions with various neighboring regions, 3) how does it impose organization on neighboring tissues, and 4) how does it coordinate morphogenesis and inductive patterning so that a normal body axis results? Our partial answers to these questions come largely from the study of abnormal embryos, ones with modified organizer function.

This review is a companion to another written recently (Gerhart *et al.* 1989) in which the reader can find a more complete discussion of pre-gastrula developmental events.

THE RANGE OF EMBRYONIC AXIAL PHENOTYPES

With suitable experimental intervention, the *Xenopus* egg can develop into a wide range of ventralized and dorsalized embryos, all abnormal but related by a systematic modification of the body axis. These may comprise a canonical set based on organizer size, as will be discussed. The antero-posterior and dorsoventral dimensions of the body axis are affected in concert, and the practice of calling these aberrant forms "ventralized" or dorsalized" gives only a partial description. In *Drosophila*, by contrast, the two dimensions of the body axis seem to develop independently (Nüsslein-Volhard *et al.* 1987). In *Xenopus*, identical axial modifications can be produced by widely different treatments applied at different stages, while different grades of the set arise following different doses of a single treatment (Table 1). The effects are not "treatment or agent specific" but instead reflect the embryo's response to inhibition or stimulation of different but interdependent developmental processes, the perturbation of any one of which is sufficient to upset the whole of dorsoventral and antero-posterior development. So frequently encountered are members of this set that, if mutants of early *Xenopus* development were available, many would surely have these phenotypes.

Ventralized Embryos

The limit-form lacks dorsal structures, and hence it lacks the main elements of the body axis (Grant and Wacaster 1972; Malacinski *et al.* 1977; Scharf and Gerhart 1983). It probably also lacks some ventral structures. It is identical to the "belly piece" obtained by Spemann (1938) from the ventral half of a pregastrula embryo, separated from its dorsal half and allowed to develop in isolation. Three germ layers form and differentiate into a short gut, abundant red blood cells, coelomic mesentery, mesenchyme, and ciliated epidermis, these cell types and tissue organization occupying the ventral midline of the normal tadpole. Although cleavage of the ventralized egg is normal, gastrulation is not. It occurs at the correct distance from the vegetal pole (40-60°) but is symmetrical and delayed, as if the entire blastopore circumference

Table 1. Treatments Leading to Ventralized or Dorsalized Development of *Xenopus laevis*. Numbers in parentheses indicate literature references given at the foot of the Table. DV and DMZ refer to "dorsal vegetal region" and "dorsal marginal zone", respectively (see Figure 1). UV indicates UV irradiation; MT, microtubules; MBT, mid-blastula transition. NT refers to "normalized time" of the first cell cycle.

Stage treated:	Ventralization	Dorsalization
Stage VI oocyte	vegetal UV (1)	
First cell cycle	prevent cortical rotation by deploymerizing MT with cold, pressure, UV, or nocodazole before 0.45 NT (2)	70% D_20 before 0.3 NT (3)
Cleavage stages	Remove all or part of DV, or add back insufficient dorsal vegetal cells to a fully ventralized embryo (4, 5)	0.1-0.3 M lithium ion before MBT (6,7)
Late Blastula	Remove DMZ, all or part, or add back an insufficient amount of DMZ to a fully ventralized embryo (8).	graft in extra DMZs (9)
Gastrula	reduce convergent extension with polysulfonated agents injected in the blastocoel (10)	

References: (1) Elinson and Pasceri 1989; (2) Scharf and Gerhart 1983; (3) Scharf *et al.* 1989; (4) Gimlich and Gerhart 1984; (5) Gimlich 1986; (6) Kao and Elinson 1988; (7) Yamaguchi and Shinagawa 1989; (8) Stewart and Gerhart 1990; (9) Stewart 1990; (10) Gerhart *et al.* 1989.

forms a ventrolateral lip. There is little or no convergent extension, as if the marginal zone is wholly an LVMZ. No organizer can be found by tissue transplantation, and neurulation does not occur. Ventral development, though, can proceed in the absence of an organizer.

Intermediate ventralized larvae (Figure 2) possess a partial body axis in which anatomical structures, especially dorsal ones, are missing from the anterior end of the axis. For example, the head can be absent while the trunk and tail are present, or the head and trunk absent while the tail is present. At gastrulation, these have an "intermediate dorsal lip", a blastopore sector appearing slightly earlier than the rest but

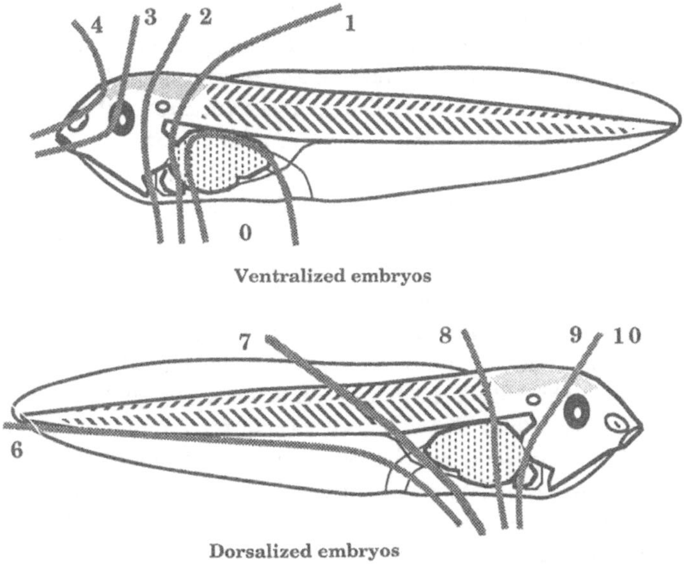

Figure 2. The range of dorsalized and ventralized phenotypes in *Xenopus*. Two normal tadpoles (stage 41) are shown in side view. Each is scored as grade 5 on the "dorsoanterior index" ("DAI"; Kao and Elinson 1988) which spans the range from complete ventralization (grade 0) to complete dorsalization (grade 10). Ventralized anatomies can be estimated by looking at the upper tadpole. A grade 4 embryo lacks those structures to the left of the line marked with a 4. A grade 3 embryo lacks those structures to the left of line 3. Grades 2 and 1 can be approximated in the same way using lines 2 and 1. A grade 0 embryo is fully ventralized. It lacks all structures outside the region contained under line 0. Dorsalized anatomies can be estimated by looking at the lower tadpole. A grade 6 embryo lacks those structures under the line marked with a 6. A grade 7 embryo lacks those structures to the left of line 7. Grades 8 and 9 can be approximated in the same way using lines 8 and 9. A grade 10 embryo is like a grade 9 case, but fully cylindrically symmetric.

not as early as a normal dorsal lip. The final differentiated pattern of the axis can be roughly predicted from the time of appearance of the blastopore and the vigor of gastrulation (especially convergent extension) at the prospective dorsal position. All aspects of weakened DMZ behavior seem monotonically correlated with the final axial pattern.

We think we now understand, in part, why this might be so (Gerhart *et al.* 1989). The cylindrically symmetric unfertilized egg has a latent capacity for dorsal, ventral, or lateral development at every meridian. After fertilization the egg generates a bilateral symmetry wherein meridians now differ. This occurs by "cortical rotation", a 30° displacement of the egg's cortical layer of cytoplasm over the deeper contents. The sperm, via its aster, can bias the direction of rotation, and therefore the meridian passing through the sperm entry point often coincides with the prospective ventral midline of the embryo (reviewed in Gerhart *et al.* 1989). The opposite meridian coincides with the midline of the neural plate, the prospective dorsal midline. Dorsal axial development depends critically on cortical rotation, which is the first in a series of indispensable developmental processes. Many ventralizing agents work simply by preventing rotation, as a result of their depolymerization of microtubules which are needed in the vegetal cortex as parallel tracks for cytoplasmic movement (Vincent and Gerhart 1987; Vincent *et al.* 1987; Elinson and Rowning 1988). If rotation is completely blocked, the final embryo is fully ventralized. The more rotation to occur (up to the 30° amount), the more anteriorly developed are dorsal structures of the body axis. We have been particularly interested in this quantitative relationship.

Rotation probably activates the egg's cytoplasm in a narrow sector (60-90° wide) straddling the prospective dorsal midline, in both the animal and vegetal hemispheres (a "dorsal sector"), enabling this region, when cellularized in the blastula stages, to use its latent capacity to initiate dorsal development, while other regions remain unactivated and take ventral development as their default option. By the early blastula stage, cells cleaved from the activated sector express special properties important for organizer formation. Cells of the dorsal sector of the vegetal hemisphere have the ability to induce nearby cells of the animal hemisphere to form the gastrula organizer (Gimlich and Gerhart 1984). These "dorsal vegetal cells" (also collectively called the dorsal vegetal base (Figure 1) or "Nieuwkoop center", to recall the fact that Pieter Nieuwkoop (1973) was the discoverer of vegetal inductions) do not contribute progeny to the organizer or to dorsal axial structures. Due to their inductive abilities, these cells have been called the "organizer of the organizer" (Nakamura 1978), although as we will soon see, the gastrula organizer needs additional steps of organization. Vegetal induction may be the second of the series of pre-gastrula dorsalizing developmental processes.

At the same time, cells originally cleaved from the animal and equatorial levels of the dorsal sector of the rotated egg may contribute progeny directly to the organizer, and they also may induce neighboring cells to join the organizer (Gimlich 1986; Kageura 1990). The relative importance of animal, equatorial, and vegetal activated regions in normal development is not known, although each is sufficient on its own to establish a body axis in transplantation situations.

We will illustrate the induction of the gastrula organizer by dorsal vegetal cells. If these cells of an early blastula are transplanted to the vegetal hemisphere of a ventralized recipient embryo of the same stage, the rescue of the host can first be seen at the early gastrula stage when it forms an early dorsal lip at the blastopore position closest to the graft (Gimlich and Gerhart 1984). This is the site of the DMZ and gastrula organizer induced by graft cells and composed wholly of host cells. Dorsal vegetal cells meet the definition of an organizer since grafted hosts form nearly complete body axes. These organizer cells are not located in the DMZ and don't function at the gastrula stage. They comprise an early blastula organizer. Any animal hemisphere cell can respond to the vegetal inducer and become part of a functional gastrula organizer, provided the cell is brought into range of the vegetal inducer by transplantation into the marginal zone.

At the late blastula stage, vegetal inductive cells lose their activity and a "late blastula organizer", located in the DMZ, now begins to function. As discussed later, there is probably at least one more step in the formation of the true gastrula organizer which directs convergent extension and signals neighboring cells of the LVMZ and animal cap to develop as particular dorsal mesodermal and neural tissues, respectively. This formation and functioning of the gastrula organizer is the third and final major process in patterning the body axis. Together, the three processes constitute a causal chain of events. If one step fails, subsequent steps fail and axial development, especially of dorsal parts, is precluded. Ventral development is left as the default option.

Regarding axial development, this interpretation suggests that cortical rotation, vegetal induction, and organizer function may be quantitatively related. That is, intermediate amounts of rotation or induction should result in intermediate amounts of organizer function. This has been tested recently by a set of operations on embryos at the late blastula stage to remove part or all of the late blastula organizer, that is, to artificially reduce its function by reducing its size (Stewart and Gerhart 1990). This stage was chosen in order to complete the operations before internal gastrulation movements begin. When as little as a 60° sector is removed spanning the prospective dorsal midline, the rest of the embryo develops as a ventralized embryo. This sector contains the organizer. Without it, gastrulation is delayed and weak; neurulation does not occur; and only ventral cell types and tissues differentiate. Thus, the late blastula embryo is unspecified for dorsal development in a full 300° of its circumference, even though the LVMZ would normally contribute most of the somites and kidney, and the animal cap would provide neural structures, if the organizer had been present. Dorsal development depends wholly on the interaction of lateroventral cells with cells of the gastrula organizer at times after the late blastula stage. In reciprocal experiments, when the 60° organizer sector is restored to a ventralized embryo at the late blastula stage, a nearly complete body axis is developed. Thus, cortical rotation and vegetal induction must at least establish an organizer if dorsal development is to ensue.

Now, when a quarter or half of the organizer is left after surgery (*i.e.* a 15° or 30° width), the remaining embryo develops a partial body axis (Stewart and Gerhart 1990). The more organizer, the more anterior is the truncation level of the body axis. The entire ventralization series (Figure 2) can be reproduced by surgically reducing the width of the organizer at the late blastula stage. It is therefore plausible that reduced cortical rotation in the first cell cycle, due to various inhibitors, or reduced vegetal induction in the early blastula, does indeed lead to a smaller or less functional population of organizer cells in the late blastula.

We will return later to the questions of why the size of the organizer might determine the anterior completeness of the body axis, and why organizer-dependent morphogenesis is related to larval axial pattern.

Dorsalized Embryos

These can be produced by a variety of treatments differing from those for ventralization, and the phenotypes stand in opposition to those of the ventralized series. The cylindrically symmetric limit-form develops an early circular blastopore and convergent extension is exaggerated, as if the entire circumference of the blastopore forms a dorsal lip and the entire marginal zone consists of DMZ cells. Convergent extension is sometimes so vigorous that the marginal zone extrudes as a tube, giving the embryo a light bulb shape. In other cases convergent extension is less vigorous and this tube is not extended. Neurulation occurs around the entire blastopore or along the tube. The cylindrically symmetric embryo differentiates a large central heart and core or tube of notochord, and a ring of eye pigment and sucker material around its circumference. Muscle and blood are absent. A portion of gut may be present. The embryo can be approximately described as developing excessive dorsal anterior structures (although anatomically its heart would be called a ventral structure) and insufficient posterior ventral structures. As a more accurate description, though unwieldy for use as a scoring scale, the dorsalized limit-form can be said to overdevelop anatomical structures derived from positions close to the prospective dorsal midline and organizer on the early gastrula fate map.

Intermediate forms (see Figure 2) include embryos which have the external appearance of lacking tails or tails and trunks. That is, posterior and ventral parts are successively deleted from the axis leaving dorsal anterior ones to fill the circumference of the embryo. As in the ventralized series, gastrulation events are coordinately affected. Embryos with the most dorsalized type of gastrulation give the most dorsalized final differentiations.

How is this series related to the ventralized series? These embryos probably have unusually large gastrula organizers and DMZ regions, and correspondingly diminished LVMZ regions. In the extreme case, the DMZ and organizer must encircle the early gastrula, and the LVMZ must be absent. Kao and Elinson (1988) have in fact shown that organizer grafts can be successfully obtained from all positions of the circumference of the marginal zone of such gastrulae.

Two effective agents for producing dorsalized embryos are D_20 (Scharf *et al.* 1989) and lithium ion (Kao and Elinson 1988), which probably work in different ways to cause an enlargement of the organizer. D_20 is effective only when applied in the first third of the first cell cycle, well before cortical rotation begins. The dorsalizing effect of D_20 is negated if "ventralizing agents" are subsequently applied to prevent formation of the vegetal microtubules normally needed for cortical rotation. D_20 perhaps allows a microtubule-dependent activation of cytoplasm in many or all sectors of the egg's circumference so that an expanded Nieuwkoop center develops, which later induces an expanded late blastula organizer. There may also be an expanded region of activated animal hemisphere cytoplasm, from which cells could contribute directly to the organizer.

Lithium, on the other hand, can be applied long after the first cell cycle, in fact at cleavage stages up to the midblastula transition (4000 cells) (Yamaguchi and Shinagawa 1989). Unlike D_20, lithium can even act on ventralized embryos, that is, ones blocked entirely in cortical rotation and thought to lack a Nieuwkoop center. This is taken to indicate that lithium ion causes animal hemisphere cells to alter their response to vegetal signals and to form an organizer even when receiving a lateroventral vegetal signal, the kind that normally would induce only the formation of the LVMZ (Slack *et al.* 1989). An oversized organizer would form because the lateroventral vegetal signal doesn't require cortical rotation for its release and occurs around most or all of the blastula circumference. It is not known whether lithium also activates the vegetal hemisphere to form a wider Nieuwkoop center with its unique inductive signal, even without cortical rotation. But in any event, these embryos acquire an abnormally wide organizer by the gastrula stage.

Taking the ventralized and dorsalized series together, we see that the egg has access to an impressively wide range of developmental possibilities. It can establish dorsal or ventral development at all meridians, any fraction of meridians, or no meridian. Normally a cortical rotation of 30° insures a balance of dorsal and ventral development by limiting the late blastula organizer width to 60° (see later), and by ensuring the development of just one organizer, hence one dorsal midline. Grades of the ventralized-dorsalized range of phenotypes are quantitatively related by the single variable of gastrula organizer size (width), no matter whether this is altered by reduced cortical rotation, reduced or enhanced activation of egg cytoplasm, addition or removal of cells of the Nieuwkoop center, or the abnormal responsiveness of animal hemisphere cells to vegetal signals.

MORPHOGENESIS AND PATTERN

As discussed above, aspects of morphogenesis in the early gastrula predict the completeness of the pattern of the final body axis: ventralized embryos lack early bottle cells of the DMZ and convergent extension is reduced; dorsalized embryos have an excess of early bottle cells and convergent extension increases. The length of the archenteron gives a rough measure of the amount of convergent extension, and this is very short in ventralized limit forms and can be very long (the length of the extruded tube) in dorsalized limit forms.

We have been interested in this correlation of early morphogenesis and late pattern, and have looked to see if interference with morphogenesis during gastrulation can result in ventralization. This is indeed the case, as shown by the use of polysulfonated dyes (trypan blue, trypan red, congo red) or the related colorless compound suramine (Figure 3). These agents cause acephaly in mammalian embryos and block sea urchin gastrulation (Beck and Lloyd 1966). For the best demonstration of these effects, the agent is injected into the blastocoel at different times during gastrulation at high doses that stop further convergent extension (Waddington and Perry 1956; Danilchik *et al.* 1991). In many cases, the blastopore still succeeds in completing its closure. As shown in Figure 4, the archenteron length found at the end of gastrulation (stage 13) correlates with the time of injection of the agent, and with the level of truncation of the final body axis of the larva (stage 41). When gastrulation is blocked early (stage 10+), the archenteron never extends beyond its initial short length, and the final embryo is largely ventralized in its pattern. When gastrulation is blocked midway (stage 11), the archenteron remains approximately half its length, and the embryo develops a tail and part of the trunk, but fails to develop head structures. When the agent is injected still later as morphogenesis nears completion (stage 12.5), the final embryo lacks only the most anterior head parts, and by the end of gastrulation the embryo is refractory to the agents. Thus, morphogenesis correlates with pattern even when inhibitors are applied during the period of organizer function. Furthermore, lithium-dorsalized embryos can be treated at the gastrula stage with injections with suramin or trypan, which partially restore normal development by causing the embryos to develop more ventral and posterior axial parts than they otherwise would. Clearly the pattern of the body axis remains in an undetermined or labile condition well into the gastrula stage (Doniach *et al.* 1991).

As a generalization, then, interference with gastrulation, particularly with convergent extension, leads to a reduction of anterior dorsal development and an increase of posterior ventral development, even though the embryo may have had a normal organizer at the start of gastrulation. The final effect on pattern is the same as found for embryos in which the organizer was surgically reduced in size, or was modified as the result of reduced cortical rotation or vegetal induction.

Perhaps these polysulfonated agents directly reduce the gastrula organizer's function, and this is effectively the same as reducing its size. Perhaps they also interfere with the antero-posterior differentiation of the gastrula organizer (the

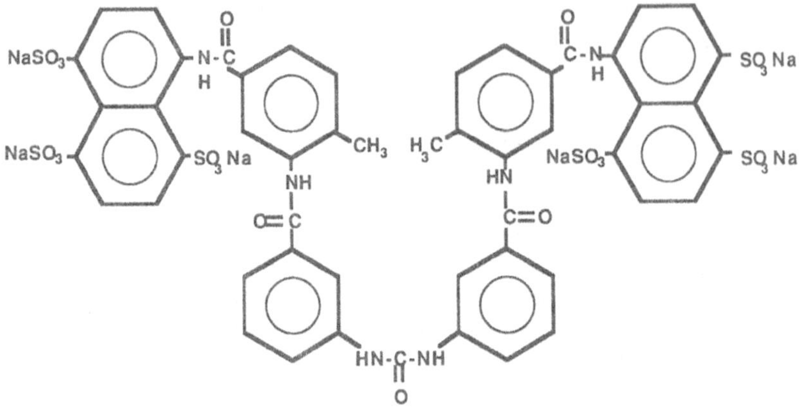

Figure 3. Suramin (Germanin), a polysulfonated agent effective at blocking *Xenopus* gastrulation movements. The same effects are obtained with trypan blue, trypan red, and congo red, polysulfonated agents also known to block sea urchin gastrulation and to cause acephaly and microcephaly in mice (Beck and Lloyd 1966). Suramin is known to dissociate ligand:receptor complexes (Peacock *et al.* 1988), to block growth factor effects on cultured cells (Coffey *et al.* 1987), and to inhibit convergent extension in Keller sandwiches (Danilchik *et al.* 1991).

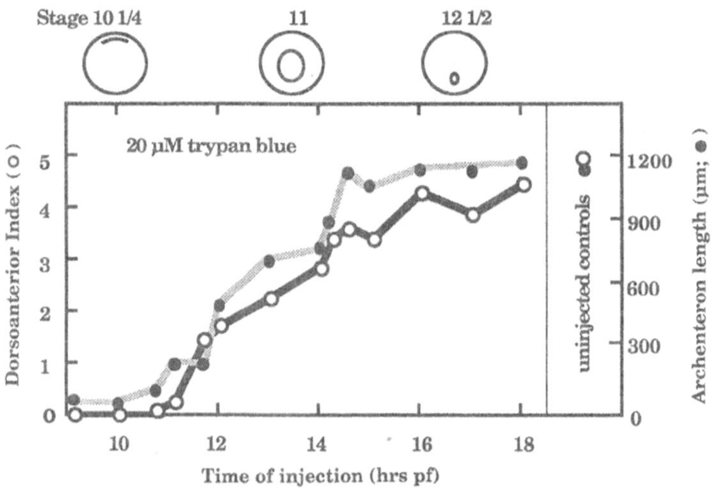

Figure 4. The correlation of morphogenesis and axial pattern in *Xenopus*. The horizontal axis gives the time post fertilization (pf) at which trypan blue was injected into the blastocoel. Approximately 30 nl of 200 μM trypan blue was injected to achieve a final blastocoel concentration of 20 μM. Gastrulation starts at 10 hr and ends at approximately 18 hr (19°C). The vertical axis shows axial pattern (dorsoanterior index; dark line) and convergent extension morphogenesis (archenteron length; stippled line), which increase together during gastrulation. Twenty embryos were injected at each time point and 10 of these were allowed to develop to stage 41 to assess their final pattern, and the remaining 10 were fixed when control embryos had reached the end of gastrulation (stage 12.5-13; 18 hr). After fixation, these were sectioned sagittally and the length of the archenteron was measured on a video monitor. Uninjected stage 13 controls are shown in the right panel of the graph.

positioning of its neuralizing and mesodermalizing parts), which may not be completed until late in gastrulation. These possibilities are hard to distinguish, especially if an aspect of the gastrula organizer's function is to complete its spatial differentiation. The agents observably inhibit the organizer's function in convergent extension (this can also be demonstrated in cultured DMZ explants ["Keller sandwiches"]; Danilchik *et al.* 1991) and may also interfere with the gastrula organizer's acquisition of its two inductive functions of neuralization and dorsalization. Preliminary results indicate that this is so: the embryo is injected with these agents at the late blastula stage, and the gastrula organizer is then removed at the early gastrula stage and tested for its neural inducing ability in a "blastocoel implantation assay", that is, by insertion into the blastocoel of a normal early gastrula. A normal gastrula organizer in such an assay induces the formation of a secondary head in the ventral midline of the host, but the organizer from a trypan-treated donor organizes only a tail or a part of the trunk as well (Doniach *et al.* 1991). As discussed later, this is the behavior of a late dorsal lip (the "trunk-tail organizer") or of parts of the LVMZ.

We conclude that convergent extension, one of the three activities of the gastrula organizer, correlates with the anterior completeness of the pattern of the body axis. This holds no matter how an agent works and when it is applied, as long as its consequence is the reduction of morphogenesis during gastrulation. Probably, though, morphogenesis does not diminish without a coordinate loss of the inductive activities of the gastrula organizer, perhaps due to an incomplete formation of the gastrula organizer, and so the true correlation may hold between organizer function and final embryonic pattern. To go further with these questions, we must look into the last step of the formation of the gastrula organizer.

PATTERN IN THE LATE BLASTULA ORGANIZER

We are still left with the question of why the anterior end of the body axis is missing when organizer size and/or function is reduced. This is paradoxical since according to the fate map, prospective anterior regions invaginate first and lead morphogenesis. And reciprocally, why is the body axis truncated from the posterior end when the organizer is too large (excessive function)? Why not the other way around, or why doesn't the entire axis proportionately enlarge or diminish? To continue this inquiry, we should ask about antero-posterior and dorsoventral organization within the organizer itself at the late blastula stage, just before gastrulation begins. Is the organizer already fully organized as a miniature mosaic of the final body axis, and are certain parts of the organizer absent from ventralized and dorsalized embryos?

Mediolateral pattern: To test for this pattern, we bisect two late blastulae 15° from the prospective dorsal midline, one to the right and one to the left side, and combine the two halves, each with one-quarter organizer (Stewart and Gerhart 1990). This recombinant embryo contains just the lateral extremities of the organizer and no medial part. In a reciprocal experiment a center wedge of cells is cut 15° to either side of the midline and is inserted in a slit in a ventralized late blastula, to give an embryo with medial but not lateral parts of the organizer. Both types of recombinants develop equally well, giving embryos with tail, trunk, and posterior head parts (DAI 2.7; see Figure 2). Thus, the edges and center of the organizer seem qualitatively the same. Either there is no mediolateral pattern in the organizer or it readily regulates its pattern after surgery.

Animal-vegetal pattern: To test for organization in this second direction, which may relate to the antero-posterior dimension of the body axis, late blastulae are cut equatorially at a level just below the blastocoel floor (Stewart 1990, 1991). In control experiments with lineage labelled halves, the cut was shown to fall in the marginal zone approximately at the boundary between the prospective head mesoderm and prospective notochord territories. If the late blastula is proportioned in the same way as the early gastrula, the cut would have separated the marginal zone into animal and vegetal halves.

We then combine a normal half with the reciprocal half from a ventralized embryo of the same age, that is, one lacking a late blastula organizer. When a normal animal half and a ventralized vegetal half are together, the recombinant occasionally develops a tail and posterior parts of the trunk (DAI 1.1), and it lacks the notochord. Axial development is poor, only slightly better than that of ventralized controls alone (DAI 0.7). Even though the normal animal half contains most or all of the notochord territory, and even though this territory possesses organizer activity in the gastrula stage, it is not autonomous for any aspect of dorsal development at this stage, any more so than is the animal half of a ventralized embryo. The upper part of the DMZ must gain its special gastrula organizer properties after the late blastula stage.

Reciprocal recombinants (normal vegetal half plus ventralized animal half) develop body axes with tail, trunk, and heads complete to the midbrain level on average (DAI 3.8). Larvae contain well formed notochords, of which much of the tissue derives from the animal half which had originated from a ventralized donor unable to form notochord on its own. The vegetal half clearly contains the organizer region at this stage (the "late blastula organizer"), and this induces a notochord territory (a gastrula organizer) directly above it, even in ventralized tissue. Experiments by us (Stewart 1990) and others (Boterenbrod and Nieuwkoop 1973) have shown that dorsal vegetal cells no longer harbor organizer activity at the late blastula stage. The lower half of the DMZ is the sole locus of organizer activity at this stage shortly before gastrulation movements begin. This rectangular region, which is approximately 5-10 cells high and 10-20 cells wide, and 2-4 cell layers deep, corresponds to head mesoderm and head endoderm territories of the fate map.

Not all of these latter recombinates develop exactly the same, though all resulting larvae are members of the standard ventralized series. Some variation may be due to the variable level of the cut since it is impossible to hit the head mesoderm/notochord boundary every time. Since recombinants probably receive different amounts of the lower DMZ organizer, due to inadvertent cuts at different animal-vegetal levels, we suggest once again that axial completeness depends on the quantity of late blastula organizer material, not its quality. We detect no antero-posterior organization within the late blastula organizer, and if it does exist, it must be convoluted in relation to the fate map, for we are obliged to preferentially remove posterior parts of the organizer (according to the fate map) and yet anterior parts of the axis are ultimately absent. The only apparent organization in the late blastula DMZ is the difference of upper and lower parts; the upper part is not yet induced to have organizer activity and autonomy for notochord differentiation. Similar conclusions apply to the DMZ of the *Cynops* (urodela) gastrula (Kaneda 1981; Kaneda and Suzuki 1983).

Taken together, the results indicate that the late blastula organizer either lacks internal organization or can re-establish it after surgical perturbations in the animal-vegetal and mediolateral directions. We prefer to think that at this stage it is just a spatially uniform population of cells. Favoring this is the finding that animal cap explants, uniformly treated with XTC-MIF, can act as organizers when grafted to the ventral marginal zone of a late blastula (Cooke, this volume). If true, the definitive axial pattern of the larva would then develop, not from a mosaic miniature contained in the organizer, but from interactions of the organizer with asymmetrically arranged neighboring cell populations of the animal cap and LVMZ. The organizer's context may provide organization.

PATTERN IN THE LATE GASTRULA ORGANIZER

As gastrulation proceeds, the gastrula organizer develops organization regarding its morphogenetic and inductive functions. For inductive functions, this is concluded from the work of O. Mangold (1933) who dissected the archenteron roof (the locus of the prospective head mesoderm and notochord after the DMZ has involuted) of an early neurula stage urodele embryo, cut it into antero-posterior pieces, and implanted individual pieces into the blastocoel of an early gastrula. Each piece induced in the ventral midline of its host a portion of secondary axis similar in character to the axial level from which it originated. In related experiments, Spemann (1931) removed the dorsal lip of urodele gastrulae of different stages and showed in blastocoel implantation assays that early lips induced secondary heads, mid-gastrula lips induced heads or trunk-like parts, and late lips induced secondary tails. Thus, by the end of gastrulation the gastrula organizer has become antero-posteriorly differentiated into head, trunk, and tail inducing subregions.

According to various interpretations of these observations, the gastrula organizer releases just two types of inducers, neuralizing and mesodermalizing types, in different amounts from its different parts. Proposed inducer distributions differ, for example: 1) inducers may be arrayed in a double gradient with neural induction strongest from the head mesoderm and/or the anterior end of the notochord territory and mesoderm induction strongest from the posterior end of the notochord territory (Hamburger 1988), or 2) neural induction may be uniformly strong throughout all territories, but mesoderm induction would be strongest from the posterior end of the notochord territory, grading off anteriorly (Saxen and Toivonen 1961), or 3) neural induction may be exclusive to the head mesoderm territory and mesoderm induction exclusive to the notochord territory (Nieuwkoop 1985). These models seek to explain antero-posterior patterning of the neural plate by the pattern of inductive signals in the underlying mesoderm (the now involuted and extended gastrula organizer), and all consider the mesodermalizing inducer to posteriorize or "caudalize" inductive effects of the neuralizing inducer as a means to get posterior neural structures. The neural signal alone is proposed to induce only forebrain and midbrain parts, and the caudalizing signal alone is proposed to induce no neural structures but only mesoderm. Thus, the posterior part of the late gastrula organizer is seen as a "caudalizing center" acting on the "anteriorizing center" located anteriorly or ubiquitously in the organizer.

These proposals of course relate to our consideration of the axial patterns of ventralized and dorsalized larvae, at least of the neural aspect of their patterns. Perhaps late gastrula organizers of modified embryos differ in their antero-posterior organization of inducers from that of normal embryos. Ventralized cases could be seen as ones resulting from too strong a caudalizing center or too weak a neuralizing center, and dorsalized cases could be seen as the opposite. As discussed later, it may be important to understand the organizer differentiation process.

These proposals are not comprehensive for they ignore the organizer's induction of the LVMZ to form somites and kidney ("dorsalization" of the LVMZ) and the early induction of the notochord territory of the DMZ by the head mesodermal part, this also being an important step of mesoderm induction. Dorsalization could be added to the proposals simply by saying that the mesodermalizing regions dorsalize the LVMZ as well as posteriorize neural induction. Neural and mesodermal patterning are coordinately modified in our abnormal cases. It is unknown whether dorsalization requires further organization of the gastrula organizer.

There are also morphogenetic differences within the gastrula organizer, as discussed by Winklbauer *et al.* and by Keller *et al.* (this volume). Prospective head mesoderm cells at the early gastrula stage act as a spreading, crawling population, whereas prospective notochord cells converge and extend as a population. Convergent extension activity is graded in an antero-posterior direction in the prospective notochord territory, being maximal at the posterior limit, which becomes the "limit of involution" of the gastrula, and the mesoderm/neural boundary thereafter. At the midgastrula stage, this behavior is also acquired by cells of the NIMZ (the dorsal animal cap above the organizer) and grades off with distance from the limit of involution. As reviewed by Keller *et al.* (this volume), there may also be differences between superficial and deep layers of the organizer, the former being an inductive source at early gastrula stages and the latter inductive later. Thus, there is ample evidence that even though the late blastula organizer may start gastrulation with little or no organization, the gastrula organizer manages to acquire regional differences of morphogenetic and inductive activities in the course of gastrulation.

ORGANIZING THE ORGANIZER

Little is known of the processes by which the gastrula organizer gains antero-posterior (and perhaps also dorsoventral) organization during gastrulation, and so our comments must be speculative. We suggest two major classes of proposals: 1) those in which the organizer "self-organizes", that is, it develops pattern within itself without the need for interactions with neighboring tissues, and 2) those in which such interactions are required. In discussing the proposals, we will assume that, as a starting state, the largely unpatterned late blastula DMZ consists of uniform late blastula organizer cells in the lower part (the prospective head mesoderm region) and uniform uninduced cells in the upper part (the prospective notochord region; the soon-to-be gastrula organizer), and that, as a final state, the well-patterned DMZ of the late gastrula contains in its upper part a gastrula organizer with a preponderance of neuralizing cells with weak convergent extension activity (close to the prospective head mesoderm), and a preponderance of mesodermalizing cells with strong convergent extension activity (close to the limit of involution).

DMZ Self-Organization

The first two proposals given below involve spreading signals within the DMZ and gain plausibility in light of the fact that, as will be discussed later, the neural plate can be patterned in part by signals spreading to it from the DMZ in the plane of the neural tissue. Perhaps similar spreading signals establish pattern in the DMZ itself. The third proposal involves interactions of parts within the DMZ.

By cellular age differences: According to this proposal, inductive signals would propagate slowly as a wave from the head mesoderm region through the prospective notochord region. Cells of this latter region would receive signals at different times depending on their distance from the head mesoderm. The older the cell at the time of signal reception, the more strongly it responds by initiating convergent extension, mesoderm induction (to become a caudalizing center), and mesoderm differentiation, up to a certain age after which these responses diminish and the NIMZ-relevant responses are given. The limit of involution would be set by cells of the most responsive age. The part close to the head mesoderm region, which receives signals very early, would become strongly neural inductive. This sort of proposal could be tested with "age chimeras".

By gradients: Prospective head mesoderm cells would initially serve as a source of a diffusible or transported inducer, and the nearest prospective notochord cells would later release neuralizing inductors by virtue of their exposure to the highest concentration of this inducer. Cells of the prospective notochord region would receive less of the initial inducer the greater their distance from the source and would become less neuralizing. Those receiving a critical concentration of inducer would give the maximal convergent extension response (thus setting the limit of involution) and the maximal mesodermalizing inductive response (thus setting the site of the caudalizing center).

New contacts made between DMZ parts: The lower DMZ may begin gastrulation as a mesoderm inducing region, acting on the posterior DMZ to mesodermalize it, and then in time the lower part (and the nearest cells of the upper part) may gain neural inducing activity as the result of involuting and making contact with the not-yet-involuted and not-yet-mesodermalized DMZ cells near the limit of involution (Kaneda and Suzuki 1983; Suzuki *et al.* 1984). This could be tested in situations where involution is prevented, as discussed later.

Context Dependence of the DMZ

According to this proposal, the true gastrula organizer is formed in the notochord territory during gastrulation. These cells are initially uniform and uninduced, but reside in a structured environment. They could receive signals from cells of lower DMZ which is the late blastula organizer (the prospective head mesoderm territory), and from the LVMZ to gain an antero-posterior pattern of release of neuralizing and mesodermalizing inducers.

A rather detailed proposal will be given for use in explaining how ventralized and dorsalized embryos, which differ only in the size of their late blastula organizers, might come to develop differently patterned organizers by the late gastrula stage. At the start of gastrulation, the lower DMZ cells (the late blastula organizer; prospective head mesoderm cells) would act as strong mesoderm inducers and would gradually mesodermalize the upper DMZ by signals spreading in the plane of the tissue (giving it autonomy for notochord development), and would confer mesodermalizing activity on the upper DMZ. The lower DMZ might then continue to induce the upper DMZ, slowly converting it to neural inducing cells. If this induction continued unopposed, cells of the upper DMZ would eventually all become neuralizing and would induce only anterior neural structures and differentiate as anterior dorsal mesoderm (anterior notochord). As an aside we suggest that the lower DMZ (prospective head mesoderm) never becomes a neuralizing inductor itself, but acts on the upper DMZ to make it a neuralizing source. It "anteriorizes" the upper DMZ.

The LVMZ might act as the true "caudalizing center" in the sense that it might release signals inhibiting the conversion of the upper DMZ from a mesodermalizing to neuralizing state. Since the anteriorizing signal would come from the lower DMZ and the caudalizing signal from the LVMZ, the upper DMZ would come to have neuralizing and mesodermalizing activities graded in opposite directions, the former strongest at the lower (anterior) end of the notochord territory and the latter strongest at the upper (posterior) end of this territory. Without LVMZ interactions, no part of the upper DMZ would become capable of inducing posterior neural structures or posterior mesoderm, or of differentiating as posterior mesoderm.

In applying this proposal to dorsalized and ventralized embryos, we assume that their final body patterns differ because their organizers, which only differed by size before gastrulation, now come to differ in their neuralizing and mesodermalizing proportions during gastrulation. D_2O and lithium treated embryos would not be able to induce posterior axial structures because there is no LVMZ to posteriorize the DMZ. DMZ cells would become entirely neuralizing, inducing anterior dorsal neural parts and differentiating as head mesoderm and anterior notochord. Yet if the DMZ of an early dorsalized gastrula is transplanted into the ventral marginal zone of a normal recipient (Kao and Elinson 1988), it will organize a full axis because it becomes posteriorized by host LVMZ cells. The model also might explain why a small and/or less functional organizer would organize only posterior parts of the embryonic axis, as in a partially ventralized embryo, since it would receive less signal from the lower DMZ cells to offset LVMZ signals; it would therefore not convert effectively to neuralizing cells and anterior notochord differentiation. Finally, in a fully ventralized gastrula, there are no lower DMZ cells to mesodermalize the upper DMZ and subsequently make it a neuralizing source, and so the LVMZ has no target for its caudalizing action. Aspects of this model could be tested by appropriate tissue grafts.

HOW DOES THE ORGANIZER ORGANIZE NEIGHBORING TISSUES?

We will address mostly the question of neural induction and omit somite mesoderm induction. Research of the 1930's showed that an explanted piece of archenteron roof (late organizer DMZ) in urodeles can induce neural differentiation in gastrula ectoderm wrapped around it (Holtfreter and Hamburger 1955). Also, if inductive tissue is implanted in the blastocoel of an early gastrula, it will induce secondary neural structures in the ventral midline of the host. This has been called "transplanar induction" or "vertical induction" by various authors (Kaneda and Suzuki 1983; Savage and Phillips 1989; Dixon and Kintner 1989), since the signal passes between two tissue layers. It has been long assumed that this is the exclusive route of transmission of a rough pattern of neuralizing and mesodermalizing inductive signals needed to map out the antero-posterior and dorsoventral dimensions of the neural plate.

More recently Phillips (London *et al.* 1988; Savage and Phillips 1989; Phillips 1991), following a suggestion of Spemann (1938), raised the possibility of "planar inductions" or "tangential inductions" (Kaneda 1981) that is, neuralizing inductions passing in the plane of the tissue. In collaboration with Phillips, we have pursued this by the study of neural induction in Keller open-faced explants (see Keller *et al.*, this volume, for similar studies). In this preparation, the DMZ and prospective neural tissue (the "NIMZ" and nearby animal cap tissue) are taken as a single piece from an early gastrula and held flat to prevent the DMZ from turning under the neural part. Thus, vertical signals are precluded but planar ones are not, since the DMZ and neural region maintain edge-on contact. In this configuration, prospective neural tissue does in fact express NCAM RNA (neural specific; Dixon and Kintner 1989). NCAM protein can be detected by antibody staining in regions almost as far from the from the mesoderm/neural boundary as would be found in the neural plate of a normal neurula embryo (Phillips and Doniach, submitted). More recently it has been found that neural tissue of the Keller explant forms a near-normal antero-posterior pattern of expression of the homeobox genes en (engrailed; for midbrain/hindbrain), Xlhbox1 (anterior spinal cord) and Xlhbox6 (posterior spinal cord) (Doniach, in preparation). Thus an elaborate antero-posterior neural pattern can form despite the fact that only planar signals are involved. This is a surprise to us. We don't know whether this planar signalling requires concentration gradients, or signal propagation and age-dependent responses, and whether neuralizing and mesodermalizing (caudalizing) signals, which must both cross the limit of involution even though the neuralizing source is presumably much farther away, pass in the plane simultaneously or in order.

Still, vertical signals can be demonstrated to exist in *Xenopus*, for Hemmati-Brivanlou *et al.* (1990) have shown that pieces of anterior notochord, taken from an early neurula, serve as effective inducers of neural tissue (NCAM expression) and en expression in gastrula ectoderm (taken from a ventralized donor). Here, at least the initiating signal must be vertical, although subsequent signals could pass within the neuroectoderm plane.

Why might there be two pathways for signals? Keller and Danilchik (1988) have noted that although their explant undergoes convergent extension in the posterior part of the neural region, as does the posterior part of the normal neural plate, it fails to display other aspects of neural morphogenesis such as tube formation and columnarization of cells. One could say that these morphogenetic aspects depend on vertical signals. In the normal embryo vertical signals might allow the nascent and labile antero-posterior pattern in the neuroectoderm (due mostly to planar signals) to be matched with the related labile pattern forming in the underlying dorsal mesoderm (due to planar signals from the lower DMZ and LVMZ), to bring the two patterns into register in the body axis.

CONCLUSIONS

Different Organizers. The organizer is usually defined as a population of cells capable of inductively establishing a body axis when transplanted into the appropriate host environment. Three cell populations may possess organizer activity at different times in early *Xenopus* development. At the early and mid blastula stages, dorsal cells of the vegetal hemisphere have this activity, as the consequence of having received maternal cytoplasmic materials activated by cortical rotation in the first cell cycle. Early blastula cells of the dorsal part of the animal hemisphere may also receive activated cytoplasm and display organizer activity upon transplantation. At the late blastula stage, dorsal vegetal cells lose activity, and cells of the lower DMZ possess it. This is the late blastula organizer. Some of these cells gain activity by an induction from the dorsal vegetal cells, while others may acquire it directly from activated animal hemisphere cytoplasm they contain. As gastrulation proceeds, cells of the upper dorsal marginal zone (the prospective notochord territory) become inductive and act as the gastrula organizer, while lower DMZ cells may lose their organizer activity. In these three steps, organizer activity spreads in a vegetal to animal direction within the plane of marginal zone cells. This continues in a modified form as upper DMZ cells induce prospective neural cells by planar signals passing in an animal direction. In the course of gastrulation, it becomes harder to obtain a single small population of cells capable of inducing a full body axis on transplantation. This is probably in part because the DMZ organizer of the gastrula differentiates into anterior (neuralizing) and posterior (mesodermalizing) inductive regions, and partly because, for technical reasons, the assay changes from one of transplanting tissue fragments into the marginal zone of a recipient embryo, to one of implanting tissue fragments into the blastocoel, which yields different results.

Axial Pattern and Organizer Size. The entire series of ventralized and dorsalized phenotypes can be related to the single variable of the width of the organizer in the marginal zone at the late blastula stage. All dorsalizing and ventralizing treatments, applied at pre-gastrula stages, are unified by their effect on organizer size, that is, on the magnitude of late blastula organizer function. Ventralizing treatments applied during the gastrula stage, as the organizer leads morphogenesis and inductively signals neighboring tissues, perhaps interfere with gastrula organizer formation and/or reduce its function. For all pre-gastrula and gastrula treatments, the extent of convergent extension during gastrulation (measured as archenteron length) correlates with the completeness of the final axial pattern.

Organizing the Organizer. There is little information on the process by which the gastrula organizer develops from a small group of apparently homogeneous cells to a differentiated population with region-specific inductive and morphogenetic properties. This occurs in the gastrula period. We propose that this differentiation requires interactions of the upper DMZ with the LVMZ (which would "caudalize" the organizer) and with lower DMZ cells (which anteriorize it). Changes in these interactions in ventralized and dorsalized embryos may lead to organizers with different axis-inducing properties.

REFERENCES

Beck, F. and J.B. Lloyd. 1966. The teratogenic effect of azo dyes. *Adv. Teratol.* 1:133-191.

Boterenbrood, E.C. and P.D. Nieuwkoop. 1973. The formation of the mesoderm in urodelian amphibians. V. Its regional induction by the endoderm. *Wilhelm Roux's Arch. Dev. Bio.* 173:319-334.

Coffey, R.J., E.B. Leof, G.D. Shipley, and H.L. Moses. 1987. Suramin inhibition of growth factor receptor binding and mitogenicity in AKR-2B cells. *J. Cell. Physiol.* 132:143-148.

Cooke, J. 1991. The Arrangement of Early Inductive Signals in Relation to Gastrulation; Results from Frog and Chick. p. 79-100. *In: Gastrulation: Movements, Patterns, and Molecules.* R. Keller, W.H. Clark, Jr., F. Griffin (Eds.). Plenum Press, New York.

Dale, L. and J.M.W. Slack. 1987a. Fate map of the 32 cell stage of *Xenopus laevis*. *Development* 99:527-551.

Dale, L. and J.M.W. Slack. 1987b. Regional specification within the mesoderm of early embryos of *Xenopus laevis*. *Development* 100:279-295.

Danilchik, M., T. Doniach, and J.C. Gerhart. 1991. Patterning of the embryonic body axis during *Xenopus* gastrulation: Experimentally reduced morphogenesis leads to anteriorally truncated embryos. In preparation.

Dixon, J.C. and C.R. Kintner. 1989. Cellular contacts required for neural induction in *Xenopus* embryos: Evidence for two signals. *Development* 106:749-757.

Doniach, T., M. Danilchik, and J.C. Gerhart. 1991. Patterning of the embryonic body axis during *Xenopus* gastrulation: The progressive anteriorization of cell fates. In preparation.

Elinson, R.P. and P. Pasceri. 1989. Two UV-sensitive targets in dorsoanterior specification in frog embryos. *Development* 106:511-518.

Elinson, R.P. and B. Rowning. 1988. A transient array of parallel microtubules in frog eggs: Potential tracks for a cytoplasmic rotation that specifies the dorso-ventral axis. *Dev. Biol.* 128:185-197.

Gerhart, J., M. Danilchik, T. Doniach, S. Roberts, B. Rowning, and R. Stewart. 1989. Cortical rotation of the *Xenopus* egg: Consequences for the anteropoasterior pattern of embryonic dorsal development. *Development* 107 (Suppl.):37-51.

Gerhart, J.C. and R.E. Keller. 1986. Region-specific cell activities in amphibian gastrulation. *Annu. Rev. Cell Biol.* 2:201-229.

Gimlich, R.L. 1986. Acquisition of developmental autonomy in the equatorial region of the *Xenopus* embryo. *Dev. Biol.* 115:340-352.

Gimlich, R.L. and J.C. Gerhart. 1984. Early cellular interactions promote embryonic axis formation in *Xenopus laevis*. *Dev. Biol.* 104:117-130.

Grant, P. and J.F. Wacaster. 1972. The amphibian grey crescent–a site of developmental information? *Dev. Biol.* 28:454-471.

Hamburger, V. 1988. *The Heritage of Experimental Embryology, Hans Spemann and the Organizer.* Oxford University Press, Oxford.

Hemmati-Brivanlou, A., R.M. Stewart, and R.M. Harland. 1990. Region-specific neural induction of an engrailed protein by anterior notochord in *Xenopus*. *Science* 250:800-802.

Holtfreter, J. and V. Hamburger. 1955. Embryogenesis: Progressive differentiation. Amphibians. p. 230-296. *In: Analysis of Development.* B.H. Willier, P.A. Weiss, and V. Hamburger (Eds.). Hafner Publishing Co., New York.

Kageura, H. 1990. Spatial distribution of the capacity to initiate a secondary embryo in the 32-cell embryo of *Xenopus laevis*. *Dev. Biol.* 142:432-438.

Kaneda, T. 1981. Studies of the formation and state of determination of the trunk organizer in the newt, *Cynops pyrrhogaster*. III. Tangential induction in the dorsal marginal zone. *Dev. Growth & Differ.* 23:553-564.

Kaneda, T. and A.S. Suzuki. 1983. Studies on the formation and state of determination of the trunk organizer in the newt, *Cynops pyrrhogaster*. IV. The association of neural inducing activity with the mesodermization of the trunk organizer. *Wilhelm Roux's Arch. Dev. Biol.* 192:8-12.

Kao, K.R. and R.P. Elinson. 1988. The entire mesodermal mantle behaves as a Spemann's organizer in dorsoanterior enhanced *Xenopus laevis* embryos. *Dev. Biol.* 127:64-77.

Keller, R.E. and M. Danilchik. 1988. Regional expression, pattern and timing of convergence and extension during gastrulation of *Xenopus laevis*. *Development* 103:193-210.

Keller, R.E., J. Shih, and P. Wilson. 1991. Cell Motility, Control and Function of Convergence and Extension During Gastrulation in *Xenopus*. p. 101-120. *In: Gastrulation: Movements, Patterns, and Molecules*. R. Keller, W.H. Clark, Jr., F. Griffin (Eds.). Plenum Press, New York.

London, C., R. Akers, and C.R. Phillips. 1988. Expression of epi 1, an epidermal specific marker, in *Xenopus laevis* embryos is specified prior to gastrulation. *Dev. Biol.* 129:380-389.

Malacinski, G.M., A.J. Brothers, and H.-M. Chung. 1977. Destruction of components of the neural induction system of the amphibian egg with ultraviolet irradiation. *Dev. Biol.* 56:24-39.

Mangold, O. 1933. Über die Inductionsfähigkeit der verschiedenen Bezirke der Neurula von Urodelen. *Naturwissenschaften* 21:761-766.

Moody, S. 1987. Fates of the blastomeres of the 16-cell stage *Xenopus* embryo. *Dev. Biol.* 119:560-578.

Nakamura, O. 1978. Epigenetic formation of the organizer. p. 179-220 *In: Organizer: A Milestone of a Half Century from Spemann*. O. Nakamura and S. Toivonen (Eds.). Elsevier/North Holland, Amsterdam.

Nieuwkoop, P.D. 1973. The "organization center" of the amphibian embryo: Its spatial organization and morphogenetic action. *Adv. Morphogen.* 10:1-39.

Nieuwkoop, P.D. 1985. Inductive interactions in early amphibian development and their general nature. *J. Embryol. Exp. Morphol.* 89 (Suppl.):333-347.

Nüsslein-Volhard, C., H.G. Frohnhöfer, and R. Lehmann. 1987. Determination of anteroposterior polarity in *Drosophila*. *Science*. 238:1675-1681.

Peacock, S.L., M.P. Bates, D.W. Russell, M.S. Brown, and J.L. Goldstein. 1988. Human low density lipoprotein receptor expressed in *Xenopus* oocytes. *J. Biol. Chem.* 263:7838-7845.

Phillips, C.R. 1991. Effects of the dorsal blastopore lip and the involuted dorsal mesoderm on neural induction in *Xenopus laevis*. *Symp. Soc. Dev. Biol.* 49:93-107.

Phillips, C.R. and Doniach, T. 1991. Effects of the dorsal blastopore lip and the involuted dorsal mesoderm on neural induction in *Xenopus laevis*. Submitted.

Savage R. and C.R. Phillips. 1989. Signals from the dorsal blastopore lip region during gastrulation bias the ectoderm toward a non-epidermal pathway of differentiation in *Xenopus laevis*. *Dev. Biol.* 133:157-168.

Saxen, L. and S. Toivonen. 1961. The two-gradient hypothesis in primary induction. The combined effect of two types of inductors mixed in different ratios. *J. Embryol. Exp. Morphol.* 9:514-528.

Scharf, S.R. and J.C. Gerhart. 1983. Axis determination in eggs of *Xenopus laevis*: A critical period before first cleavage, identified by the common effects of cold, pressure, and ultraviolet irradiation. *Dev. Biol.* 99:75-87.

Scharf, S.R., B. Rowning, M. Wu, and J.C. Gerhart. 1989. Hyperdorsoanterior embryos from *Xenopus* eggs treated with D_2O. *Dev. Biol.* 134:175-188.

Slack, J.M.W., B.G. Darlington, L.L. Gillespie, S.F. Godsave, H.V. Isaacs, and G.D. Paterno. 1989. The role of fibroblast growth factor in early *Xenopus* development. *Development* 107 (Suppl.):141-148.

Smith, J.C. and J.M.W. Slack. 1983. Dorsalization and neural induction: Properties of the organizer in *Xenopus laevis*. *J. Embryol. Exp. Morphol.* 78:299-317.

Spemann, H. 1931. Über den Anteil von Implantat und Wirtskeim an der Orientierung und Beschaffenheit der Induzierten Embryonalanlage. *Wilhelm Roux's Arch. Dev. Biol.* 123:390-517.

Spemann, H. 1938. *Embryonic Development and Induction*. Yale University Press, New Haven. (reprinted by Hafner Publishing Co., New York, 1967).

Stewart, R.M. 1990. *The active inducing center of the embryonic body axis in Xenopus*. 183 pp. Ph.D. Dissertation, University of California, Berkeley.

Stewart, R.M. 1991. Dorsal mesoderm induction during normal *Xenopus* development and its relationship to anteroposterior patterning. Submitted.

Stewart, R.M. and J.C. Gerhart. 1990. The anterior extent of dorsal development of the *Xenopus* embryonic axis depends on the quantity of organizer in the late blastula. *Development* 109:363-372.

Toivonen, S. 1978. Regionalization of the embryo. p. 119-156. *In: Organizer: A Milestone of a Half Century from Spemann*. O. Nakamura and S. Toivonen (Eds.). Elsevier/North-Holland, Amsterdam.

Suzuki, A.S., Y. Mifune, and T. Kaneda. 1984. Germ layer interactions in pattern formation of amphibian mesoderm during primary induction. *Dev. Growth & Differ.* 26:81-94.

Vincent, J.-P. and J.C. Gerhart. 1987. Subcortical rotation in *Xenopus* eggs: An early step in embryonic axis specification. *Dev. Biol.* 123:526-539.

Vincent, J.-P., S.R. Scharf, and J.C. Gerhart. 1987. Subcortical rotation in *Xenopus* eggs: A preliminary study of its mechanochemical basis. *Cell Motil. Cytoskeleton* 8:143-154.

Waddington, C.H. and M.M. Perry. 1956. Teratogenic effects of trypan blue on amphibian embryos. *J. Embryol. Exp. Morphol.* 4:110-119.

Winklbauer, R., A. Selchow, M. Nagel, C. Stoltz, and B. Angres. 1991. Mesoderm Cell Migration in the *Xenopus* Gastrula. p. 147-168. *In: Gastrulation: Movements, Patterns, and Molecules*. R. Keller, W.H. Clark, Jr., F. Griffin (Eds.). Plenum Press, New York.

Yamada, T. 1950. Dorsalization of the ventral marginal zone of the *Triturus* gastrula. I. Ammonia treatment of the medio-ventral marginal zone. *Biol. Bull.* 47:98-121.

Yamaguchi, Y. and A. Shinagawa. 1989. Marked alteration at the midblastula transition of the effect of lithium on the formation of the larval body pattern of *Xenopus laevis*. *Dev. Growth & Differ.* 31:531-541.

THE ARRANGEMENT OF EARLY INDUCTIVE SIGNALS IN RELATION TO GASTRULATION; RESULTS FROM FROG AND CHICK

Jonathan Cooke

Laboratory of Embryogenesis
National Institute for Medical Research
Mill Hill, London, NW7 1AA, U.K.

EARLY REGIONALIZATION FOR THE BODY PATTERN IN AMPHIBIAN DEVELOPMENT

Nieuwkoop and his students first showed clearly that specification of presumptive mesodermal territories in the amphibian embryo, and of their overall orientation, takes place by agency of signals deriving from the yolky vegetal zone (Nieuwkoop 1977, review). This process begins during (possibly early) blastula stages, and is progressive so that by onset of gastrulation, when the first movements begin to produce rearrangement in the induced territories, there is a significant geographical pattern and differential time schedule to these movements, as well as a pattern of differentiation capacities in the tissue when cultured in isolation (Keller *et al.* 1985; Dale and Slack 1987b). This pattern relates to the subsequent axes of organization of the body. Geographical regionalization on a finer scale is most advanced in a relatively narrow (*ca.* 90°) sector around the future dorsal midline, and is related to deep *vs.* superficial position within the blastula wall as well as to cells' distances from initial sources of induction, *i.e.*, to 'height' in the marginal zone towards the animal pole. As described by Keller and his associates (op. cit. and Keller 1986; Wilson *et al.* 1989) the role of this sector in gastrulation and neurulation and its capacities when developing in isolation entitle it to the designation 'morphogenetic organ'. Recent results of various recombination experiments (Dale and Slack 1987b; Gimlich 1986; Stewart and Gerhart 1990) suggest that until the start of gastrulation the 'induced' zone exists as the above- mentioned minor, dorsal axial (DA) sector, abutting against the remaining much larger non-axial, ventrolateral (VL) sector, with a relatively sharp transition between them. This is because the more general 'mesoderm inducing' signals from all around the vegetal region are replaced by, or have superimposed upon them, a specifically dorsal axial induction in a narrow (possibly *ca.* 60°) dorsal sector. Presence of both sectors, and in roughly normal proportionate extents, appears to be required for production of normal body pattern *in vivo* (Figure 1a).

Nieuwkoop had originally noted that the distinctively suprablastoporal, superficial layer of the gastrula fated to be definitive body endoderm, *i.e.* gut lining, was also part of the dorsal induced zone. The distinctive role of this in axial morphogenesis in conjunction with its underlying tissue is now recognized (Keller *et al.* 1985). Indeed, definitive body endoderm may originate in the induced zone all around the gastrula (see later discussion of *in vitro* inducers), whereas the mass of large yolky cells, the initiating source of inducing signals (Gimlich and Gerhart 1984), may make no long term structural contribution in the amphibian body. Such an arrangement would bring the inductive sequence in relation to fate map, for amphibian development, into line with that believed to operate in 'blastoderm-type' vertebrate embryos (see final section). In view of this, the term 'mesoderm induction' is an improperly restrictive one for this first major intercellular signalling system in the vertebrate embryo. Furthermore, it has recently become apparent from a number of observations that a significant priming signal, determining the location of the nervous system territory within the tissue lying animalwards of the mes-endodermal one, is part of the same early patterning episode occurring within the plane of the blastula and gastrula wall (London *et al.* 1988; Dixon and Kintner 1989; Sharpe *et al.* 1987; Ruiz-Altaba 1990). Thus one signalling episode, in the hours leading up to gastrulation, achieves a broad regionalization that marks out the major regions and polarities of the body.

However many molecular components are involved in the cascade of signals and in the feedback, or regulative responses that must be involved to effect patterning (see later discussion), the process seems to be initiated by two contrasting signals from the vegetal zone. These experimental embryological data and anatomical observations relating to the organization of gastrulation, that have been briefly introduced here will be laid out in detail and hopefully illustrated by their authors in other contributions to this volume. Figure 1 gives an impression of the geometry of the blastular inductive system as so far understood. In what follows I shall describe various observations, concerning the responses of *Xenopus* and *Rana* animal cap tissue to purified proteins from certain 'growth factor' families, and relate each observation to a documented aspect of the natural organization during blastula, gastrula and neurula phases. Together, these observations form impressive evidence that molecules identical with these ones, or close congeners of them, are the natural vegetally arising VL and DA inducing agents that appear to be required by the embryological data (Cooke *et al.* 1987; Smith *et al.* 1988, 1990a; Cooke 1989b; Slack *et al.* 1987; Kimelman *et al.* 1988). In a final section I shall draw attention to features that must differ from the organization

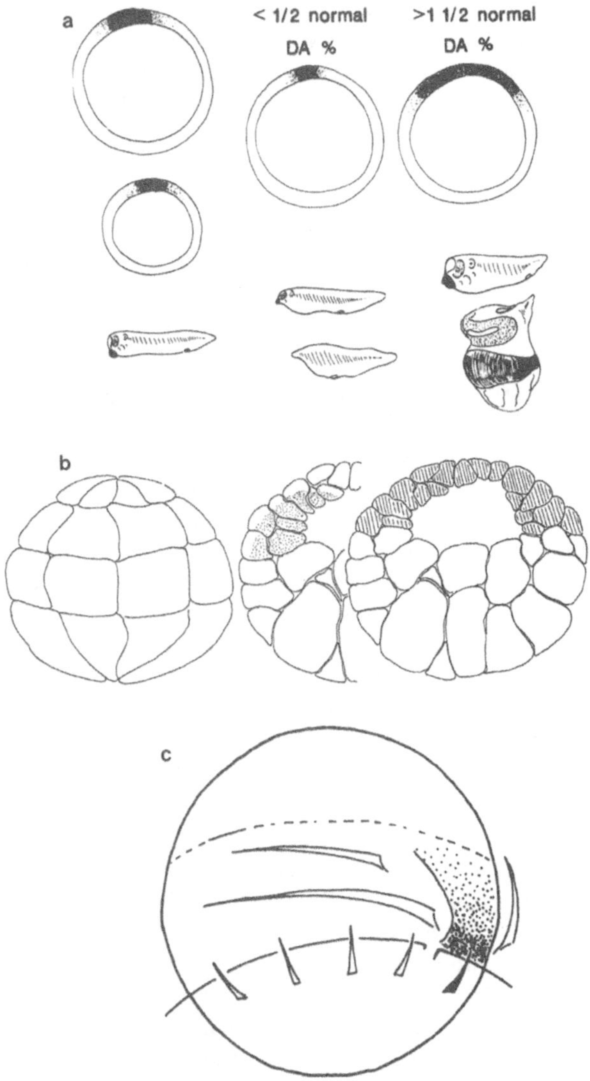

Figure 1. Overall arrangement of the blastular inductive system. **(a)** Three situations are shown, with the initial configurations in the egg viewed from one pole in the upper line, and sketches of the larval body plans that finally result in the lower. From left to right are shown respectively the normal balance between Dorsal Axial (stippled) and Ventro-Lateral (clear) inductive sectors, which is achieved over a range of natural egg sizes in *Xenopus* and is installed by the 2-cell stage, and then abnormally restricted or wide DA sectors which can be produced by experimental interference prior to that stage, and are not subsequently corrected. Anterior deficiency, or dorso-anterior over-representation in the body pattern respectively, result from these two situations. **(b)** The early system in lateral view. At left is shown a schematic 32-cell pattern of cleavage in surface view, and at right an early blastula two rounds of cleavage later, in section. The shaded blastomeres, descendant from the animal most two tiers of the 32 cell, usually differentiate only as abnormal epidermis when cultured in isolation from this stage unless soluble inducing factors are added (see text). But in the central half-replica of the blastular structure, those regions are stippled which we estimate, in normal development, to be ultimately recruited into mesodermal structure by induction initiated vegetally. The more surface parts of some of the same blastomeres will become definitive body endoderm, also induced. **(c)** The signalling system by onset of gastrulation. The vegetally arising inductions of two characters (short spearheads) have established their zones of DA and VL character in the marginal region. As these begin the mechanical activities that constitute gastrulation, the DA zone in particular reveals a degree of pre-organization important in establishing head to tail sequence at involution (stipple gradation and long spearhead off outline). During and perhaps before gastrulation an extensive signalling system whereby DA tissue modifies the state of more lateral sectors, further refines the specification map for mesodermal pattern (long spearheads).

in amphibians in any comparable inductive signalling phase of 'higher' vertebrate embryos, even if homologous molecules are in play. I then outline recent evidence, necessarily less direct than that coming from amphibian experiments, that homologues of such factors indeed bring about comparable steps of specification in blastula phases of bird development.

COMPETENCE, NATURE OF SIGNAL AND TEMPORAL CHARACTERISTICS OF RESPONSE IN BLASTULA INDUCTIVE EVENTS; EPIBOLY

All tissue of the animal hemisphere is competent (though not necessarily equally responsive) to induction by the vegetal signals from around the 64 cell stage until early in gastrulation itself. The signals do not depend upon intact gap junctional function (Warner and Gurdon 1987), and can cross interposed filters revealing that direct cell contact is not required (Slack *et al.* 1987). The first dramatic response to induction is mechanical participation in gastrulation by individual cells, consisting *in vivo* either of acquisition of the ability to adhere and locomote on 'basement membrane' extracellular matrix or of the commencement of the force-producing interaction among like cells, known as convergent extension (Keller *et al.* 1985). Particularly in the dorsal sector where the latter activity dominates, the time- or 'age'- schedule of this response to induction is quite constant among embryos in relation to the original time of fertilization. Experiments with animal/vegetal recombinates, in which the relatively distant response marker of muscle actin synthesis is monitored, suggest that this constancy of the age of cells at response is inherent in the responding cells, and not controlled by the exactitude with which the earlier intercellular signal has been timed (*e.g.* Gurdon *et al.* 1985).

Only a highly defined group of small proteins, known to be active as small secreted products and either identical with or very closely related to the activins (Smith *et al.* 1990a; Sokol *et al.* 1990), or perhaps 'bone morphogenic protein', mimic DA vegetal inductive activity when used *in vitro* as purified inducers of animal cap tissue, or when injected into amphibian blastocoels (Cooke *et al.* 1987; Smith *et al.* 1988; Cooke and Smith 1989; Symes and Smith 1987). Various members of the heparin-binding, fibroblast growth factor (FGF) family, small extracellularly active proteins whose mechanisms of secretion are unclear, as well as certain β-transforming growth factors which are related to the activin group in structure, mechanism of secretion and action, act as potent experimental inducers in the above ways but give rise to VL, or non-axial cell states. They are thus candidates for identity or relatedness with the natural VL type signal assumed to emanate from the major vegetal sector (or perhaps the whole vegetal zone) in normal development. It should be stressed that while RNA sequences coding for all these protein subfamilies are found at appropriate early stages of *Xenopus* eggs and embryos, the identities and numbers of proteins that actually mediate *in vivo* induction are currently unknown.

The competent period for response of animal cap cells to the XTC cell derived activin-like factor, that I call XTC-MIF pending its complete characterization, extends into beginning gastrula stages though it diminishes an hour or so before this. Competence to FGF ends earlier, slightly before the normal onset time of gastrulation (Slack *et al.* 1989). Experimentally, XTC-MIF will act only extracellularly or *via* intracellular injection of RNA into cells not normally being sources of inducer.

Intracellular injection of the protein is ineffective. The natural signal is therefore assumed to travel in the intercellular spaces, with extents of travel and effective concentrations possibly controlled by interaction with extracellular materials in the restricted diffusion space normal between cells of the blastular wall. Significant free diffusion in the greater blastocoel space would be incompatible with the rather restricted origin of mesoderm in the embryo, and particularly with its considerable spatial pre-organization in the marginal zone by the time of gastrulation (see below). Also, embryos with collapsed blastocoels achieve pretty normal morphogenesis.

FGF is stated not to be able to induce cells of the outer blastula layer that inherit original egg surface membrane (Slack *et al* 1989), whereas XTC-MIF can definitely recruit such cells from animal caps into either axial mesodermal or endodermal specifications (J. Cooke, L. Jones, and H. Woodland, unpublished observations). The latter in particular is consistent with normal fate maps, in that anterior and dorsal definitive endoderm is a crucial part of the 'organizer' region, mapping to the surface layer just above the dorsal blastoporal lip. But FGF too is reported to induce an endodermal along with mesodermal specifications, in animal caps *in vitro*, and the normal induced zone *in vivo* would include a surface, presumptive endodermal territory in 'ventral' as well as dorsal sectors.

Subtle effects are seen with the purified factors before the 'age'-point of true gastrulation in responding tissues, which may relate to mechanisms of the normal pre-gastrulation movements, epiboly (Keller 1986). When relatively early blastulae of *Xenopus* or *Rana* are injected with inducing concentrations of amphibian FGF or XTC-MIF into the blastocoel, a dramatic departure from normality occurs at gastrulation because an ectopic sheet of mesoderm has been specified from the deep layers of the entire animal hemisphere (Cooke and Smith 1989). But more subtle anatomical abnormality precedes this. The normal thinning of the animal pole region as cell layers reduce to two there, with corresponding build up of cell numbers displaced to the marginal zone, does not occur. This abrogation of the normal cell redistribution in epiboly is more pronounced when FGF has been the ectopic inducer. Normal epiboly may therefore occur because entry to the induced state, even ahead of the dramatic new cell behavior at gastrulation, inhibits the active process of interdigitation (Keller 1980) which otherwise thins the still ectodermally specified tissue near the animal pole. Perhaps significantly in view of the more prominent inhibition of epiboly by FGF, anatomical and cell lineage studies suggest that the normal ventrolateral marginal zone actually incorporates more cells than an equivalent sector of DMZ, despite the more dramatic mechanical role subsequently to be played by cells of the latter (Cooke and Webber 1985; Cooke 1987; Cooke and E.J. Smith 1988).

At onset of the gastrulation time-schedule, cells of blastocoel roofs ectopically induced with XTC-MIF or bFGF respectively pursue schedules and modes of new activity that correspond with those of the DA and of the VL non-axial marginal zones. XTC-MIF induced roofs segregate abruptly into a more or less massive inner tissue of induced cells, dependent upon injected concentration, and a correspondingly depleted outer layer of cells still behaving normally for ectoderm. The new layer of inner cells emerges and comes to sit, within a few minutes, upon a clean interface formed by basal surfaces of the cells left behind. This segregation can be seen as an abnormal spatially synchronized version of the events that occur in the form of a 'wavefront' (Cooke and Zeeman 1976) at the normal 'inner' dorsal blastoporal lip. There, cells from the deep layer of the marginal zone alter their properties in vegetal-to-animal sequence and leave that layer, to sit upon the basement surface of the original blastocoel wall as members of an emerging new layer, the definitive mesoderm. Rapid fix-dissection of large samples of injected and of synchronous control embryos reveals that this abnormal analogue of involution behavior occurs over only a few minutes within each subinjected blastocoel roof, and always at a time very shortly *after* the beginning of the much more extended schedule of normal involution within the DA type, convergently extending marginal zone which starts during stage 10 (Nieuwkoop and Faber 1967). The ectopic layer further reveals its DA status by beginning to undergo, after about a further 30 minutes, convergence or densification that produces considerable contractile force, stiffening and sometimes extension of the tissue, with gathering and columnarization of the animal pole surface layer that superficially resembles (but is probably not) an episode of neurulation.

In bFGF subinjected blastocoel roofs the reorganization to give a new ectopic layer is less abrupt when it comes, and less quickly forms a definite basal interface. But there tends to be a more massive recruitment into the induced state, from which a coherent remaining surface layer does not survive at the higher inducing doses. The blastocoel roof then disaggregates entirely. Even after lower injected doses histological recording of what has occurred is very difficult, since the ectopic new tissue has no mechanical integrity but consists of a population of quasi-independent, loosely adhering and locomoting cells (J. Cooke, unpublished observations), reminiscent of the normal gastrula's VL marginal zone sector after the latter has begun ingression. Most strikingly, the new behavior corresponding to ingression postdates that caused by XTC-MIF by around an hour in *Xenopus*, and 3 hours in *Rana* at 20°C. This again corresponds to normal VL mesoderm ingression, and suggests that we are dealing with two, effectively qualitatively different new cell states just as the overall results suggest a dichotomy of character for the natural inducing signals before gastrulation. Whatever the ultimate mechanism of these differences and their significance for normal gastrular organization, they are remarkable as cell biology. They suggest an inherent time schedule with an independent expression in cells that are competent to respond to either of the two signal categories *via* (presumably) separate receptor systems. According to the category of the signal, but independently of its concentration or of their own earlier 'age' at the time they received it, cells instigate particular new activities at either of two different 'ages', counting from fertilization (see also Symes and Smith 1987; Cooke and Smith 1990). Some anatomical details of ectopic induction *in vivo* after blastocoel injections are shown in Figures 2 and 3.

The ultimate halting of the normal gastrulation activity in experimentally injected embryos, or transformation of their own body pattern after later recovery of gastrulation, are complex but expected consequences of the presence of large populations of abnormally organized ectopic mesoderm. These details are less relevant to the topic of this symposium—the normal organization of gastrulation. Also less relevant is the evidence from the tissue types and structures ultimately differentiated in animal cap tissue *in vitro*, in response to the respective classes of factor, that the states of specification indeed correspond with those normal to early DA and VL marginal zone sectors (see Green *et al.* 1990). Thus extensively segmenting somite, often with accompanying notochord rod and capable of extensive neural induction, is the normal product of XTC-MIF induction, while non-axial mesenchymal and

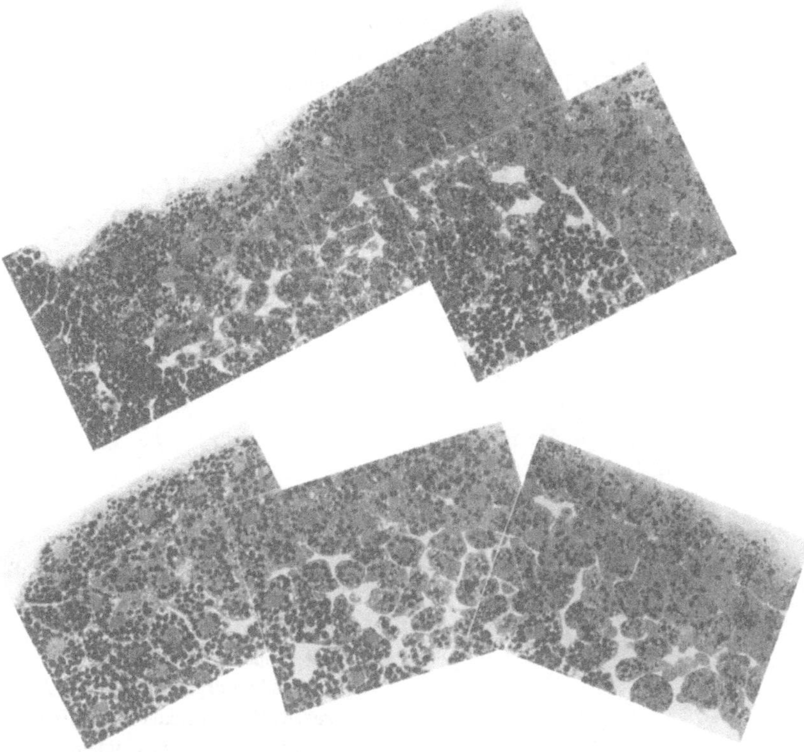

Figure 2. Blastocoel injection with XTC MIF; effects at an early gastrula stage. **Upper and lower** photomontages show longitudinal semithin sectional appearances of, respectively, a distorted but normally organized dorsal marginal zone (stage 8 injection saline only) and a disorganized marginal zone in a sibling that had been synchronously injected with XTC-MIF, *ca.*20 ng/ml. Vegetal region is to the left. The experimental embryo has been unable to form a stage 10 external lip. Relative mixing of cells from different neighborhoods is revealed by yolk platelet character. The normal interfaces between the basal surface of the marginal zone, the recently emerged, orderly mesoderm and the endoderm are obscured, and there are early signs of loosening cell adhesions in the deeper layers of the blastular wall. The latter will give rise to ectopic, synchronously emerging mesoderm and a relict monolayer of 'ectoderm'.

mesothelial tissue and an overall morphology like that of natural ventral marginal zone explants characterize bFGF (and TGFβ2) -induced animal caps. Complexities to this interpretation will be mentioned later in discussing overall spatial arrangement of the natural signals.

Before passing to consideration of the normal pre-organization of the DMZ for gastrulation, one observation relating to mechanisms of the different cell activities that cause involution and ingression is pertinent. Disaggregated *Xenopus* animal cap cells exposed to either bFGF or XTC-MIF, when they reach gastrula 'ages', acquire sharply enhanced ability to flatten and locomote on fibronectin as substrate. The convergent extension movements undergone by XTC-MIF induced whole tissue, however, which involve active cell rearrangement without reference to any extraneous substrate, may be independent of the RGD (cellular FN receptor) -mediated mechanism of cell adhesion (Smith *et al.* 1990b). Since convergent extension dominates the behavior of the DMZ from early stages in *Xenopus*, it appears that one activity, perhaps independent of the RGD/FN system, masks an altered propensity for interaction with FN that DA induced cells in fact share with VL induced ones. In the latter cells, and perhaps in all gastrulating mesodermal cells in embryos of other gastrulation types, that alteration remains of importance to gastrulation (Boucaut *et al.* 1984).

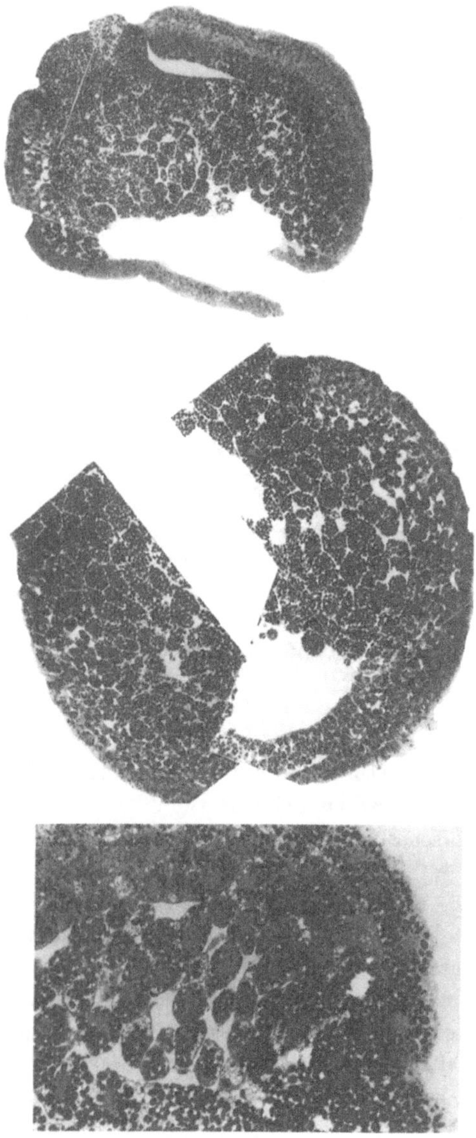

Figure 3. Blastocoel injections with XTC MIF; effects by the mid-gastrula stage. Upper and middle photomontages show longitudinal semithin sectional appearances of, respectively, normal and XTC-MIF injected embryos, as in Figure 2 but having reached midgastrula age. Yolk plug region is uppermost and original dorsal midline to right. Note in the normal case the orderly mesoderm involution of two different (DA and VL) characters, the formation of an archenteron, the control of the anterior most position of endoderm by the extent of mesoderm migration and the absence of mesoderm over the anterior blastocoel roof. In the experimental embryo the marginal zones are disrupted by chaotic, largely ectopic mesoderm, while the layer structure is not clear and archenteron formation has not been possible. The marginal mesoderm is more or less continuous with the palisade-like layer of ectopic, DA type mesoderm which is contracting (= converging) after recruitment out of the blastocoel roof. Endoderm is encroaching upon this. The lower photomicrograph is of the normal appearance at the 'inner' dorsal lip, at somewhat higher magnification and earlier gastrula stage. Compare the appearance of normal newly emerged DA type mesoderm with that lining the blastocoel roof in the experimental whole embryo.

EARLY ORGANIZATION OF THE DMZ IN RELATION TO THE AXIAL BODY PLAN AND THE PATTERN OF DORSAL MESODERMAL TISSUES; RESULTS WITH XTC-MIF

The normal DMZ reveals its pre-organization as a 'morphogenetic organ' during gastrulation by the wavefront-like ordering, in space and time, with which successive groups of cells perform repacking and begin the development of force by alterations of adhesions, stiffening, etc., during lip formation, invagination and involution and the beginning of segmentation (Keller *et al.* 1985; Wilson *et al.* 1989). Massive re-organization of cells' short range neighbor relationships must accompany the great shape change that transforms the squat broad pre-involution axial material into the long axial column of notochord and segmenting somite (without corresponding individual cell shape change). Nevertheless, there is correlation between cells' relative positions in pre- and post-involuted conditions. Cells near the blastula dorsal midline and low (vegetally situated) in the induced marginal zone begin their active participation in involution early and come to occupy prechordal or relatively anterior chordamesodermal sites in the larval plan, whereas those initially more lateral and further from the vegetal sources of DA induction change their physical properties later, and come to occupy successively more posterior positions. That this actively graded schedule or 'wavefront' of involution activity is integral to axial organization is evident both from normal anatomical and from experimental observations. Cells in intact gastrula DMZ and in explants of the entire 'organ' are observed to complete the phases of involution-related change as a succession of wavefronts for each new phase sweep through the tissue. Part of the organization sequence *in vivo* must be brought about because, by later gastrula stages, signalling interactions have recruited into a DA state cells from much more lateral zones of tissue, which were not originally induced to that state by the narrow (60-90°) DA vegetal inductive zone of the beginning gastrula (Keller 1976; Slack and Forman 1980 and see below). This helps to complete the late–involuting posterior axial pattern.

Individual embryos perturbed in a variety of experimental ways such that their inductive systems will not, ultimately, produce a complete and balanced set of mesoderm specifications, all reveal this in correspondingly altered and incomplete time-schedules for involution activity during gastrulation (Cooke 1985 review). The correlation holds at the local level within involuting tissue; early performance of movements predicts anterior/dorsal pattern contribution, while progressively later movements go with the normally more posterior and laterally derived pattern contributions. Embryos failing to specify major sectors of the body pattern are simply missing the mesoderm that performs involution at times appropriate for those sectors. The mechanism of the correlation may be only indirectly causal (see J. Gerhart contribution, this volume), but the pre-organization of the DMZ very soon after its initial induction does mediate normal pattern after gastrulation. What then is the mechanism of this pre-organization?

The experiments involving injection of XTC-MIF into blastocoels, during the time period within which normal marginal zone induction is thought to be proceeding, have enabled us to reject in their simplest forms two possible principles of organization that might underlie the time sequence of involution behavior. A simple gradient of effective signal concentration, correlated with distance in the marginal zone from a vegetally localized source of the natural inducer, might be transduced into the observed 'wavefront' of the onset time of response, even if the effective signalling had occurred altogether earlier (see Cooke and Zeeman 1976). An alternative to this gradient principle is one of signal timing. Successively later times of signal receipt by cells at more distant positions in the field of induction might be expressed as a gradation of timings (again, a wavefront) for the onset of response. Both principles are conceivable because we already know that these cells can measure elapsed developmental time as such, *via* as yet unknown molecular mechanisms. In fact, neither principle appears to dominate the establishment of an involution schedule. When the ectopic response to signalling that was abnormally homogeneous (in space and time) is being made across a wide sheet of tissue in the blastocoel roof after factor injection, the rapid and dramatic change that corresponds to involution in the normal gastrulation sequence is *un*affected, in its timing, by variation across wide ranges in the prior timing or concentration of the injected signal molecule. This ectopic 'involution' behavior always occurs near the beginning of the extended sequence of normal DA involution (- a sequence which last some hours even in *Xenopus*). This does not mean that either timing or effective concentration of the natural signal are altogether irrelevant in the mechanism of normal axial sequencing, but that if they are involved it is in a more subtle way, perhaps as triggers for other mechanisms of self-organization.

Reaction-diffusion schemes for morphogenetic intercellular signalling that could lead to a series of cell states laid out in space, *i.e.* to a morphogenetic field, have been widely modelled. These require a relatively slowly spreading but self-enhancing 'activated' cell state (*i.e.* one incorporating autocatalysis or positive feedback) and then relatively more diffusible (or at least further-reaching) inhibitory or modulatory signals that are produced in cells as consequences of high local levels of the 'activating' one, but that then prevent activation from spreading to distant sites. For a detailed understanding of how such schemes could create positional information and a pattern of states in tissue, the theoretical literature should be read (*e.g.* Gierer and Meinhardt 1972; Meinhardt 1982; see also Cooke 1989a). Reaction-diffusion systems are considered usually to require an initiating reference point, as in the vegetal origin of induction, for reliably timed and polarized operation. Their dynamics are however such that completely spontaneous, if delayed and partial patterns, could result from homogeneous starting conditions. The sheet of ectopic 'DA' specified tissue that lines the antero-ventral region at postgastrular stages, following blastocoelic injection of the lower concentrations of XTC-MIF, tends to be self-organizing (Cooke and Smith 1989). In such sheets of mesoderm a significant axial formation, with convergent extension that generates a somite segment series and organizes neural induction and a tail, frequently develops. Such formations are positioned without reference to the injection position in the original blastocoel roof, and the inducing signal concentration that initiated them cannot have been spatially graded on any large scale within the tissue. Some signal molecules whose functions in the later body are thought to involve modes of secretion (-and perhaps autocrine or paracrine stimulation within tissue) similar to mesoderm inducers, *e.g.* TGFβ, can exhibit autocatalytic dynamics of synthesis (van-Obberghen-Schilling *et al.* 1988). In this they resemble the more locally spreading, auto-activating component required by the above reaction-diffusion schemes. A relatively more successful experimental simulation of normal DMZ organization, involving grafting of newly induced tissue that aids polarization, is discussed in a following section on 'larger scale patterning during the gastrula stage'.

It would be consistent with the known organization of the DMZ at onset of gastrulation if the initiation of induction (=activation) at its vegetal (presumptive anterior) end were found to be coupled to later production of modulating signals that specifically interacted with slightly later-induced cells to delay their time of involution response. Such a cascade of signals could also be involved in initial establishment of a pattern of activation of the earliest position-specific genes, such as certain homeobox-containing genes (*e.g.* Rosa 1989; Ruiz-Altaba and Melton 1989a), whose transcription may begin very early and be instrumental in further elaboration of the gastrulation and differentiation pattern.

Cells disaggregated and treated with soluble XTC-MIF, then reaggregated, do not create an episode of convergent extension such as is seen *in vivo* or when whole animal cap tissue is treated *in vitro*, even though as discussed next they differentiate as dorsal axial cell types. Once again, a systematic gradation of cell states may be required for convergent extension to become organized; cells need labels with which to order their rearrangements to produce co-herent shape change. An alternative explanation is that an adhesive interaction between cells of two specific parts in the normal structure, *viz* deep and superficial marginal zone cells, is required for force production (Keller *et al.* 1985), and cannot be set up in such reaggregates.

Specification of the various tissue types is related to, but not co-incident with, the axial positional system that gives head-to-tail character to the induced mesoderm. Thus perturbations that truncate axis specification first ablate the formation of prechordal, forebrain-inducing territories, followed by notochord and then by somite in a progressive erosion of the front-to-back segment series as the severity increases. This, together with the blastular fate map dispositions of territories for these cell types (see Figure 4a), suggests that both kinds of specification could be partly controlled in relation to proximity to the initiating source for DA induction. Recent experiments in which cells are treated with XTC-MIF as suspensions in CaMg free saline, then reaggregated and allowed to differentiate (Green and Smith 1990), have offered more evidence of relationship between signal concentration and *tissue type* specification than could be obtained for head-to-tail specification by experiments at the whole embryo level (see above). But the relationship is not simple enough to imply that the *in vivo* inducer is a classical morphogen, forming a concentration gradient whose levels are directly read to give the pattern. As the MIF concentration in these experiments is raised above low threshold (around 0.2 ng/ml), the result is the switching off of epidermal differentiation among the reaggregated cells and the final generation of at least two cells types, somite muscle and notochord. These are found in adjoining cell groups, that are scattered within a background of undefinable cells that may be inviable but may represent a third specified cell state. With very modest increase in the initiating MIF level these defined cell types disappear, to be replaced by one that is a powerful neuralizing inducer of any further added cells that have not been directly exposed to the MIF. This is suggested to represent the prechordal tissue of the whole embryo which is the distinctive inducer of anterior brain regions. bFGF behaves very differently in such experiments, as muscle is only produced after a large concentration 'gap' above threshold for suppression of epidermis, and is not itself suppressed above a further threshold.

A fascinating aspect of these results is that the whole response spectrum occurs across a MIF concentration range that is orders of magnitude narrower than, and at the lower end of, the extended range which results in a general progression of tissue types and morphologies in whole *in vitro* induced animal caps (Smith *et al.* 1988; Cooke *et al.* 1989). The latter produce structures ranging from mesenchymal non-axial, through segmenting, notochord-containing axes like isolated whole DMZs (Keller *et al.* 1985), to pre-axial, non-elongating head-like formations, in a way loosely controlled by XTC-MIF concentration over a 200-fold concentration range. Based on these earlier results, it was proposed that *in vivo* an XTC-MIF like signal might specify all regions of the induced zone including the VL, but the new information, especially the large concentration disparity between whole tissue and single cell responsiveness, makes any simple interpretation of the natural mechanism of patterning unlikely. When the response is initiated in whole tissue, *no* MIF concentration will recruit more than about 50% of the final cleavage products into 'induced' pathways. Experience in handling explants suggests that inadequate concentration and/or time of exposure of cells'

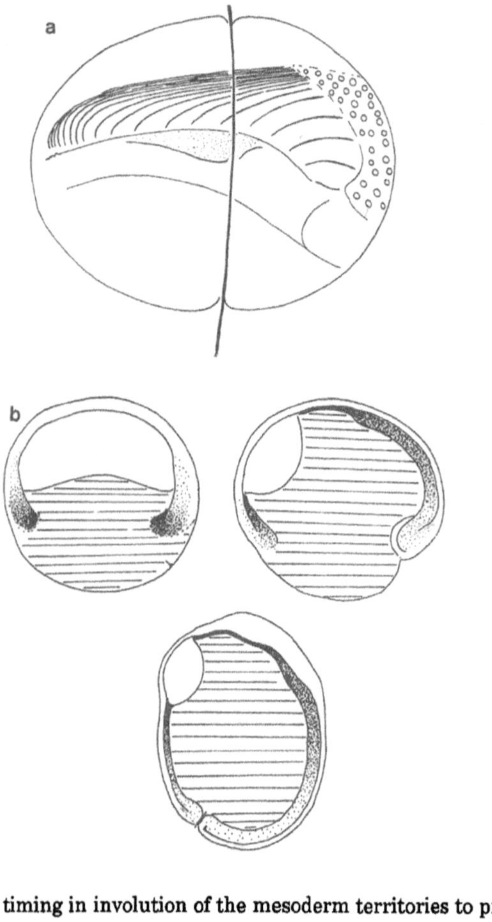

Figure 4. The fate-map for the body axis and the sequence of mesoderm involution. **(a)** Impression of the fate map for deployment of the marginal zone material in constructing the normal axial plan, as depicted in early blastula stages (*i.e.*, before epiboly). Heavy vertical line marks typical position for the second cleavage plane within this map, which has been assembled from *in situ* tracing of descendants of blastomeres labelled between 4 and 64-cell stages. The notochord territory (circles) with prechordal head territory at its vegetal end, is at right. The somite territory overlies the lateral mesodermal one around most of the circumference, with the restricted pronephric territory between them tending to lie athwart the second cleavage plane. Normal dispositions for the material of future medial *vs.* lateral parts of relatively anterior (right and low) and more posterior (high in the marginal zone) somite segments can be deduced, and justifies the orientated lines shown, though segment boundaries are of course not literally mappable. **(b)** Schematic of longitudinal sections of beginning, mid- and late gastrulae, oriented as in (a), with stippling and its density representing the disposition and relative timing in involution of the mesoderm territories to produce the head-to-tail, and dorsal-to-ventral organization of the body axis. From relating parts (a) and (b), it can be seen how both height in the marginal zone above the original source of inducers, and distance from the dorsal midline, are components influencing the final positions of involuted cells in the axial plan, and how this is organized *via* the variable timing of participation in involution it can also be seen how later in gastrulation the 'signal 3' system, of intramesodermal communication (see Figure 1c), must cause a considerable sector of originally 'VL' specified tissue to enter DA-type development posteriorly. This contributes massively to lateral somite in trunk and tail but not, in our belief, to notochord.

surfaces in the whole tissue situation is a most unlikely explanation of the effective dose disparity. There is a strong tendency for whole treated tissue to self-organize, giving a co-herent set of sequential convergent extension movements and an axis as in the gastrula. The fact that treated reaggregated cells cannot do this suggests that spatial inhomogeneity in intensity of the initiating inducer, however slight, is important for organization but does not directly dictate pattern parts.

The whole assemblage of results suggests an *in vivo* reality where a cascade of signals accomplishes patterning, with the XTC-MIF equivalent perhaps being an initial activator (or perhaps one early in the chain of synthesis but *not* the initial, maternally supplied one) but with downstream inhibitory or regulatory signals also playing a prominent role (Cooke 1989b). The activins in their already established roles in cell-cell communication (*e.g.* Vale *et al.* 1986) are paired with known inhibitory molecules, though these may act more by directly preventing response to signal than by actually modifying that response as is required in a mechanism for generating diversity. Any successful model for the patterning of blastular induction must account for the fact that only a fixed *proportion* of the competent animal hemisphere cells are diverted from ectodermal specification, and that this is due to active regulation because it is maintained if the inducible field is experimentally modified in size (Cooke 1989c).

A general model has been proposed for the process creating spatial patterns of specified states on a large scale in early development (Cooke 1983), in which certain intercellular signals act as an orientating or ranking principle by becoming graded or arising at one edge of the system, but are *not* true morphogens whose local levels directly give positional information (Wolpert 1971). An ordered multiplicity of cell states is brought about by a cascade of domain- or cell type-specific modulatory signals that successively divert the pathways of development in 'younger', *i.e.* less advanced cells distant from their site of synthesis. Detailed exegesis of this model would be out of place here, but interested readers of the original presentation will perceive that many dynamic aspects of pattern formation in *Xenopus*, and the experimental responses to purified mesoderm inducing factor, are in accord with its general predictions.

LARGER SCALE PATTERNING IN LATE GASTRULA/BLASTULA STAGES; THE 3 SIGNAL MODEL AND THE SIMULATION OF DORSAL LIP GRAFTING USING XTC-MIF

Slack and his colleagues first formalized a hypothesis of step-by-step patterning of the marginal zone as a whole, to initiate the body pattern, which they termed the '3 signal model' (Dale and Slack 1987b). Certain elements of this model now correspond with secure observations about organization of induction from the egg vegetal region, and about subsequent interaction around the gastrular marginal zone. That is, a major sector of 'VL', or non-axial inductive character appears to abut against a minor (*ca.* 60°) sector of DA inductive character, with rather little interaction between these sectors until near the onset of gastrulation (*e.g.* Dale and Slack 1987; Stewart and Gerhart 1990). I have tried to indicate the remarkable extent to which induction by bFGF (or by TGFβ2 - unpublished observations) on the one hand, and by XTC-MIF (or its mammal-derived congeners WEHI-MIF and 'PIF') on the other, specify cell states corresponding to those specified by the natural VL and DA vegetal initiating signals of the model. Although it has no equivalently strong candidates among known molecules, several observations support the existence of a subsequent signalling system across from DA into VL *induced* sectors during gastrulation, that integrates these to give the specification map for axial pattern (see Figure 1c). Dorsal marginal tissue will 'dorsalize' originally VL specified tissue after surgical recombination during gastrula stages *in vivo* (Smith and Slack 1983) and *in vitro* (Slack and Forman 1980). Extensive posterior axial contributions derive, finally, from initially VL specified marginal zone tissue from the 'sperm entry' side of the gastrula (Cooke and Webber 1985; Dale and Slack 1987a; Keller 1976). Formally, this 'signalling system 3' could utilize the same intercellular signal as has mediated dorsal axial induction, *via* a homoiogenetic process in which DA tissue produced more of the agent that had specified it. This would still constitute a third signalling step, since it occurs largely when the periods of competence for induction of ectodermally specified cells by signals one or two is over, and the target cells are now not uninduced, but rather already induced ones of a different category.

Organization of a new axial plan *de novo* can be initiated when new interfaces are made between DA- and VL-specified marginal zone tissue in the embryo, provided that certain quantitative conditions are met. Thus Stewart and Gerhart (1990) conclude that at an early stage there is little or no spatial heterogeneity in 'intensity' within the normal DA sector, and that the precise range of specification during axis formation depends rather upon the extent of the DA organizing tissue in relation to its VL surroundings. Certainly, neither DA nor VL states should be thought of as direct specifications for particular structures or cell types, but rather as intermediary states whose specific function is to interact spatially, across an interface between territories, in creating a signalling system (see previous section) that *does* more directly specify cell types or axial levels.

The grafting of small panels of XTC-MIF induced animal cap tissue to the ventral margins of late blastula hosts has been revealing about the natural mechanism in both positive and negative ways (Cooke 1989b). The results generally support the idea that a 'third signal system', with dynamics like a diffusion-controlled gradient from DA source into VL sink that operates largely once gastrulation is underway, sets up and orders the axial specifications. Such grafting, of homogeneously induced tissue from blastocoel-injected donors, is often a remarkably good but only occasionally a perfect simulation of the natural dorsal lip or 'organizer' graft first perforated by H. Mangold (Spemann and Mangold 1924). The results are outlined in Figure 5. It is in keeping with our own results using bFGF as an inducer that comparable bFGF-induced grafts do not give second organizations, as if they created no new interface between sectors of contrasting cell state. Other authors however, using somewhat different procedures (Ruiz-Altaba and Melton 1989b), do provoke accessory tail axial formations with such grafts. Significantly, with XTC-MIF, the set of new body territories organized is maximized by grafting small panels of a particular shape that have been exposed to high inducer concentrations. Delay of grafting until near host gastrulation then has little attenuating effect on the outcome. This suggests that a coherent and very localized territory of DA-specified cells offers the best chance of a good patterning interaction, and that such interaction can occur successfully during gastrulation even if it normally begins before this.

Second patterns which are complete, including notochord and prechordal region with its associated anterior brain parts, are rare after these 'experimental organizer' grafts, though regularly seen after grafts of *appropriately orientated* dorsal

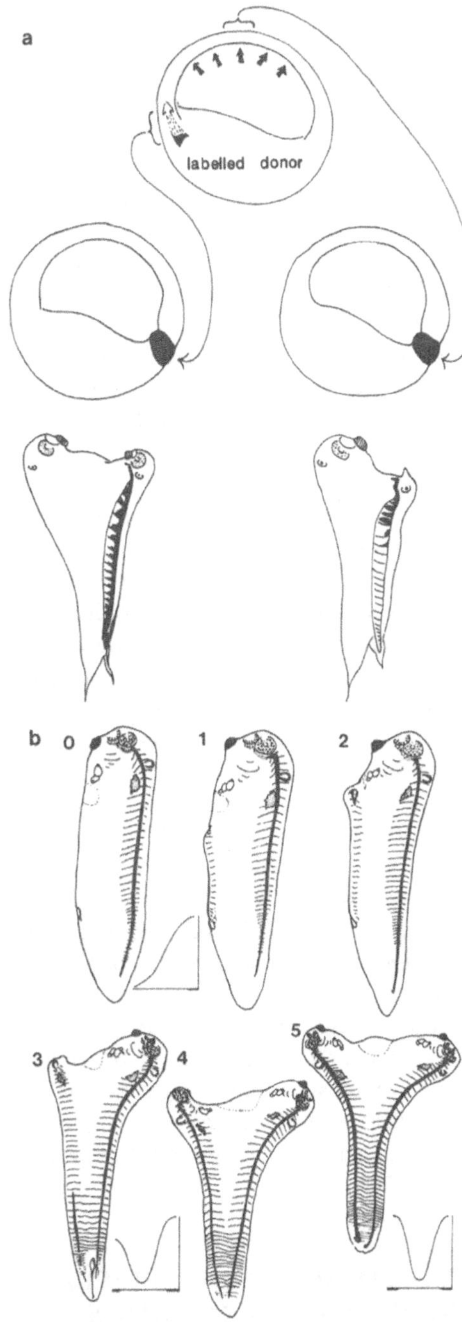

Figure 5. The production of new axial patterns by grafting of DA-specified tissue. **(a)** The lineage contributions to new body plans, made by natural dorsal lip grafts and by pieces of XTC-MIF induced blastocoel roof when transferred to the ventral margins of host blastulae, are compared. Both graft types contribute to the anteriormost meso- and endodermal structure in the new patterns they cause, but the natural graft which usually organizes a more complete pattern also contributes to the dorsal mesodermal midline along much of its length, The 'XTC-MIF' graft, by contrast, tends not itself to populate the more posterior axial regions (including segmenting somites) that it has organized. **(b)** Side views of the *Xenopus* larval body (0), and of bodies with various grades (1-5) of second pattern to give twin structure. Patterns of grades 4-1 are missing increasing subsets of anterior mesodermal and correlated neural structure, as can be seen from comparison with the complete 'host' pattern in 0. Even optimally sized and shaped 'XTC-MIF' grafts tend to result in patterns of grades 3 or 4 (although 5 has been recorded), whereas grade 5 is commonly obtained with the dorsal lip graft. Small gradient profiles over baselines with thick and thin sectors, inset next to particular diagrams, represent one way of conceiving how the action of a morphogen signalling system with DA tissue as a source and other tissue a 'sink', operating mainly during gastrulation, may underlie such results.

suprablastoporal tissue. Enough full second patterns have nevertheless occurred, after the XTC-MIF grafting procedure, to indicate that the commoner incomplete pattern is not caused by lack of any specific head or prechordal inducer with properties over and above those of the purified inducer at hand. Instead, the different effectiveness seems to be accounted for by differing degrees of internal organization, at time of grafting, between recently, homogeneously XTC-MIF induced tissue and the future anterior part of the natural DA induced sector. Between one and three hours after grafting, as their embryos of origin reach the age of gastrulation, grafts of the two types behave differently. Natural organizer tissue adheres to its original animal-to-vegetal polarity, expressed as a timed sequence or wavefront of involution activity, thus creating a 'lip' and localized invagination at the graft vegetal edge and giving rise to a convergently extending column of tissue moving with the host's ventral mesodermal mantle. As observed previously (Cooke 1972; Keller 1986), this pre-organization is obstinate and obstructive, preventing graft-host integration if the animal/vegetal axis is mismatched with the host's 'Experimental organizer' grafts on the other hand, while most effective if a particular *shape*, show little evidence of anisotropy at this early time. Synchronous involution behavior by their deep cells leads to formation of a shallow placode-like pit, with little elongation of ingressing tissue until much later stages.

When the embryos with their extra axial patterns have finally differentiated, the different degrees of pre-organization within the two types of graft are also revealed in their lineage contributions to the new axes (Figure 5a). Natural organizer grafts in *Xenopus* typically contribute extensively to notochord and to a medial strip of the somite column along the axis, as well as to prechordal mesoderm and to pharyngeal and archenteron roof endoderm. This suggests that they had already generated a considerable series of provisional antero-posterior 'values' or codings among their cells, whatever the nature of such codings at the early time of grafting. XTC-MIF treated animal pole grafts, in contrast, tend to contribute more extensively and massively to the cross-sectional structure of axes, but only at the anterior end of whatever series of structures develops. Axial patterning of mesoderm posterior to this, organized by presence of the graft, is nevertheless not populated by its cells. Thus these cells can instigate a new morphogenetic field (hypothetically, a new signal 3 gradient system) that specifies trunk and tail levels of segmentation in host tissue, without themselves diversifying so as to populate any of those 'positional addresses'. One can imagine that the (belated) setting-up of a new gradient field, by interaction of a patch of homogeneous DA tissue with VL surroundings, is a delayed process in relation to the normal one. The latter is controlled by initial induction at the vegetal edge, and early organization of DA tissue in relation to this to give the preliminary sequence of states that organizes at least the time course of gastrulation. If a substantial part of a normal DA territory is transplanted, intact in its organization as in the classical dorsal lip graft, the process of second axial patterning is considerably more effective.

AMNIOTE (BLASTODERM TYPE) VERTEBRATE DEVELOPMENT; EVIDENCE FOR COMPARABLE ROLES OF HOMOLOGOUS SIGNAL MOLECULES

A major difference between typical amphibian and the 'blastoderm type' versions of vertebrate development lies in the more epigenetic nature of early stages in the latter. There is no mechanism localizing the formation of the equivalent of a DA inducing sector within the single-celled egg. Emergence of a plane of symmetry, that defines the orientation with which tissue will be allocated to parts of the body pattern, is delayed until there are *ca.* 10^3 cells in the embryo-forming tissue. In addition we suspect that cell sorting mechanisms around this stage can play more pronounced roles than in amphibians, so that at least some early cell specification steps need not await the definition of an overall axis, but could diversify cells that would then migrate under the influence of a special signalling region (see Stern and Canning 1990). It is unclear whether truly inducing (*i.e.* respecifying) signals ever come from regions that are clearly spatially segregated from those to be induced. Thus, even if there are homologies within a cascade of intercellular signals involved in the respective developments, the cellular arrangements for their deployment might be quite different, for instance involving autocrine or paracrine mechanisms of self- or cell neighbor transformation rather than action at a distance. Certainly the experimental and observational embryology of birds suggests that time is an important agent of diversification before gastrulation, in that successive populations of cells leave a blastoderm, each in a 'pepper and salt' manner, perhaps then to act back on it in bringing about subsequent changes (Kochav *et al.* 1980).

I have explored the effects of peptides, certified to be potent inducers for frog animal cap cells, on cells from the early (pre-primitive streak) epiblast of bird embryos. Evidence that significant respecifications are brought about by brief (2-4 hrs) incubation with picomolar concentrations of all these factors derives from three kinds of observation.

1) Cells having been treated with the factors, then disaggregated and plated on a fibronectin substrate at times when their embryos of origin would be forming streaks (showing ingressed or involuting cells), adhere and spread strongly on this substrate within an hour. Epiblast similarly cultured and treated but without exposure to the active factors does not interact with FN in this way. Mammalian and amphibian bFGF, mammalian TGFβ2, and the activin-like factors XTC-MIF (amphibian-derived) and WEHI-MIF (mammal-derived) are all potent in causing this behavior, whereas platelet-derived growth-factor, inactive as a frog 'mesodermal inducer', is ineffective,

2) Similarly treated cells, seeded down to form a layer covering the denuded basal epiblast surface of a host embryo inverted as in New (1955) ring culture, actively adhere to that surface and interfere strikingly with subsequent gastrulation and organization of pattern using the chick/quail nucleolar marker system to analyze contributions, it has been found that while bFGF treatment gives cells which may colonize the relatively disorganized body pattern of the host in mesoderm and endoderm, treatment with the amphibian DA inducers seems to give cells which disrupt the embryo-forming area but are finally found only in peripheral hypoblast, *i.e.*, extra embryonically. Control epiblast cells, disaggregated and added in this way, show little tendency to stick to the host basal surface or to integrate into the cell layers, and do not disrupt the host morphogenesis.

3) Whole blastoderms from pre-streak stages can be incubated for a few hours in very low concentrations of the factors in shallow layers of medium, before washing and setting up in New ring culture. Such pre-treatment with the factors active as inducers in frogs, though not with platelet-derived growth-factor, causes highly abnormal, thin layered expanded blastoderms to develop. Examples are seen in Figure 6. These show little or no sign of axial patterning and great reduction of cell numbers in the epiblast (ectodermal) layer when controls have reached headfold stages. They are thus reminiscent of comparably staged amphibian embryos after blastocoelic injection of low

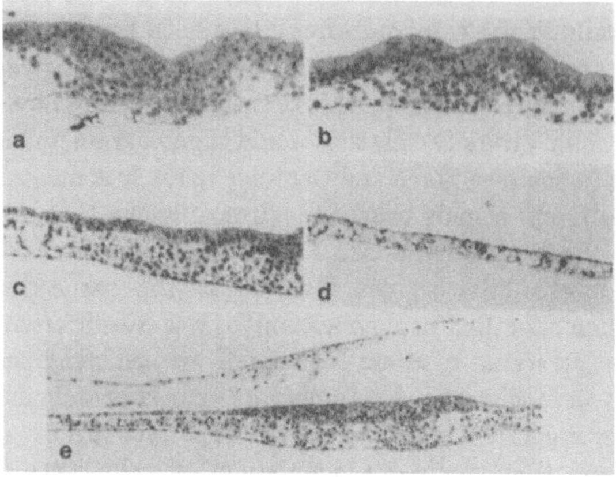

Figure 6. Effects of pre-incubating the bird blastoderm in low concentrations of an amphibian inducing factor. **(a)**, **(b)**, Semithin transverse sections at different post-nodal levels, during streak regression, of normal chick development occurring in New culture after pre-incubation in saline alone. Note organized involution of mesoderm from a restricted site, and the cell-rich epiblast layer with an organized streak thickening, **(c)**,**(d)**, higher magnification and **(e)**, lower magnification views of comparable sections, where blastoderms had been pre-incubated for 3 hours in *ca.* 2 ng/ml XTC-MIF before the 10 hrs of culture that led in controls to the appearance in **(a)** **(b)**. Note that very large areas are reduced to thin, stretched layers in the epiblast, and that areas most resembling axes are nevertheless wide and chaotic in their organization of involution. Higher concentrations of the factor do not allow survival of the integrity of blastoderms.

to moderate doses of inducing factor. Control pre-incubated examples are of largely normal morphogenesis.

These observations, together with other recent ones (Mitrani and Shimoni 1990), indicate that although we have much to learn of the arrangements for morphogenesis in this embryo type, specification steps corresponding to those of amphibian blastular induction do occur and can be specifically mediated by peptides of 'growth-factor' families homologous to those active in amphibians.

While the descriptions, conclusions and speculations in this particular paper are my own, it is a pleasure to acknowledge the many discussions and collaborations over the last five years, with members of the Laboratory of Embryogenesis at Mill Hill, that have led to them.

REFERENCES

Boucaut, J.C., T. Darribère, H. Boulekbache, and J.-P. Thiery. 1984. Prevention of gastrulation but not neurulation by antibodies to fibronectin in amphibian embryos. *Nature* 307:364-367.

Cooke, J. 1972. Properties of the primary organization field in the embryo of *Xenopus laevis*. I. Cell autonomy and behavior at the site of the organizer. *J. Embryol. Exp. Morphol.* 28:13-26.

Cooke, J. 1983. Evidence for specific feedback signals underlying pattern control during vertebrate embryogenesis. *J. Embryol. Exp. Morphol.* 76:95-114.

Cooke, J. 1985. The system specifying body position in the early development of *Xenopus*, and its response to perturbations. *J. Embryol. Exp. Morphol.* 89 (Suppl.):69-87.

Cooke, J. 1987. Dynamics of the control of body pattern in the development of *Xenopus laevis*. IV. Timing and pattern in the development of twinned bodies after re-orientation of eggs in gravity. *Development* 99:417-427.

Cooke, J. 1989a. The early amphibian embryo: Evidence for activating and for modulating or self-limiting components in a signalling system that underlies pattern formation. p. 145-158. *In: Cell to Cell Signalling*. A. Goldbeter (Ed.). Academic Press, New York.

Cooke, J. 1989b. Mesoderm-inducing factors and Spemann's organiser phenomenon in amphibian development. *Development* 107:229-241.

Cooke, J. 1989c. *Xenopus* mesoderm induction: Evidence for early size control and partial autonomy for pattern development by onset of gastrulation. *Development* 106:519-529.

Cooke, J. and E.J. Smith. 1988. The restrictive effect of early exposure to lithium upon body pattern in *Xenopus* development, studied by quantitative anatomy and immunofluorescence. *Development* 102:85-99.

Cooke, J. and J.C. Smith. 1989. Gastrulation and larval pattern in *Xenopus* after blastocoelic injection of a *Xenopus*-derived inducing factor: Experiments testing models for the normal organisation of mesoderm. *Dev. Biol.* 131:383-400.

Cooke, J. and J.C. Smith. 1990. Measurement of developmental time by cells of early embryos. *Cell* 10:891-894.

Cooke, J., J.C. Smith, E.J. Smith, and M. Yaqoob. 1987. The organisation of mesodermal pattern in *Xenopus laevis*: Experiments using a *Xenopus* mesoderm-inducing factor. *Development* 101:893-908.

Cooke, J., K. Symes, and E.J. Smith. 1989. Potentiation by the lithium ion of morphogenetic responses to a *Xenopus* inducing factor. *Development* 105:549-588.

Cooke, J. and J.A. Webber. 1985. Dynamics of the control of body pattern in the development of *Xenopus laevis*. I. Timing and pattern in the development of dorso-anterior and of posterior blastomere pairs, isolated at the 4-cell stage. *J. Embryol. Exp. Morphol.* 88:85-112.

Cooke, J. and E.C. Zeeman. 1976. A clock and wavefront model for the control of the number of repeated structures during animal morphogenesis. *J. Theor. Biol.* 58:455-476.

Dale, L. and J.M.W. Slack. 1987a. Fate map for the 32-cell stage of *Xenopus laevis*. *Development* 99:197-210.

Dale, L. and J.M.W. Slack. 1987b. Regional specification within the mesoderm of early embryos of *Xenopus laevis*. *Development* 100:279-295.

Dixon, J.E. and C.R. Kintner. 1989. Cellular contacts required for neural induction in *Xenopus* embryos: Evidence for two signals. *Development* 106:749-757.

Gerhart, J., T. Doniach, and R. Stewart. 1991. Organizing the *Xenopus* Organizer. p. 57-78. *In: Gastrulation: Movements, Patterns, and Molecules*. R. Keller, W.H. Clark, Jr., F. Griffin (Eds.). Plenum Press, New York.

Gierer, A. and H. Meinhardt. 1972. A theory of biological pattern formation. *Kybernetik* 12:30-39.

Gimlich, R.L. 1986. Acquisition of developmental autonomy in the equatorial region of the *Xenopus* embryo. *Dev. Biol.* 115:340-352.

Gimlich, R.L. and J. Gerhart. 1984. Early cellular interactions promote embryonic axis formation in *Xenopus laevis*. *Dev. Biol.* 104:117-130.

Green, J.B.A., G. Howes, K. Symes, J. Cooke, and J.C. Smith. 1990. The biological effects of XTC-MIF: Quantitative comparison with *Xenopus*. *Development* 108:173-183.

Green, J.B.A. and J.C. Smith. 1990. Graded changes in dose of a *Xenopus* activin A homologue elicit stepwise transitions in embryonic cell fate. *Nature* 347:391-394.

Gurdon, J.B., S. Fairman, T.J. Mohun, and S. Brennan. 1985. Activation of muscle-specific actin genes in *Xenopus* development by an induction between animal and vegetal cells of a blastula. *Cell* 41:913-922.

Keller, R.E. 1976. Vital dye mapping of the gastrula and neurula of *Xenopus laevis*. II. Prospective areas and morphogenetic movements of the deep region. *Dev. Biol.* 51:118-137.

Keller, R.E. 1980. The cellular basis of epiboly: An SEM study of deep cell rearrangement during gastrulation in *Xenopus laevis*. *J. Embryol. Exp. Morphol.* 60:201-234.

Keller, R.E. 1986. The cellular basis of amphibian gastrulation. p. 241-328. *In: The Cellular Basis of Morphogenesis*. L.B. (Ed.). Plenum Press, New York.

Keller, R.E., M. Danilchik, R.L. Gimlich, and J. Shih. 1985. The function and mechanism of convergent extension during gastrulation of *Xenopus laevis*. *J. Embryol. Exp. Morphol.* 89 (Suppl.):185-209.

Kimelman, D., J.A. Abraham, T. Haaparanta, T.M. Palisi, and M.W. Kirschner. 1988. The presence of fibroblast growth factor in the frog egg, its role as a natural mesoderm inducer. *Science* 242:1053-1056.

Kochav, S., M. Ginsburg, and H. Eyal-Giladi. 1980. From cleavage to primitive streak formation: A complementary normal table and a new look at the first stages of the development of the chick. II. Microscopic anatomy and cell population dynamics. *Dev. Biol.* 79:296-308.

London, C., R. Akers, and C. Phillips. 1988. Expression of Epi-1, an epidermis-specific marker in *Xenopus laevis* embryos, is specified prior to gastrulation. *Dev. Biol.* 129:380-389.

Meinhardt, H. 1982. *Models of Biological Pattern Formation*. Academic Press, London.

Mitrani, E. and Y. Shimoni. 1990. Induction by soluble factors of organised axial structures in chick epiblasts. *Science* 247:1092-1094.

New, D.A.T. 1955. A new technique for the cultivation of the chick embryo *in vitro*. *J. Embryol. Exp. Morph.* 3:326-331.

Nieuwkoop, P.D. 1977. Origin and establishment of embryonic polar axes in amphibian development. *Curr. Top. Dev.* 11:115-132.

Nieuwkoop, P.D. and J. Faber. 1967. *Normal table of Xenopus laevis (Daudin)*. 2nd edition. North Holland, Amsterdam.

Rosa, F.M. 1989. Mix.1, a homeobox RNA inducible by mesoderm inducers, is expressed mostly in the presumptive endodermal cells of *Xenopus* embryos. *Cell* 57:965-974.

Ruiz-Altaba, A. 1990. Neural expression of the *Xenopus* homeobox gene Xhox3: Evidence for a patterning neural signal that spreads through ectoderm. *Development* 108:595-604.

Ruiz-Altaba, A. and D.A. Melton. 1989a. Bimodal and graded expression of the *Xenopus* homeobox gene Xhox3 during embryonic development. *Development* 106:173-183.

Ruiz-Altaba, A. and D.A. Melton. 1989b. Interaction between peptide growth factors and homeobox genes in the establishment of antero-posterior polarity in frog embryos. *Nature* 341:33-38.

Sharpe, C.R., A. Fritz, E.M. De Robertis, and J.B. Gurdon. 1987. A homoebox-containing marker of posterior neural differentiation shows the importance of predetermination in neural induction. *Cell* 50:749-758.

Slack, J.M.W. and D. Forman. 1980. An interaction between dorsal and ventral regions of the marginal zone in amphibian embryos. *J. Embryol. Exp. Morphol.* 56:283-299.

Slack, J.M.W., B.G. Darlington, L.L. Gillespie, S.F. Godsave, H.V. Isaacs, and G.D. Paterno. 1989. The role of fibroblast growth factor in early *Xenopus* development. *Development* 107 (Suppl.):141-148.

Slack, J.M.W., B.G. Darlington, J.K. Heath, and S.F. Godsave. 1987. Mesoderm induction in early *Xenopus* embryos by heparin binding growth factors. *Nature* 326:197-200.

Smith, J.C., B.M.J. Price, K. Van Nimmen, and D. Huylebroeck. 1990a. Identification of a potent *Xenopus* mesoderm-inducing factor, as a homologue of activin A. *Nature* 354:729-731.

Smith, J.C. and J.M.W. Slack. 1983. Dorsalization and neural induction: Properties of the organizer in *Xenopus laevis*. *J. Embryol. Exp. Morphol.* 78:299-317.

Smith, J.C., K. Symes, R.O. Hynes, and D.W. DeSimone. 1990b. Mesoderm induction and the control of gastrulation in *Xenopus laevis*: The roles of fibronectin and integrins. *Development* 108:229-238.

Smith, J.C., M. Yaqoob, and K. Symes. 1988. Purification, partial characterisation and biological effects of the XTC mesoderm-inducing factor. *Development* 103:591-600.

Sokol, S., A.A. Wong, and D.A. Melton. 1990. A mouse macrophage factor induces head structure and organises a body axis in *Xenopus*. *Science* 249:561-564.

Spemann, H. and H. Mangold. 1924. Uber Induktion von Embryonenanlagen durch Implantation Artfremder Organisatoren. *Wilhelm Roux's Arch. Dev. Biol.* 100:599-638.

Stern, C.D. and D.R. Canning. 1990. Origin of cells giving rise to mesoderm and endoderm in chick embryo. *Nature* 343:273-275.

Stewart, R.M. and J.C. Gerhart. 1990. The anterior extent of dorsal development of the *Xenopus* embryonic axis depends on the quantity of organizer in the later blastula. *Development* 109:363-372.

Symes, K. and J.C. Smith. 1987. Gastrulation movements provide an early marker of mesoderm induction in *Xenopus laevis*. *Development* 101:339-350.

Vale, W., J. River, J. Vaughan, R. McClintock, A. Corrigan, W. Woo, D. Karr, and J. Spiess. 1986. Purification and characterisation of a FSH-releasing protein from porcine ovarian follicular fluid. *Nature* 321:776-779.

Van-Obberghen-Schilling, E., N.S. Roche, K.C. Flanders, M.B. Sporn, and A.B. Roberts. 1988. Transforming growth factor β1 Positively regulates its own expression in normal and transformed cells. *J. Biol. Chem.* 263:7741-7746.

Warner, A. and J.B. Gurdon. 1987. Functional gap junctions are not required for muscle gene activation by induction in *Xenopus* embryos. *J. Cell Biol.* 104:557-564.

Wilson, P.A., G. Oster, and R. Keller. 1989. Cell rearrangement and segmentation in *Xenopus*: Direct observation of cultured explants. *Development* 105:155-166.

Wolpert, L. 1971. Positional information and pattern formation. *Curr. Top. Dev.* 6:183-223.

CELL MOTILITY, CONTROL AND FUNCTION OF CONVERGENCE AND EXTENSION DURING GASTRULATION IN *XENOPUS*

Ray Keller[1], John Shih[1],and Paul Wilson[2]

[1] Department of Molecular and Cell Biology
University of California
Berkeley, CA 94720
[2] Department of Biochemistry and Molecular Biology
Harvard University
Cambridge, MA 02138

INTRODUCTION

In this paper we will discuss some recent work on the cell motility underlying the convergence and extension movements during gastrulation and neurulation of *Xenopus laevis*. We will also discuss some of the tissue interactions controlling this motility, and we will refine our previous ideas on how convergence and extension functions in gastrulation of *Xenopus*.

Gastrulation, Edited by R. Keller *et al.*
Plenum Press, New York, 1991

A Brief History of Studies on Convergence and Extension

The morphogenetic movements of convergence and extension have recently gained the attention of developmental biologists after 50 years of neglect. The vital dye mapping of Vogt (1929) showed that narrowing (convergence) and elongation (extension) of the dorsal part of the embryo comprise a major part of the tissue movements of amphibian gastrulation and neurulation. Work by Vogt, Spemann, Schechtman, Holtfreter, and others on microsurgically altered gastrulae and explanted tissues led to the conclusion that these were active, autonomous, and region-specific movements, rather than ones passively produced by forces generated elsewhere (reviewed in Spemann 1938 and Keller 1986). Schechtman (1942) demonstrated autonomous convergence and extension in gastrulation of the California tree frog, *Hyla regilla*, and he also suggested how these processes might be integrated with others during gastrulation.

Despite their importance, convergence and extension were ignored in favor of phenomena that better fit the notions at the time about how cells function in morphogenesis. These phenomena include the dramatic change in shape of the bottle cells (Holtfreter 1943a,b; Baker 1965), adhesion-guided movement of tissue layers (Holtfreter 1939; Townes and Holtfreter 1955; Phillips and Davis 1984), and the crawling of cells on the roof of the blastocoel (Holtfreter 1944; Nakatsuji 1975; Kubota and Durston 1978). In contrast, convergence and extension were mysterious "mass movements" in which deformation of the cell population seemed molded by an unseen hand, and there was no paradigm explaining how the motility or adhesive properties of individual cells could produce the movements of the population. Waddington (1940, page 109) insightfully suggested that these movements must occur by cell rearrangement since they could not be accounted for by oriented cell division or by change in cell shape. But this idea, which turned out to be correct, did not take root and no observations or experimental work followed. Likewise little effort was made to explain how these movements functioned in gastrulation, despite a good start on the problem with Schechtman's paper on the subject, cited above.

Our first attempt to demonstrate that convergence and extension were autonomous processes in *Xenopus* failed. Extension of explants of the dorsal marginal zone was poor and occurred rather late in development (Keller, unpublished results 1973). These results were similar to those of others (Ikushima and Maruyama 1971). Thus it appeared that the convergence and extension movements *in vivo* might be passive responses to forces generated by other processes, and accordingly we began investigations of some of these processes, including cell migration (Keller and Schoenwolf 1977), epiboly (Keller 1980), and bottle cell formation (Keller 1981). But much of gastrulation could not be explained by these processes. Removal of the bottle cells did not stop gastrulation but only truncated the periphery of the archenteron, which is the tissue normally formed by the respread bottle cells (Keller 1981; Hardin and Keller 1988). When the entire blastocoel roof (animal cap) was removed, which disrupts both migration of mesodermal cells on the roof and the epiboly of the animal cap, most of the gastrulation occurred, including involution of the marginal zone, closure of the blastopore, and extension of the axial mesodermal tissues in their proper pattern (Keller *et al.* 1985a,b; Keller 1991; see Holtfreter 1933). Thus the marginal zone alone could do much of the movements of gastrulation without the bottle cells or the blastocoel roof. Since the marginal zone could do so much alone, it was of interest to learn if the superficial endodermal epithelium or the deep, nonepithelial mesoderm of the marginal zone was primarily responsible for this remarkable property. Several manipulations of the deep mesodermal cells of the dorsal marginal zone, designed to stop or misdirect convergence and extension, if they exist as autonomous processes, resulted in failure of involution and blastopore closure (Keller 1981, 1984; Keller *et al.* 1985a,b). Similar manipulations of the superficial, epithelial layer of the same region had little or no effect. Therefore it appeared that the deep cells were the active, force-producing cell population and that the superficial cells were moved along passively on the outer surfaces of the deep cells. Moreover, these results suggested that the mysterious convergence and extension movements described by the early workers might be the movements generated by these deep marginal zone cells, since by a process of elimination, this was the only one of the regional, putatively autonomous processes left for consideration. We could not, however, demonstrate conclusively that convergence and extension occurred autonomously in the complex mechanical environment of the whole gastrula. Thus we returned to explants.

Demonstration of Autonomy in Sandwich and Open-faced Explants

Demonstration of the function of convergence and extension in gastrulation hinges on being able to show that these tissue movements are locally autonomous and active rather than being passively generated by forces produced elsewhere. In fact, defining the regional, autonomous processes in gastrulation and how they interact has been one of the most difficult stepping-stones in finally making a meaningful analysis of the process at the cellular level (see Spemann 1938; Keller 1986; Gerhart and Keller 1986) or at the molecular level (see Keller and Winklbauer 1990). Thus we returned to the problem of directly demonstrating local autonomy of convergence and extension in explants, this time with improved techniques. We made "sandwich" explants by apposing the inner surfaces of two marginal zones and allowing them to heal. The method of construction, the structure, and the prospective fates of these explants are shown in Figure 1. Surprisingly, these explants showed convergence and extension in not one but *two* regions--in the involuting marginal zone (IMZ), which differentiated into a normal array of somites and notochord, and in the noninvoluting marginal zone (NIMZ), which differentiated into neural tissue (Keller *et al.* 1985a,b; Keller and Danilchik 1988) [Figure 1A]. By assaying the capacity of other regions of the gastrula to extend and converge we found that these movements are local, region-specific and autonomous process associated with the notochordal and somitic mesoderm in the IMZ and with the posterior nervous system in the NIMZ (Keller and Danilchik 1988). Below, in the section on "Function of Convergence and Extension", we will describe how we think these two regions of convergence and extension act coordinately to produce involution and blastopore closure.

The convergence and extension shown in both regions are powerful, robust movements when the conditions for their expression are met. However, they are complex processes and a variety of trivial factors will prevent their expression. Our success in demonstrating these movements after an initial failure was a matter of developing the right technique, and failure to demonstrate them in other systems should not necessarily be taken to mean that they do not occur in those systems.

The sandwich method did not allow observation of deep cell behavior, and since the evidence recounted above pointed to the deep cells as the active, force-generating

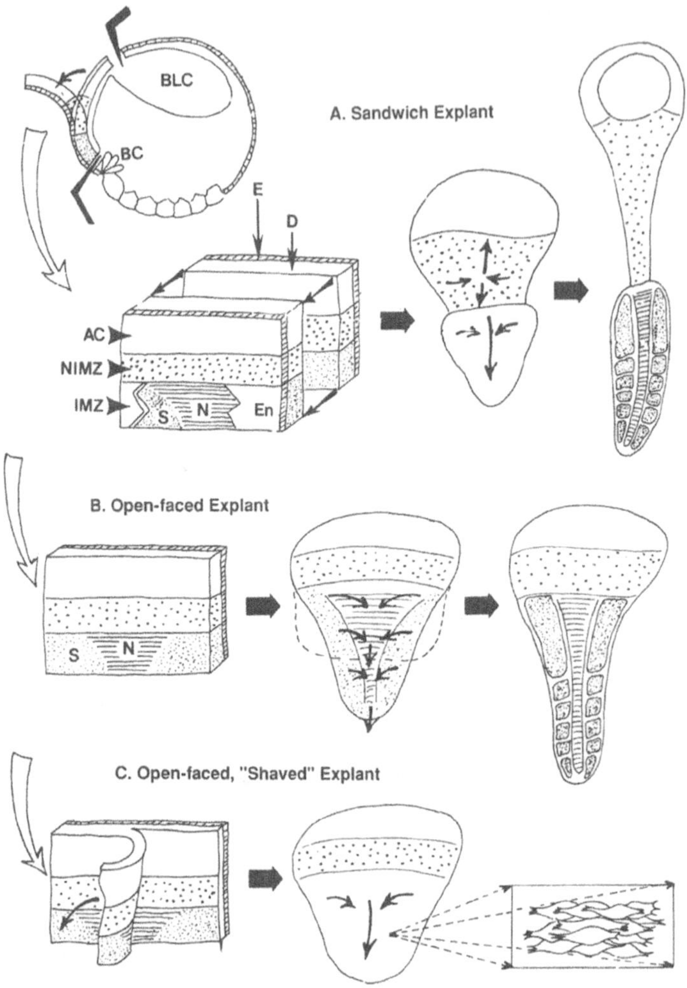

Figure 1. A schematic diagram shows the development of sandwich explants (**A**), open-faced explants (**B**), and shaved open-faced explants (**C**). The dorsal sector of the early gastrula, depicted in sagittal section at the upper left, is excised on both sides about 60 degrees from the midline and extending from the bottle cells (BC) to near the animal pole. This explant consists of a superficial epithelial region (E) and a deep, nonepithelial (mesenchymal) region (D). The explants consist of an animal cap region (AC), a noninvoluting marginal zone (NIMZ), and an involuting marginal zone (IMZ). The AC and NIMZ have a uniform tissue fate in both layers, which is neural ectoderm, the AC making the brain and the NIMZ making the ventral part of the spinal cord. The IMZ differs in that the superficial and deep regions have different fates; the superficial layer forms endoderm (En) and the deep region forms mesoderm, specifically notochordal (N) and somatic (S) mesoderm. The sandwich explants show two regions of convergence and extension, one in the NIMZ and one in the IMZ (arrows in **A**). The NIMZ develops into a tangle of neurons and the IMZ develops into notochord and somites. The open-faced explants are cultured in Danilchik's solution under a coverslip conditions under which the IMZ will converge, extend, and differentiate into somites and notochord (**B**). In open-faced explants, the NIMZ does not converge or extend (Wilson and Keller 1991). Shaved explants (**C**) are made and cultured like open-faced explants except that the inner most layers of deep cells are shaved or peeled off with an eyebrow knife. Exposed deep cells, next to the epithelium, are observed and recorded on video for analysis of cell behavior (box, lower right).

population of cells, we developed a culture solution and a method of explanation that allows normal development and morphogenesis of deep cells in view of the microscope. A single explant of the dorsal sector of the gastrula is cultured between a coverslip and a culture dish in Danilchik's solution (Keller *et al.* 1985a,b). These are called "open-faced" explants because the deep cells are exposed to the medium and to the microscope objective [Figure 1B]. Danilchik's solution is a saline that mimics the salts in blastocoel fluid (Gillespie 1983) and is characterized by high pH and low chloride concentration. The original formula (Keller *et al.* 1985a,b) has been revised twice–once to increase its stability (Wilson and Keller 1991) and again to improve its capacity to support deep cell behavior (Shih and Keller 1991a). We have used the open-faced explant method to resolve the cellular behavior of convergence and extension in the axial mesoderm in the neurula (Wilson *et al.* 1989; Keller *et al.* 1989a) and in the gastrula stages of development (Wilson and Keller 1991; Shih and Keller 1991b).

MECHANISM OF CONVERGENCE AND EXTENSION IN THE IMZ

Probable Causes

What causes this tissue to get longer and narrower? We suspected from indirect evidence that cell rearrangement was involved. Scanning electron microscopic studies suggested that area increase in the marginal zone and in the animal cap probably involved radial intercalation of cells, to produce a thinner tissue of larger area (Keller 1980). Time-lapse cinemicrographic tracing showed that superficial epithelial cells accommodated convergence and extension movements by *mediolateral intercalation*–they intercalated between one another along the mediolateral axis to form a longer but narrower array (Keller 1978). Tracing cells with the lineage tracer, fluorescein dextran amine (FDA), in whole embryos showed that the deep cells in the axial mesoderm also undergo mediolateral intercalation during convergence and extension (Keller and Tibbetts 1989). Since these were the cells that microsurgical alteration experiments had suggested were the active, force-producing cell population, this indirect evidence suggested that deep cell intercalations were involved in producing convergence and extension, but the details were hidden from view in the opaque embryo.

Direct Observation and Analysis of Deep Cell Intercalation

To resolve the details of deep cell intercalation, we used the "open-faced" explant system with high resolution time-lapse videomicrography (Wilson *et al.* 1989; Keller *et al.* 1989a; Wilson 1990; Wilson and Keller 1991; Shih and Keller 1991b). At explant stages equivalent to the early gastrula, the explant extends but does not converge significantly [Figure 2]. High resolution recordings show more cells entering than leaving the inner most layer of deep cells. As they do so, the explant increases in area. These observations reflect the fact that deep cells intercalate between one another along the radial aspect of the embryo to form fewer layers of greater area [Figure 2] (Wilson 1990; Wilson and Keller 1991). The area increase does not result in extension in all directions but only in the antero-posterior direction [Figure 2]. This bias may be

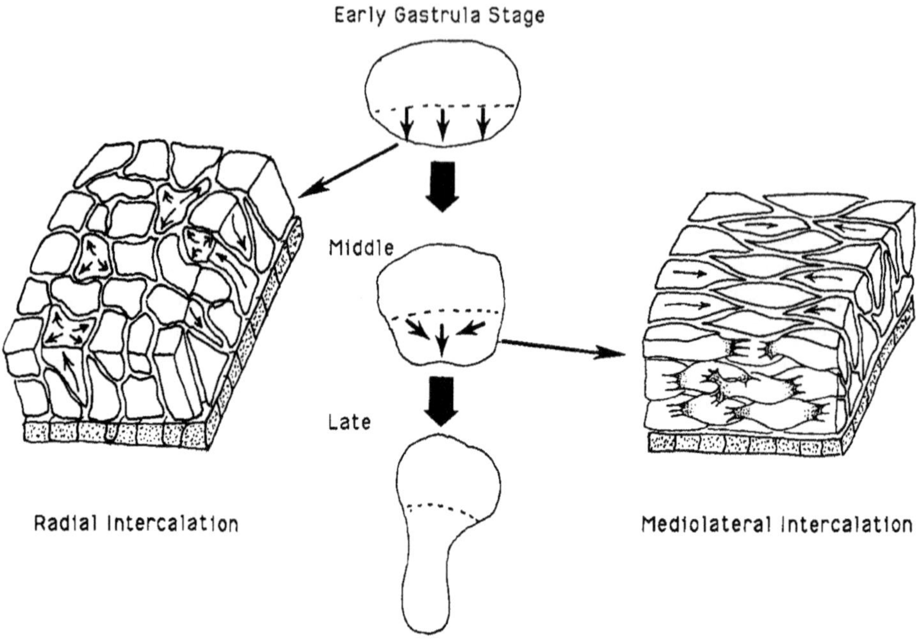

Figure 2. A schematic diagram shows the behavior of cells during convergence and extension of the involuting marginal zone (IMZ) in open-faced explants (see Wilson and Keller 1991). In the center column, outlines of open-faced explants are shown at early, middle and late gastrula-neurula stages. At the sides, deep cell behaviors are shown; the deep cells are unshaded and the superficial epithelium is shaded. Between the early and middle gastrula stages, the IMZ (area below the dashed lines) extends considerably but converges very little (arrows). High-resolution video recordings show that this extension is produced by radial intercalation (left) in which several layers of deep cells intercalate along the radius of the embryo to form a greater area. From the middle gastrula stage onward, the IMZ continues to extend but also converges. Recordings of cell behavior show that this occurs by mediolateral intercalation in which individual cells move between one another along the mediolateral axis to form a longer, narrower array (right). (modified from Wilson and Keller 1991)

because deep cells tend to intercalate between antero-posterior neighbors more often than between mediolateral neighbors. The bias could also arise if intercalation was unbiased but the restraining epithelium, which all the evidence suggests must be passively stretched by the deep cells, is easier to stretch in the antero-posterior direction. Thus any area generated by intercalation would be channeled into length. Lastly, mediolateral intercalation could be occurring in layers not seen and thus biasing the event of radial intercalation (see Wilson 1990; Wilson and Keller 1991).

Following radial intercalation, at about the midgastrula stage, the deep cells undergo *mediolateral intercalation*, moving between one another along the mediolateral axis to produce a narrower but longer array (Wilson and Keller 1991) [Figure 2]. Unlike radial intercalation, which produces extension by thinning, mediolateral intercalation produces extension by convergence (Wilson and Keller 1991). Radial intercalation ceases by the midgastrula stage, whereas mediolateral intercalation continues through neurulation, having become obvious after radial intercalation stops. As noted above, mediolateral intercalation may have been occurring during radial intercalation as well. When the entire circumblastoporal region is considered, progressively more lateral and ventral tissue is recruited into the sequence of behaviors--radial intercalation followed by mediolateral intercalation--and the ventral-most axial mesoderm is passing through this sequence of behaviors in the neurula stage (Wilson *et al.* 1989).

Convergence and Extension Are Separable Processes

Convergence and extension very often occur together and in these cases we have referred to the pair of processes as "convergent extension" (see Keller *et al.* 1985a,b). But the two movements need not occur together. The volume of tissue seems to be conserved during these movements, and extension can be produced by thinning or by convergence. Radial intercalation produces *extension* by *thinning*, whereas mediolateral intercalation produces *extension* by *converging*. It is only in this case that the term convergent extension is appropriate. Not all of the converging movements are necessarily translated into extension; in the somitic mesoderm, convergence is accompanied by thickening as well as extension (see Wilson *et al.* 1989).

PROTRUSIVE ACTIVITY UNDERLYING CELL INTERCALATION

Radial Intercalation

Although little is known about the mechanism, the superficial epithelium is necessary for radial intercalation to be successful (Wilson 1990). One possible mechanism is that deep cells are stabilized in their movements when they encounter the inner surface of the epithelium, either because their protrusive activity is stifled on contact with the epithelium or because this is a very adhesive surface from which deep cells can not de-adhere (Keller 1980). Thus deep cells would move about between several layers, not necessarily directed toward the epithelium, but they would still gradually accumulate as a monolayer at the inner surface of the epithelium, since once they attached there they could not leave.

Mediolateral Intercalation

More is known about the mechanism of mediolateral intercalation. The deep cells next to the epithelium are the ones expressing most strongly the protrusive activity that causes mediolateral intercalation (see "Tissue Interactions Controlling Cell Intercalation" below), and thus we have analyzed these cells extensively.

To observe the deep cells next to the epithelium, John Shih developed a method of making "shaved" explants, in which the deepest cell layers of the IMZ were removed until just one or in some cases two layers of deep cells, depending on the experiment, remained on the inner surface of the epithelium [Figure 1C]. These explants were cultured using modifications of procedures developed previously (see Shih and Keller 1991b). Radial intercalation does not occur, since in most cases there is only one layer of deep cells. The cells divide and oscillate about in place until the midgastrula stage when convergence and extension occurs [Figure 3a]. Tracings of individual cells shows that convergence and extension involves mediolateral intercalation [Figure 3b].

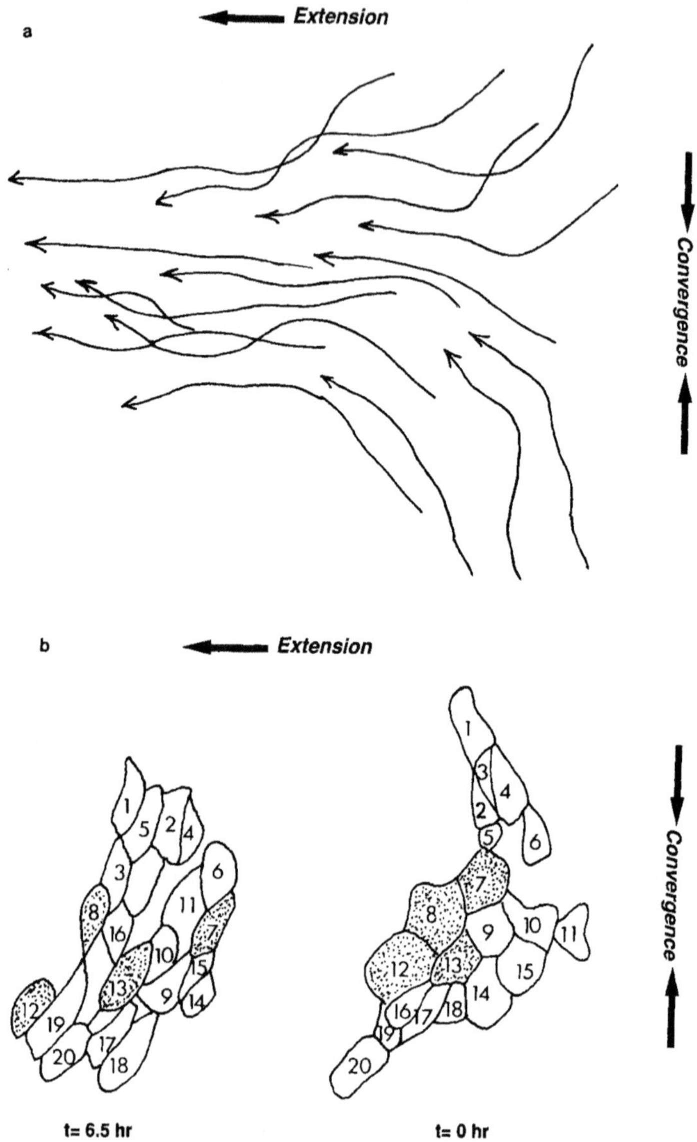

Figure 3. Tracings of individual cells from time-lapse recordings show the movements of cells during convergence and extension of an open-faced, shaved explant (**a**). Tracings of outlines of individual cells over the same interval shows the mediolateral intercalation that results in the explant becoming longer and narrower (**b**). Note the separation of the shaded pairs of cells, cells #7 and #8 and cells #12 and #13, along the axis of extension. See Shih and Keller (1991b) for details.

Mediolateral intercalation occurs in a transverse band of deep cells [see pointers, Figure 4a], characterized by a unique morphology; they become elongated in the mediolateral direction, aligned parallel to one another, and oriented perpendicular to the axis of extension. Under epi-illumination and at relatively low magnification, these cells appear

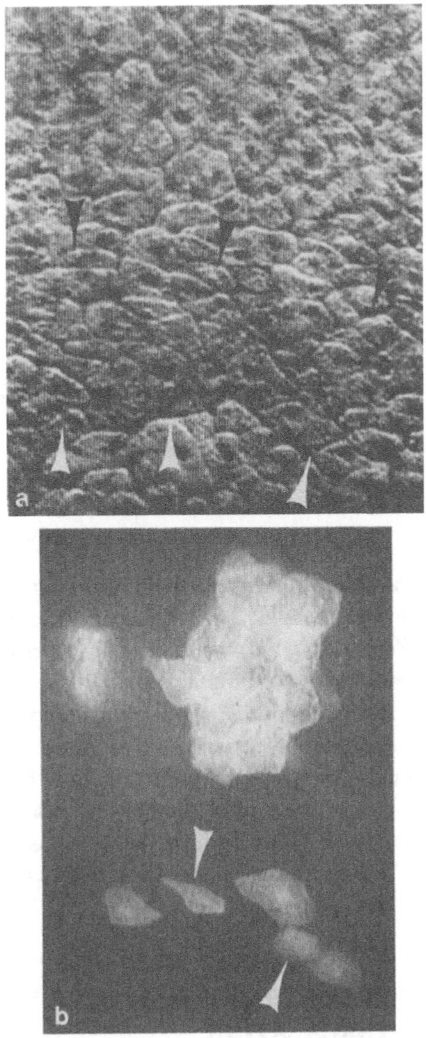

Figure 4. A light micrograph taken from a time-lapse video-recording (**a**) shows a band of mediolaterally elongate, aligned cells that develop during convergence and extension of shaved explants. The dark pointers approximate its posterior boundary and the white pointers approximate the anterior boundary. A fluorescence micrograph taken from a time-lapse recording of another explant (**b**) shows elongated cells in this band, labelled with fluorescein dextran amine (pointers), undergoing mediolateral intercalation. The axis of extension is approximately vertical in both micrographs. See Shih and Keller (1991b) for details.

to show pulsatile protrusive activity at their medial and lateral ends but little or none at their anterior and posterior sides. Mediolateral intercalation occurs during this activity, as the cells shuffle together along the mediolatateral axis.

The detailed morphology and protrusive activity of these cells is better visualized by time-lapse recording of fluorescein dextran amine labelled deep cells with fluorescence microscopy and image processing [Figure 4b] (see Shih and Keller 1991b for details). The deep cells show lamelliform protrusive activity predominately in the medial and lateral directions, and they become progressively more elongate in the mediolateral dimension and aligned parallel to this axis [Figure 5]. A plot of the protrusive activity of several labelled deep cells shows this dramatic polarization [Figure 6] (Shih and Keller 1991b). In contrast, the anterior and posterior sides of the cells occasionally extend small, filiform protrusions but these margins of the cells rarely, if ever advance across one another. Meanwhile the lamelliform protrusions at the medial

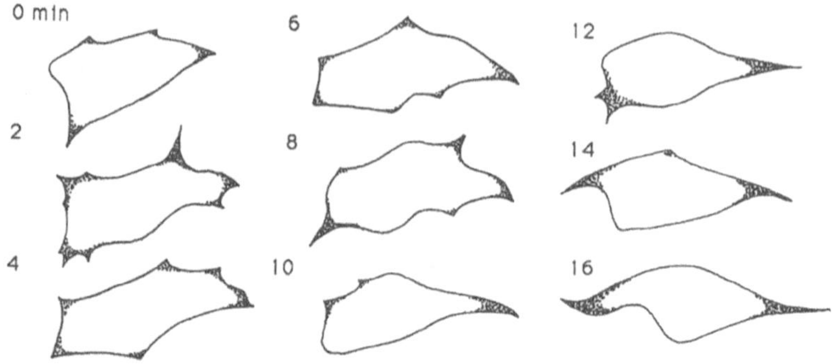

Figure 5. A series of tracings of time-lapse recordings of cells labelled with fluorescein dextran amine (FDA) shows the fine protrusive activity (shaded areas) of one the cells shown in Figure 4b during the mediolateral elongation, alignment, and intercalation (see Shih and Keller 1991b). The axis of extension is approximately vertical.

and lateral ends appear to exert traction on adjacent cells and thus move the cells between one another in the mediolateral direction, producing mediolateral intercalation [Figure 2]. The morphological polarization of these cells is further characterized by scanning electron microscopy of the cells immediately under the dorsal epithelium of whole embryos. They are elongate and aligned parallel to one another and parallel to the circumference of the blastopore. They bear large lamelliform protrusions on their medial and lateral ends and fine filiform protrusions on their long anterior and posterior sides hold the cells "cheek by jowl" in a parallel array (Keller *et al.* 1989b) [Figure 7]. The filiform protrusions may not exist as such but may be formed by shrinkage; nonetheless, they reflect contact points at the anterior and posterior sides of the cells.

HOW POLARIZED PROTRUSIVE ACTIVITY PRODUCES MEDIOLATERAL INTERCALATION

Based on the data from the shaved explants (Shih and Keller 1991b) we will refine our previous notions about how mediolateral cell intercalation works (see Keller *et al.* 1985a,b; Keller and Hardin 1987). Our observations show that the deep cells become specialized in their protrusive activity such that the small filiform protrusions on the anterior and posterior sides can not advance across their neighbors, whereas the large lamelliform protrusions on the medial and lateral ends can advance across their neighbors. The cells become highly elongated and aligned parallel to the mediolateral axis, probably as a result of this specialized protrusive activity, since the elongation and alignment develop slowly in conjunction with the exercise of this polarized protrusive activity. We propose that the two types of protrusions have different contact behavior. The small filiform ones touch, adhere and hold, but do not extend further or support advancement of the cell in that direction. In addition these small protrusions must also be dynamic, to allow sliding of cells past one another. Presence of contacts need not be an obstacle to sliding past one another; epithelial cells tightly bound at their apical

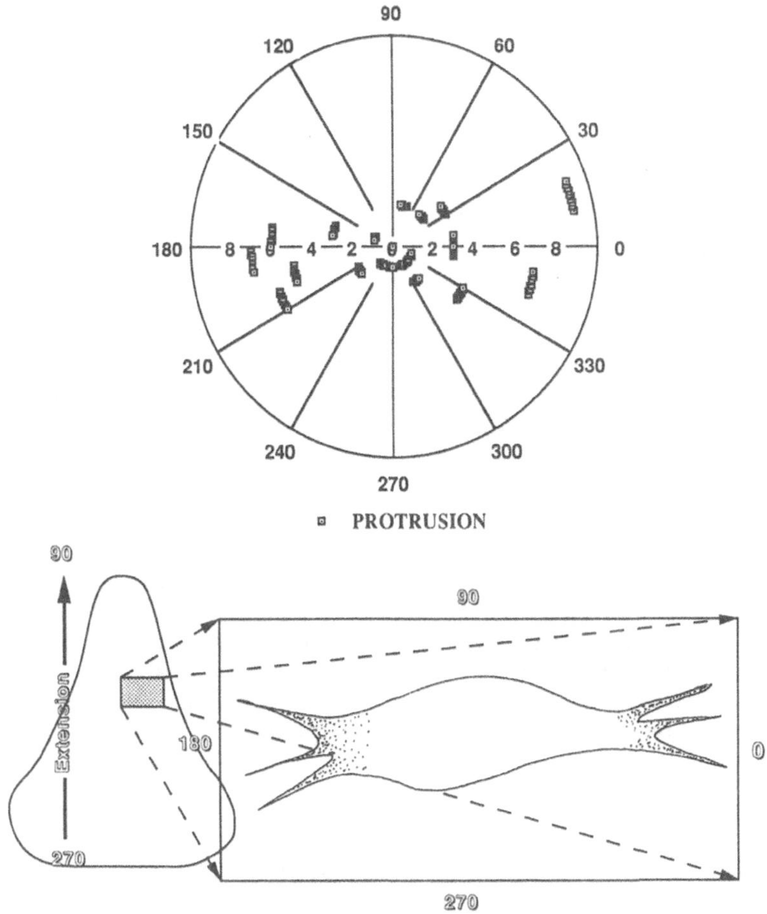

Figure 6. The mediolateral bias of protrusive activity of the deep cells during mediolateral intercalation is shown by a plot of frequency of protrusions on a circle, oriented such that the axis of extension is 90-270 degrees (see Shih and Keller 1991b).

margins by junctional complexes still rearrange (Keller 1978; Keller and Trinkaus 1987). In contrast, the large lamelliform ones extend, adhere, and pull on adjacent cells, in both directions. The result is that the cells elongate, align themselves, and pull themselves between one another in the mediolateral direction (Shih and Keller 1991b). We believe that they must exert traction in both directions, medially and laterally, to intercalate, since if they pull only one way or the other, all cells would be attempting to use cells as substrates that were also trying to move in that direction and the net movement might be zero.

It is clear that they are using one another as substrates for movement; in both SEMs and video recordings, the cells are applied to the surfaces of other deep cells (Keller *et al.* 1989b; Shih and Keller 1991b). It is also possible that they adhere to the undersurface of the epithelium, and that traction on this surface may also play a role

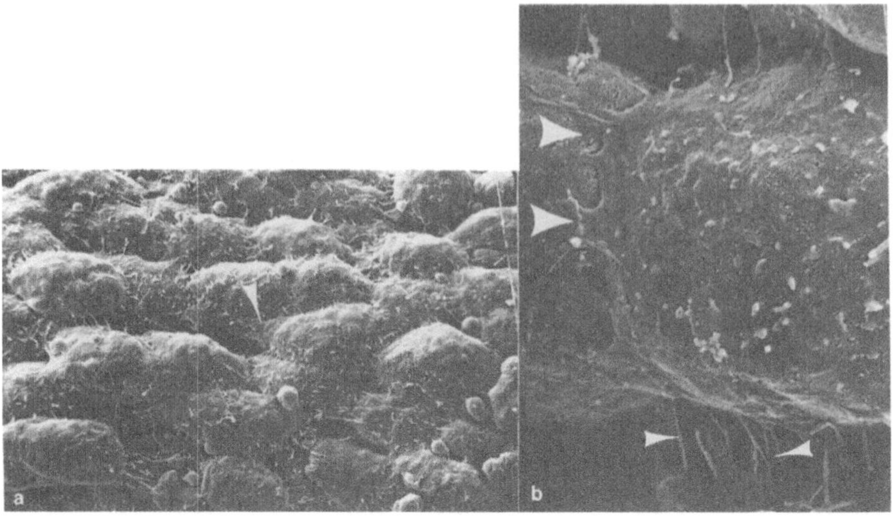

Figure 7. Scanning electron micrographs show the morphological polarization of the deep cells immediately adjacent to the endodermal epithelium during convergence and extension of the IMZ in whole embryos. Note their elongation and parallel alignment perpendicular to the axis of extension, which is vertical in the micrograph (**a**). A higher magnification (**b**) shows the large lamelliform protrusions characteristic of the medial and lateral ends (large pointers), and the fine filiform protrusions characteristic of their anterior and posterior sides (small pointers). Other examples of the large medial and lateral protrusions and their relation to the cells can also be seen at lower magnification (pointer in a) (from Keller *et al.* 1989b).

in mediolateral intercalation. We would like to observe deep cell interaction with the under side of the epithelium, but we have not yet solved the technical problems in doing this experiment.

If the explant extends by cells wedging between one another, then the greater the resistance to extension, the more difficult it would be for the cells to intercalate. If the polarized protrusive activity at the medial and lateral ends produces this wedging, then cells frustrated in this behavior should elongate mediolaterally in their unsuccessful attempt to intercalate. We plan to test this notion by offering greater resistance to extension and measuring cell elongation and intercalation. Greater resistance should produce greater elongation and less intercalation.

The deep cells should be under tension in the mediolateral direction and under compression in the antero-posterior direction if our ideas about mechanism are correct. This we are testing using micromanipulation to break the protrusions on the sides of the cells and measuring retraction on medial and lateral surfaces and on anterior and posterior surfaces.

Tissue Interactions Controlling Cell Intercalation

The convergence and extension and underlying polarized protrusive activity described above is controlled in part by the epithelial endodermal layer (Shih and Keller 1991a). An explant made of ventral deep marginal zone cells and the dorsal endodermal epithelium converges, extends and forms dorsal tissues, whereas normal ventral explants do neither (Shih and Keller 1991a). Moreover, the epithelial layer can direct the axis of extension. Epithelial sheets rotated 90 degrees in the early gastrula stage redirects the axis of extension, and high resolution recordings of cell behavior in such explants, shaved so that the deep cell behavior can be observed in great detail, shows that, in fact, protrusive activity of the deep cells is redirected by the reoriented epithelium (Shih and Keller, unpublished data). Later, in the neurula stage (Wilson *et al.* 1989), and perhaps earlier (Wilson 1990), the deep cells can converge and extend without the endodermal epithelium. Thus, it appears that early in gastrulation the epithelial layer induces the deep cells to display the polarized protrusive activity described above, and thus the direction of intercalation. Later they become independent in these respects and can continue intercalation on their own.

It could be that the epithelium stabilizes the polarized protrusive activity of deep cells and once this is done, they can continue intercalating without further instructive influence from the epithelium. Alternatively, other factors may substitute and indeed change the character of the protrusive activity later in development. When the notochord-somite boundary forms, the notochord cells adjacent to the boundary become monopolar in their protrusive activity (Keller *et al.* 1989a). When the boundary first forms, the notochord cells may show blebbing activity on their sides adjacent to the boundary, but soon the boundary ends of the cells cease all protrusive activity, and as a result become monopolar, with aggressive protrusive activity only at their inner ends. Aside from this boundary effect, the notochord cells show the same configuration of protrusive activity as the deep cells prior to boundary formation–they are mediolaterally elongated, aligned parallel to one another and perpendicular to the axis of extension, and they show translocating protrusive activity only on their narrow medial and lateral ends (Keller *et al.* 1989a). Weliky and Oster (this volume) have simulated intercalation of notochord cells with a computer, using these rules of protrusive activity. This is a powerful new technique of evaluating whether or not a given regimen of protrusive activity would be expected to produce a specific cell population behavior.

MECHANISM OF CONVERGENCE AND EXTENSION OF THE NIMZ

Before integrating the function of cell intercalation in the IMZ with the rest of gastrulation, we must return for a moment to the NIMZ, since the convergence and extension of this region comprises the other half of our mechanism of how cell intercalations function in gastrulation. The dorsal NIMZ converges and extends in the embryo to form the neural plate, and the degree to which this behavior is autonomous and independent of the involuted mesoderm has been analyzed in urodeles (Jacobson and Gordon 1976), which appear to differ somewhat from *Xenopus*. In sandwich explants of *Xenopus*, this region converges and extends autonomously to form neural tissue. The cells do not columnarize and thus it does not thicken like a normal neural plate but forms a tangle of neurons (Keller and Danilchik 1988). Convergence and extension of the NIMZ in explants occurs by radial intercalation and mediolateral intercalation (Shih and Keller, unpublished results), overall similar to what is seen in the IMZ. We have not gotten the NIMZ to routinely extend in the open-faced explants, and thus we do not know what protrusive activity occurs in this region. When we get them to extend in open-faced explants, we will be able to determine whether they show the same protrusive activity and mechanism of intercalation as IMZ cells or different ones.

We have no direct evidence that convergence and extension of the NIMZ is autonomous in the embryo, because these movements in the NIMZ occur coincident with the movements of the IMZ, which has involuted beneath NIMZ and may be dragging the NIMZ passively along. Since convergence and extension of the NIMZ do occur in sandwich explants but not in open-faced explants, these processes are likely to require some sort of physiological or mechanical interaction with another tissue at its basal surface. In the explants the basal surface of another explant just like it will suffice (see Keller and Danilchik 1988). Whether basal contact with the involuted IMZ in the embryos serves this role is not known. Indirect evidence suggests that this is the case and that, in fact, the NIMZ does actively extend in the embryo (Keller and Danilchik 1988).

FUNCTION OF CONVERGENCE AND EXTENSION IN GASTRULATION

Convergence and extension are made effective in gastrulation by virtue of the geometric and biomechanical context in which they act. In cultured sandwich explants, the two convergence and extension machines are displayed serially, one in the IMZ and the other in the NIMZ. But *in vivo*, they act in parallel and in concert—the IMZ converges and extends primarily in the postinvolution position and the NIMZ converges and extends primarily on the outside, preinvolution position [Figure 8]. The IMZ is turned inward by the formation of bottle cells early in gastrulation (Hardin and Keller 1988). Simultaneously, radial intercalation occurs, producing extension, and only extension, in the IMZ, which aids its involution (Wilson and Keller 1991). Only after it has extended somewhat and its leading edge has involuted, does the IMZ show mediolateral intercalation, which produces *convergence* and extension at the lip and in the postinvolution position. The increased length of the tissue is taken up in two ways—it is pushed backward across the vegetal region and anteriorly where its movement toward the animal pole is aided by the migration of the head mesodermal cells [Figure 8]. Meanwhile, preinvolution IMZ and the NIMZ are continuing to extend, by virtue of their continued radial intercalation, which tends to push the remaining uninvoluted IMZ material over the blastoporal lip [Figure 8]. There it begins mediolateral intercalation and convergence as well as extension. Convergence narrows the circumference of the marginal zone, putting a constriction force on the marginal zone just outside of the blastoporal lip, aiding involution and squeezing the blastopore shut (see Wilson and Keller 1991).

The process of extension by radial intercalation and of convergence and extension by mediolateral intercalation progresses from the dorsal sector of the marginal zone to the lateral sector and finally to the ventral sector, occurring in all the notochordal and somatic mesodermal tissues (see Keller and Danilchik 1988; Keller and Tibbetts 1989). Radial intercalation is still occurring in the ventral, prospective somatic mesoderm at the neurula stage, and its is followed by mediolateral intercalation (see Wilson *et al.* 1989).

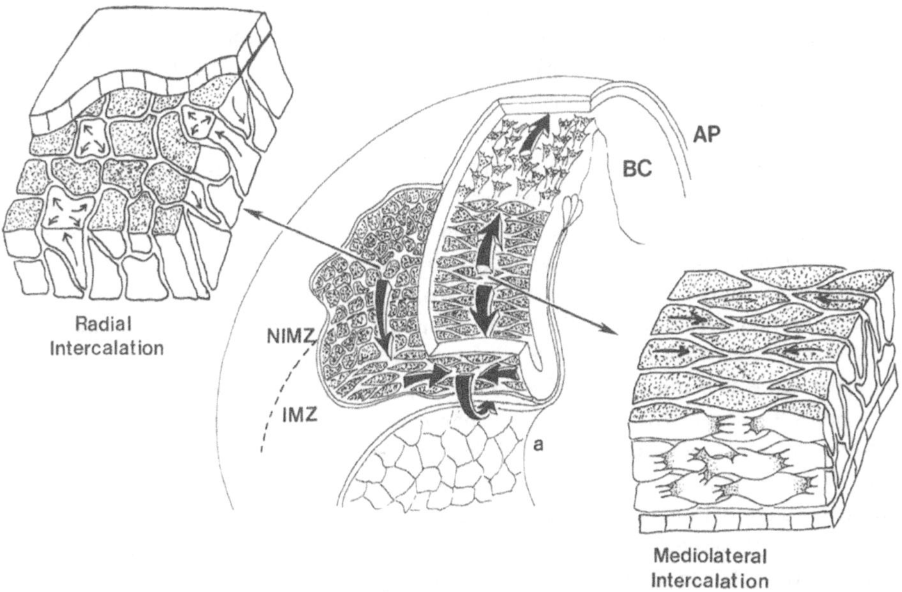

Figure 8. A schematic, cut-away diagram of the dorsal aspect of an early mid-gastrula shows how cell intercalation in the IMZ and NIMZ regions and involution are integrated in the marginal zone. In the central drawing, shaded isodiametric cells are intercalating radially, while transversely elongated cells are involved in mediolateral intercalation. Early in gastrulation, radial intercalation occurs in both the NIMZ and IMZ, causing extension toward the blastopore lip. Just outside the lip, mediolateral intercalation begins in the IMZ. The resulting convergence generates a hoop stress that tends to pull the IMZ over the lip. After involution, the cells of the IMZ continue to intercalate mediolaterally, elongating and narrowing the axial mesoderm. This elongation is accommodated in two ways: migrating cells at the leading edge of the mesodermal mantle (shingled cells beneath topmost arrow) advance toward the animal cap, and the blastopore lip moves vegetally across the yolk plug. Involution continues, at least in part because material not yet involuted (NIMZ and remaining IMZ) extends more rapidly than involuted tissue. Convergence and extension in both layers eventually squeezes the blastopore shut. The two types of intercalation are illustrated in the accompanying smaller diagrams. At left, cells move between layers to thin and expand the tissue during radial intercalation. At right, transversely elongated and aligned cells intercalate mediolaterally, displaying protrusions at their narrow ends. This characterization is based on the detailed morphology (Keller *et al.* 1989b) and protrusive activity (Shih and Keller 1990b) of deep cells close to the epithelium. BC: blastocoel; AP: animal pole. (Figure and legend from Wilson and Keller 1991).

CONVERGENCE AND EXTENSION AND CELL INTERCALATION IN OTHER SYSTEMS

Convergence and extension occur in many vertebrate and invertebrate developing systems. These movements occur during neurulation of the amphibian (Jacobson and Gordon 1976), and the chick (Schoenwolf and Alvarez 1989, and Schoenwolf, this volume), and in gastrulation of the teleost fish (Kimmel *et al.* 1990; Warga and Kimmel 1990), the sturgeon (Bolker 1989), the sea urchin (Ettensohn 1985; Hardin and Cheng 1986; Hardin 1989) and *Drosophila* (see article by Wieschaus and others, this volume).

In the sea urchin (Hardin 1989) and in the fish (Kageyama 1982; Keller and Trinkaus,1987; Kimmel *et al.* 1990; Warga and Kimmel 1990), and in the ascidian notochord (Miyamoto and Crowther 1985) there is evidence from direct observations that cell intercalation occurs during these movements, and there is compelling indirect evidence that the same occurs in *Drosophila* (Wieschaus and others, this volume). In the case of the elongation of the sea urchin archenteron, there is compelling evidence that the convergence and extension is a regionally autonomous process, at least in its middle phase, and that cell intercalation plays an active role (Hardin 1988). Cell intercalation in various patterns appears to be a universal morphogenetic process in the development and regeneration of metazoans and may occur among epithelial or nonepithelial cells (Keller 1987). Some intercalations appear to follow passively from forces generated elsewhere; others appear to be active, force-producing processes, and in other cases it is not known whether they are active or passive events.

Since we know little about the mechanisms of these behaviors, and what little we do know is only from a few systems, it is not known whether all instances of active or passive cell intercalation are based on a single mechanism or whether there are several, or perhaps many fundamentally different mechanisms. However, a detailed analysis of the protrusive activity of the actively intercalating epithelial cells of the sea urchin archenteron (Hardin 1989) shows little similarity to the behavior of the mesenchymal amphibian deep cells. This suggests that perhaps there are at least two mechanisms, one for epithelial cells and one for deep cells. The polarized, pulsatile protrusive activity seen in the amphibian gastrula mesoderm (Shih and Keller 1991b) and in the amphibian notochord (Keller *et al.* 1989a) also occurs in the ascidian notochord (Miyamoto and Crowther 1985), so there is conservation of cell behavior among mesenchyme cells of these species. Whether this extends to other mesenchymal cell populations undergoing cell rearrangement is not known. Investigations of cell intercalations in a variety of new systems, and in more depth for those that have been studied thus far, should be done to resolve these issues.

ACKNOWLEDGEMENTS

We thank former members of the laboratory, particularly Jeff Hardin, Rudolf Winklbauer, and Mark Cooper, and visitors to the laboratory, Antone Jacobson and J.P. Trinkaus, who have had a hand in developing our ideas and experiments on cell intercalation, above and beyond what they have published. The work described here was supported by National Institutes of Health grants HD18979 and HD25594 and National Science Foundation grant DCB89052 to Ray Keller, NSF grant DMS 8618975 to Paul Wilson and George Oster; Paul Wilson was supported in part by National Institutes of Health Training Grant HD7375. We thank Paul Tibbetts for technical assistance.

REFERENCES

Baker, P. 1965. Fine structure and morphogenetic movements in the gastrula of the tree frog, *Hyla regilla*. *J. Cell Biol.* 24:95-116.

Bolker, J. 1989. Gastrulation in the white sturgeon, *Acipenser transmontanus*. *Am. Zool*. 29:387.

Ettensohn, C. 1985. Gastrulation in the sea urchin is accompanied by the rearrangement of invaginating epithelial cells. *Dev. Biol.* 112:383-390.

Gerhart, J. and R.E. Keller. 1986. Region-specific cell activities in amphibian gastrulation. *Annu. Rev. Cell Biol.* 2:201-229.

Gillespie, J.I. 1983. The distribution of small ions during the early development of *Xenopus laevis* and *Ambystoma mexicanum* embryos. *J. Physiol.* 344:359-377.

Hardin, J. 1988. The role of secondary mesenchyme cells during sea urchin gastrulation studied by laser ablation. *Development* 103:317-324.

Hardin, J. 1989. Local shifts in position and polarized motility drive cell rearrangement during sea urchin gastrulation. *Dev. Biol.* 136:430-445.

Hardin, J. and L.Y. Cheng. 1986. The mechanisms and mechanics of archenteron elongation during sea urchin gastrulation. *Dev. Biol.* 115:490-501.

Hardin, J. and R. Keller. 1988. The behavior and function of bottle cells during gastrulation of *Xenopus laevis*. *Development* 103:211-230.

Holtfreter, J. 1933. Die totale Exogastrulation eine Selbstablosung Ektoderm von Entomesoderm. *Wilhelm Roux' Arch. Entwicklungsmech. Org.* 129:669-793.

Holtfreter, J. 1939. Gewebeaffinität, ein Mittel der Embryonalen Formbildung. *Arch. Exp. Zellforsch. Besonders Gewebezuecht* 23:169-209.

Holtfreter, J. 1943a. Properties and function of the surface coat in amphibian embryos. *J. Exp. Zool.* 93:251-323.

Holtfreter, J. 1943b. A study of the mechanics of gastrulation. Part I. *J. Exp. Zool.* 94:261-318.

Holtfreter, J. 1944. A study of the mechanics of gastrulation. Part II. *J. Exp. Zool.* 95:171-212.

Ikushima, N. and S. Maruyama. 1971. Structure and developmental tendency of the dorsal marginal zone in the early amphibian gastrula. *J. Embryol. Exp. Morphol.* 25:263-276.

Jacobson, A. and R. Gordon. 1976. Changes in the shape of the developing vertebrate nervous system analyzed experimentally, mathematically, and by computer simulation. *J. Exp. Zool.* 197:191-246.

Kageyama, T. 1982. Cellular basis of epiboly of the enveloping layer in the embryos of the Medaka, *Oriyzias latipes*. II. Evidence for cell rearrangement. *J. Exp. Zool.* 219:241-256.

Keller, R.E. 1978. Time-lapse cinemicrographic analysis of superficial cell behavior during and prior to gastrulation in *Xenopus laevis*. *J. Morphol.* 157:223-248.

Keller, R.E. 1980. The cellular basis of epiboly: An SEM study of deep cell rearrangement during gastrulation in *Xenopus laevis*. *J. Embryol. Exp. Morphol.* 60:201-234.

Keller, R.E. 1981. An experimental analysis of the role of bottle cells and the deep marginal zone in gastrulation of *Xenopus laevis*. *J. Exp. Zool.* 216:81-101.

Keller, R.E. 1984. The cellular basis of gastrulation in *Xenopus laevis*: Active post-involution convergence and extension by medio-lateral interdigitation. *Am. Zool.* 24:589-603.

Keller, R.E. 1986. The cellular basis of amphibian gastrulation. p. 241-327 *In: Developmental Biology: A Comprehensive Synthesis. Vol. 2. The Cellular Basis of Morphogenesis*. L.W. Browder (Ed.). Plenum Press, New York.

Keller, R.E. 1987. Cell rearrangement in morphogenesis. *Zool. Sci.* 4:763-779.

Keller, R.E. 1991. Gastrulation in *Xenopus* embryos without a blastocoel roof. In preparation.

Keller, R., M.S. Cooper, M. Danilchik, P. Tibbetts, and P.A. Wilson. 1989a. Cell intercalation during notochord development in *Xenopus laevis. J. Exp. Zool.* 251:134-154.

Keller, R.E. and M. Danilchik. 1988. Regional expression, pattern and timing of convergence and extension during gastrulation of *Xenopus laevis. Development* 103:193-210.

Keller, R.E., M. Danilchik, R. Gimlich, and J. Shih. 1985a. Convergent extension by cell intercalation during gastrulation of *Xenopus laevis*. p. 111-141. *In: Molecular Determinants of Animal Form.* G.M. Edelman (Ed.). Alan R. Liss, New York.

Keller, R.E., M. Danilchik, R. Gimlich, and J. Shih. 1985b. The function of convergent extension during gastrulation of *Xenopus laevis. J. Embryol. Exp. Morphol.* 89 (Suppl.):185-209.

Keller, R.E. and J. Hardin. 1987. Cell behavior during active cell rearrangement: Evidence and speculation. *J. Cell Sci. Suppl.* 8:369-393.

Keller, R.E. and G. Schoenwolf. 1977. An SEM study of cellular morphology, contact, and arrangement, as related to gastrulation in *Xenopus laevis. Wilhelm Roux's Arch. Dev. Biol.* 182:165-182.

Keller, R.E., J. Shih, and P.A. Wilson. 1989b. Morphological polarity of intercalating deep mesodermal cells in the organizer of *Xenopus laevis* gastrulae. p. 840. *In: Proceedings of the 47th Annual Meeting of the Electron Microscopy Society of America.* San Francisco Press, San Francisco.

Keller, R.E. and P. Tibbetts. 1989. Mediolateral cell intercalation is a property of the dorsal, axial mesoderm of *Xenopus laevis. Dev. Biol.* 131:539-549.

Keller, R. and J.P. Trinkaus. 1987. Rearrangement of enveloping layer cells without disruption of the epithelial permeability barrier as a factor in *Fundulus* epiboly. *Dev. Biol.* 120:12-24.

Kimmel, C., R. Warga, and T. Schilling. 1990. Origin and organization of the zebra fish fate map. *Development* 108:581-594.

Kubota, H. and A. Durston. 1978. Cinematographical study of cell migration in the opened gastrula of *Ambystoma mexicanum. J. Embryol. Exp. Morphol.* 44:71-80.

Nakatsuji, N. 1975. Studies on the gastrulation of amphibian embryos: Cell movement during gastrulation in *Xenopus laevis* embryos. *Wilhelm Roux's Arch. Dev. Biol.* 178:1-14.

Miyamoto, D.M. and R. Crowther. 1985. Formation of the notochord in living ascidian embryos. *J. Embryol. Exp. Morphol.* 86:1-17.

Phillips, H. and G. Davis. 1984. Liquid tissue mechanics in amphibian gastrulation: Germ-layer assembly in *Rana pipiens. Am. Zool.* 18:81-93.

Schechtman, A.M. 1942. The mechanics of amphibian gastrulation. I. Gastrulation-producing interactions between various regions of an anuran egg (*Hyla regilia*). *Univ. Calif. Publ. Zool.* 51:1-39.

Schoenwolf, G.C. Cell Movements in the Epiblast During Gastrulation and Neurulation in Avian Embryos. p. 1-28. *In: Gastrulation: Movements, Patterns, and Molecules.* R. Keller, W.H. Clark, Jr., F. Griffin (Eds.). Plenum Press, New York.

Schoenwolf, G.C. and I.S. Alvarez. 1989. Roles of neuroepithelial cell rearrangement and division in shaping of the avian neural plate. *Development* 106:427-439.

Shih, J. and R.E. Keller. 1991a. The epithelium of the dorsal marginal zone of *Xenopus* has organizer activity. Submitted.

Shih, J. and R.E. Keller. 1991b. The mechanism of mediolateral intercalation during *Xenopus* gastrulation: Directed protrusive activity and cell alignment. In preparation.

Spemann, H. 1938. *Embryonic Development and Induction.* Yale University Press, New York.

Townes, P. and J. Holtfreter. 1955. Directed movements and selective adhesion of embryonic amphibian cells. *J. Exp. Zool.* 128:53-120.

Vogt, W. 1929. Gestaltanalyse am Amphibienkeim mit Örtlicher Vitalfärbung. II. Teil. Gastrulation und Mesodermbildung bei Urodelen und Anuren. *Wilhelm Roux'Arch. Entwicklungsmech. Org.* 120:384-706.

Waddington, C.H. 1940. *Organizers and Genes.* Cambridge University Press, Cambridge.

Warga, R. and C. Kimmel. 1990. Cell movements during epiboly and gastrulation in zebra fish. *Development* 108:569-580.

Weiliky, M. and G. Oster. 1991. Dynamical Models for Cell Rearrangement During Morphogenesis. p. 135-146. *In: Gastrulation: Movements, Patterns, and Molecules.* R. Keller, W.H. Clark, Jr., F. Griffin (Eds.). Plenum Press, New York.

Wieschaus, E., D. Sweeton, and M. Costa. 1991. Convergence and Extension During Germband Elongation in *Drosophila* Embryos. p. 213-224. *In: Gastrulation: Movements, Patterns, and Molecules.* R. Keller, W.H. Clark, Jr., F. Griffin (Eds.). Plenum Press, New York.

Wilson, P.A. 1990. *The Development of the Axial Mesoderm in Xenopus laevis.* Ph.D. dissertation, University of California, Berkeley.

Wilson, P.A. and R.E. Keller. 1991. Cell rearrangement during gastrulation of *Xenopus*: Direct observation of cultured explants. *Development.* 105:155-166.

Wilson, P.A., G. Oster, and R.E. Keller. 1989. Cell rearrangement and segmentation in *Xenopus*: Direct observation of cultured explants. *Development* 105:155-166.

Winklbauer, R. and R. Keller. 1990. The role of extracellular matrix in amphibian gastrulation. *Semin. Dev. Biol.* 1:25-33.

IN VIVO ANALYSIS OF CONVERGENT CELL MOVEMENTS IN THE GERM RING OF *FUNDULUS*

J.P. Trinkaus[1,2] Madeleine Trinkaus[1,2]
and Rachel D. Fink[2,3]

[1] Department of Biology
Yale University
New Haven, Connecticut 06511
[2] Marine Biological Laboratory
Woods Hole, Massachusetts 02543
[3] Department of Biological Sciences
Mount Holyoke College
South Hadley, Massachusetts 01075

ABSTRACT

Carbon marking shows that the cells of the germ ring (GR) converge toward and enter the embryonic shield (ES). Interestingly, marks closer to the ES move faster. Cells lateral to the ES (in the prospective proximal yolk sac) also move into the ES.

Analysis of video tapes made of converging cells with DIC optics leads to the following conclusions. All cells of the GR engage in translocation almost all the time and their net directionality is always toward the ES, which they eventually join, contributing to its steady augmentation. But no cells move in a direct line toward the ES. All meander considerably. Their mean net rate during one 80 min period was 1.4

Gastrulation, Edited by R. Keller *et al.*
Plenum Press, New York, 1991

$\mu m/min \pm 0.36$ and their mean total rate was 1.9 $\mu m/min \pm 0.35$. GR cells that are near the ES move toward it at a higher net rate than those farther away. In consequence, their net trajectories are significantly longer. Since total cell movement and amount of motile quiescence are the same near and far, the greater net trajectory of the nearer cells must be due to less meandering. This suggests that there are exogenous factors that promote directionality toward the ES and that they operate more efficiently close to it. Cells in the prospective yolk sac adjacent to the ES also show net movement toward the ES. However, this movement is much less efficient than in the GR proper. Directional forces are apparently stronger in the marginal region of the blastoderm, namely the germ ring.

Converging deep cells in the GR move by both filolamellipodia and, much less frequently, by blebs. There is very little individual cell movement, and, at any moment, almost all cells are in contact with other cells in moving cell clusters. This is by far the dominant mode of movement. Clusters vary constantly in size, continually aggregating with other cells and other clusters and splitting, but always showing net movement toward the ES. Both moving cell clusters and individual filolamellipodial cells show protrusive activity solely on their free borders, which ceases whenever their surface makes contact with another cell surface. Clearly, they show contact paralysis or contact inhibition of cell movement. Nevertheless, they move and do so directionally to boot. The reason for this, we think, is that they are almost always members of moving cell clusters. As such, much of their movement is passive. Movement of clusters toward the ES is presumably due to directional factors in their environment which favor protrusive activity at their proximal margins. Both kinds of cells show intercalation or invasive activity frequently during convergence. But, consistent with their contact-inhibiting properties, filolamellipodial cells intercalate only when neighboring cells have separated enough to provide free space into which they can move. Cells that move by blebbing locomotion are not contact inhibiting and do not require free space for intercalation. Cells continue to divide during convergence at an apparently steady rate–e.g., 12% of cells observed during one 80 minute period. Although this temporarily arrests their movement, the daughter cells soon join in the mass convergent movement, usually in 2 min or less after completion of cytokinesis.

INTRODUCTION

Much of morphogenesis during early development depends on the directional movements of cells as individuals and in loosely organized cell streams (Trinkaus 1984a). Good examples are the prospective mesoderm of amphibian (Keller 1986) and chick gastrulae and the neural crest (Erickson 1985). Since the cells involved in these particular cell movements and others like them are hidden from the eye, either because of the opacity of the embryos or because they are deeply embedded within the embryo, direct study of the motile behavior of individual cells and small groups of cells in the living embryo has been precluded. This has been a frustrating and exasperating aspect of much otherwise really beautiful research on such systems. With the latest work of Keller, Shih, and Wilson (this volume), however, it seems that a ray of hope has penetrated this obscurity and direct observation of cell behavior in the prospective mesoderm of the *Xenopus* embryo has become possible.

This is an encouraging development, but neither this system nor any other system that we know of offers the supreme advantage presented by the convergent movements of deep cells of the germ ring toward the embryonic shield in certain teleost embryos. Because of the extraordinary lucidity of some of these eggs, cell surface behavior and other aspects of motility can be observed in detail during gastrulation, as individual cells and small groups of cells wend their preordained way toward their morphogenetic target (Figure 2). The developing gastrula of *Fundulus heteroclitus* presents just such material: the egg is highly transparent and cell movements in the germ ring take place near the surface. Moreover, the egg is fairly large (1.8 mm in diameter). The latter characteristic is important because the germ ring is necessarily more extended and its cells are less densely packed than in smaller embryos. In consequence, individual cells are more accessible for viewing.

It has been established for years that accumulation of cells on the dorsal side of the teleost blastoderm during gastrulation to form the embryonic shield, out of which the embryo proper is fashioned, is largely due to dorsad convergence of a particular population of cells near the margin of the blastoderm, the so-called germ ring (Figure 1). The main studies responsible for this important discovery were by Oppenheimer (1936) on *Fundulus* and by Pasteels (1936) and Ballard (1966, 1973) on the trout. Important though these investigations were, however, certain details were not provided. Determination of the extent and rate of cell movements in the germ ring and their precise relation to epiboly of the blastoderm needed a less diffuse label than vital dyes or clouds of chalk particles. Moreover, the early studies could reveal nothing of the detailed movements of the cells: whether they move as individuals or in clusters or both, their trajectories, their rates of movement, their modes of locomotion, and their contact relations with each other. Information on these important matters and others depended on following individual cells minute by minute, even second by second, within the germ ring (Figure 2).

We have attempted to gain answers to these questions by marking the blastoderm with discrete carbon particles and by time-lapse video filming of directional cell movement in the germ ring of *Fundulus*. The purpose of this communication is to describe some of the results of these studies and our attempts at analysis.

Although each system no doubt has peculiarities unto itself, it seems inevitable that detailed observations of cell motile behavior in one system are bound not only to settle important questions for that system, but in addition to suggest ways in which other similar movements are effected in other organisms. For this reason, it has occurred to us that the results of our studies of cell motile behavior in the *Fundulus* germ ring might be of general interest.

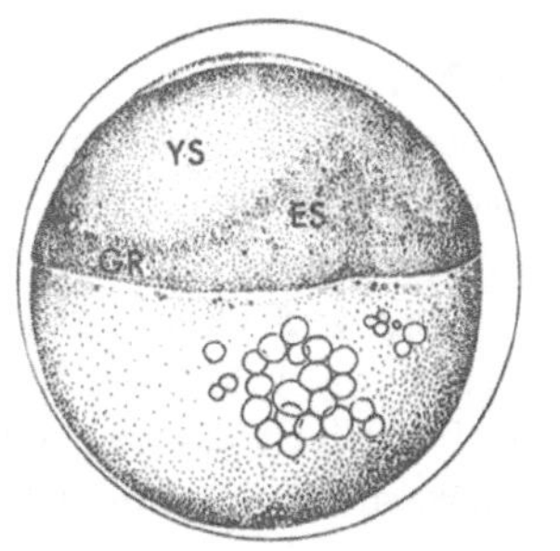

Figure 1. An early mid-gastrula of *Fundulus heteroclitus* [after stage 16 of Armstrong and Child (1965)], showing the embryonic shield (**ES**) and the germ ring (**GR**) in the marginal region of the blastoderm and the prospective yolk sac (**YS**). The blastoderm has expanded in epiboly to cover a little less than half of the yolk sphere. Cells of the **GR** and the part of the **YS** nearest the **ES** are converging toward the **ES**, contributing to its augmentation as it extends in epiboly (see text). The spheres in the yolk are lipid droplets. The egg is approximately 1.8 mm in diameter.

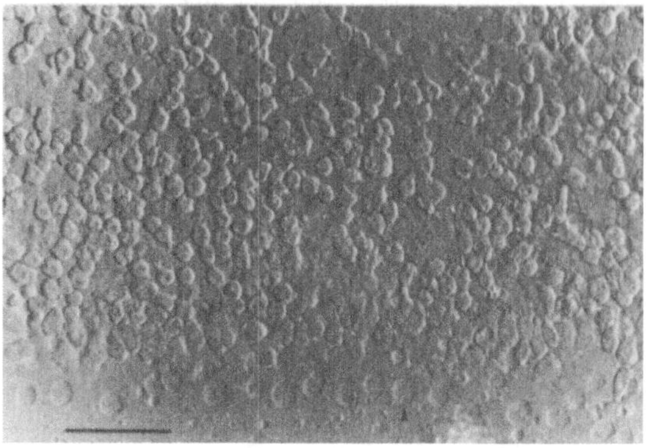

Figure 2. A light micrograph showing a small sector of a living germ ring of a mid-gastrula (stage 16½), somewhat lateral to the embryonic shield (see Figure 1). The embryonic shield is off the field at the right. The margin of the blastoderm runs roughly parallel to the bottom of the micrograph and is marked by an irregular row of large nuclei of the yolk syncytial layer (YSL) (arrowhead). Note that almost all cells in the GR are in contact with other cells in more or less loosely packed cell clusters. Many cells show protrusive activity. All of these cells and cell clusters show net convergent movement toward the ES. At its inner region, the upper 1/4 of this micrograph, where the GR is grading into the yolk sac (see Figure 1), the cells are more sparsely distributed. Scale bar equals 100 μm.

A complete, fully-illustrated paper, with quantitative details of our observations, is in press (1992).

MATERIAL AND METHODS

The materials for this study were developing gastrulae of *Fundulus heteroclitus*. Gametes were stripped from ripe adults and the eggs were artificially fertilized in sea water (Trinkaus 1967). Embryos were raised in sea water until late blastula or early gastrula stages, Armstrong and Child (1965) stages 13-15, and then manually dechorionated (Trinkaus 1967) and cultured in double-strength Holtfreter's solution. Room temperature was kept at approximately 21-22°C for all observations.

In order to obtain a global picture of the cell movements of gastrulation, certain regions of the germ ring and the non-germ ring regions near its inner margin were marked with particles of blood carbon (Trinkaus 1951) by inserting them into the deep cell population at the desired site with a sharpened tungsten needle. The embryos were then allowed to develop normally. The positions of the marks in successive observations were recorded with the aid of a camera lucida.

Cell movements were followed in detail with a Nikon inverted Diaphot microscope and Nomarski differential interference contrast optics. Time-lapse filming was performed with a Hamamatsu video-camera (Model C 2400) and an RCA time-lapse video recorder (Model TC 3920). Tapes of moving cells were supplemented by 35 mm photomicrography.

Filming chambers consisted of drilled plexiglass slides fitted with Number 0 coverslips by a thin film of silicon grease. Embryos were placed in these chambers in 2 × Holtfreter's solution and a second coverslip was added. The weight of the embryo compressed the region of the gastrula lying on the lower coverslip, providing a flat surface ideal for following cell movements. In this study, eggs were oriented so that the region of the germ ring to be observed rested squarely on the bottom coverslip. Embryos develop normally in such sealed chambers up to 4 days.

Quantitative data for analysis of cell movements were obtained from printouts of the videotapes provided by a Mitsubishi Video Copy Processor (Model P7OU). The trajectory of each cell's movement was determined from tracings of printouts made at five minute intervals from the original time-lapse sequences and its length was measured by means of a Sigma Scan Program (Jendel Scientific). For detailed observations of individual cell behavior, printouts were made at minute or even at several second intervals, as required.

Since each gastrula was left absolutely undisturbed during observation so as to avoid confusion when following cell movements, the duration of each sequence was limited. As cells converge in the germ ring toward the embryonic shield, the germ ring at the same time is moving passively toward the vegetal pole, due to the expansion of the blastoderm in epiboly. At the magnification used and at 21-22° C, this means that the margin of the germ ring moves downward and out of the field in 60-80 minutes. Also, since we wished to study directional movements of individual cells and cell clusters during the entire period of filming, we were compelled to select a region of the germ ring well lateral to the embryonic shield, where the cells are more loosely arranged. As cells approach the embryonic shield, they become so densely packed as to preclude observation of individual cell movement. For this reason, moreover, no observations of cell movements within the very early germ ring (stage 15) or within the embryonic shield proper were possible.

RESULTS

Cell Fate During Gastrulation as Studied by Carbon Marking

When the germ ring (GR) and the marginal prospective yolk sac (YS) bordering the GR (Figure 1) are marked with carbon particles in early mid-epiboly (stage 16), all marks move dorsad toward the embryonic shield (ES), except those in the ventral prospective YS approximately 180° from the future mid-dorsal line. These latter marks remain in the extraembryonic mesoderm and eventually come to reside in the ventral region of the YS. All other marks move dorsad toward the ES and those in the GR do so quite rapidly, so that all of them have moved into the lateral region of the ES by 3/4 epiboly (stage 18-18½). Significantly, marks in the GR closest to the ES at stages 16-17 move more rapidly than those farther away. This interesting observation will be considered in some detail below. Between stage 18 and 19 (5/6 of epiboly), the distance between the most anterior marks, now in the ES, and those posterior to them increases dramatically, indicating considerable antero-posterior extension of the shield. As the embryo develops and somites form, all of these marks come to lie in the lateral region of the embryo proper. Clearly, cells in the GR and adjacent to it in the early gastrula undergo massive convergent movements toward the ES and extension within the ES during epiboly. This is a striking example of the convergence extension movements that are a cardinal feature of the gastrulation of all vertebrates.

Other marks placed at the anterior edge of the embryonic shield and anterior to the shield itself at a still earlier stage (stage 15½) yield further information. All come to be embedded in the anterior region of the ES, as gastrulation progresses to late mid epiboly (stage 17½), and eventually are found in the prospective brain and optic vesicles of the embryo at the conclusion of epiboly (stage 20). These results are consistent with the discovery of Oppenheimer (1936) and Ballard (1966, 1973) that the anterior part of the embryo arises from cells anterior to the early embryonic shield.

Behavior of the Deep Cell Population in the Germ Ring

Time-lapse video taping of the motile behavior of deep cells in the GR lateral to the ES in the dorsal half of the blastoderm during both early mid and mid-epiboly (stage 16½-17) and late epiboly (stages 18½-19) shows that <u>all</u> deep cells in this region of the GR are motile and engage in translocation almost all the time. Of course, whenever a cell enters mitosis it ceases movement until cytokinesis is complete (see detailed discussion of this below) and, also, all non-mitotic cells cease moving actively for short periods from time to time for unknown reasons. But in both cases the cells are immobile only for a matter of minutes. We have not followed the motile behavior of cells of the overlying enveloping layer during this study. A previous investigation has shown that they slowly rearrange in the plane of the egg surface during epiboly (Keller and Trinkaus 1987). However, this motile process appears to have no obvious relation to the truly spectacular motility of the deep cells beneath.

Not only are the deep cells of the GR highly motile, but, as predicted by the marking studies, their movements are highly directional as well. In three sequences, at stages 16½-17, 17½, and 18-19, <u>every</u> cell whose movement could be followed throughout the sequence (*i.e.*, for an hour or more) showed net movement toward the ES. In other words, in so far as our observations go, this is a 100% efficient morphogenetic cell movement. Deep cells in the GR invariably move toward and into the ES in a massive movement of convergence. That all cells do this is a little surprising because theoretically such impressive efficiency would not be necessary to produce the end result. As this cell movement continues, accompanied by steady epiboly of the entire blastoderm, the embryonic shield continually increases in size and number of cells and gradually extends in its antero-posterior axis. As noted in MATERIALS AND METHODS, these cell movements can be observed in detail only in a sector of the GR distinctly away from its junction with the ES and in the prospective YS. Some understanding of cell movements within the more densely packed proximal region of the GR and in the ES itself awaits the application of modern cell marking techniques.

Although all cells of the GR that we have studied eventually move toward the ES, it must be emphasized that the directionality of these cell movements is net; no cells move directionally in a straight line. On the contrary, all cells meander considerably, frequently veering to the left and to the right, and even reversing direction to move away from the ES for short periods. Moreover, most cells have periods of immobility in the course of their movement, sometimes for as long as 25 min. Nevertheless, they all soon return to their proper directionality and thus eventually show net progress toward the shield.

In spite of these meanderings and brief periods of immobility, the mean rates of movement of these cells are substantial. For example, in one 80 min sequence at stage 16½-17 the mean net rate of the 60 cells whose progress could be followed throughout was 1.4 ± 0.36 μm/min at 21° and the mean total rate was 1.9 ± 0.35 μm/min. Because of the considerable variation, we also measured the minimum and maximum rates of movement of each of these cells. The mean minimum rate was 0.66 ± 0.43 μm/min and the mean maximum rate was an impressive 3.9 ± 0.94 μm/min. Clearly, this is a motile cell population, which, though variable in its rate of movement, is capable of a high rate of translocation.

The marking studies have shown that although the major source of cells for the ES is the GR, some additional cells also move into the shield from the inner region of the blastoderm that borders on the GR, just lateral and anterior to the ES. We have studied the movements of these cells, which we call yolk sac (YS) cells, in both a mid-gastrula (stage 16½-17) and a late gastrula (stage 18-19). Since this region of the blastoderm is much more sparsely populated with cells than the GR, the trajectory of every cell in the field can be readily followed throughout the sequence. As expected from the marking studies, there is a decided tendency of these cells to show net movement toward the shield. Indeed, in our observations all cells near the shield eventually move toward it. However, they meander much more than cells in the GR. Hence their net directional movement toward the ES is less than that of cells of the adjacent GR and far less than their own total movement. For example, the mean net trajectory toward the ES of 22 YS cells at stage 18-19 during 65 minutes at 21° was found to be 47.6 ± 19.8 μm, as compared to 66.6 ± 14.7 μm for 26 cells in the adjacent GR proper. And the mean total trajectory for these YS cells was 107.8 ± 31.1 μm, more than double their net trajectory, representing a much greater difference between their net and total trajectory than for the adjacent GR cells, whose total trajectory was only about 50% greater than their net trajectory (M = 94.5 μm, ± 9.6). Obviously, YS cells meander much more. In addition, the looser population of cells at the inner border of region of the GR, which we have called the "transition zone", also move toward the ES more efficiently than the more scattered cells of the YS that are farther away from the margin of the blastoderm. The mean of the net trajectory of 17 cells in this transitional zone was 63.7 ± 15.4 μm and the mean of the total trajectory was 90.7 ± 14.9 μm, essentially the same as for cells of the GR proper. From all this, it seems that there are forces in the marginal region of the blastoderm, primarily in its dorsal aspect just lateral to the ES, that direct deep cells toward the embryonic shield and that these forces are stronger in the most marginal region, the germ ring.

The carbon marking studies have demonstrated that an important feature of convergence is that particles implanted in the germ ring close to the embryonic shield move faster toward the ES than marks farther away. Our direct observations of individual cell motile behavior have confirmed this. During a given period, cells nearer the ES tend to move more toward the ES than cells farther away. Their net movement is faster and their net trajectories are longer. In one 80-minute period from stage 16½-17, the 35 cells in the proximal half of the field (closer to the ES) had a mean trajectory of 117.74 ± 25.92 μm, whereas the 25 cells in the distal half of the field had a mean trajectory of 93.2 ± 17.02 μm. The difference is highly significant; $t = 4.1369$ and $P \leq 0.01$. Three possible explanations present themselves. 1) Cells nearer the ES may have longer total trajectories; i.e., they have higher rates of movement. 2) Also, they may spend less time in an immobile state. And, finally, 3) they may meander less, as suggested by inspection of tracings of their trajectories. Comparison of the mean total trajectories of cells from both sectors reveals no difference. Nor do cells nearer the ES spend less time immobile. It seems, therefore, that cells nearer the ES do not move more efficiently toward it because of an augmentation of their motile activity *per se*. Rather, they meander less; they move more directly toward the ES. This conclusion is of considerable interest because it indicates that some extrinsic factor(s), present in the GR, directly affects the directionality of the cells, as distinct from their motility, and that this factor(s) is more concentrated nearer the ES. If, for example, it is a diffusible chemical, its action must be chemotactic, not chemokinetic.

As pointed out in MATERIALS and METHODS, epiboly of the blastoderm continues throughout convergence. In addition to limiting the duration of each period of observation, this affects the apparent trajectory of the cells. As they move toward the ES roughly parallel to the blastoderm margin, they are carried passively toward the vegetal pole by epiboly. In an egg whose position is stabilized during each period of observation this gives a downward (vegetalward) bias to each cell's apparent trajectory. Consistent with this, during one sequence, deep cells of the GR moved vegetally at average rates varying from 0.6 μm/min to 1.6 μm/min and peripheral nuclei of the underlying yolk syncytial layer, which also engages in epiboly, moved vegetally at rates varying from 0.8 μm/min to 1.2 μm/min. The wider variance of the deep cells is no doubt due to their considerable meandering.

Movements of Individual Cells and Cell Clusters

Because of the confined space in which they move, between the overlying transparent enveloping layer (EVL) and the underlying internal yolk syncytial layer (I-YSL), and because of their tendency to move in monolayers, much of the motile surface activity of deep cells in the GR is readily observed. As determined by careful focusing, they apparently use both the undersurface of the stellate layer and the EVL and the upper surface of the I-YSL as substrata for their movements.

The majority of deep cells moving in the GR show frequent protrusive activity wherever they have a free surface, forming blebs or extending filopodia and lamellipodia, or both. Most of these are "filolamellipodial", in that they are thick elongate "filopodia" that we know from SEM frequently spread out as lamellipodia at their tips (Trinkaus and Erickson 1983). Movement by means of these filolamellipodia is either quite steady and continuous or rather jerky, as when a tip adheres to another cell and contracts, pulling the cell body dramatically forward.

Mixed with these filolamellipodial cells is a small minority of cells that move by so-called blebbing locomotion (Trinkaus 1973; Trinkaus and Erickson 1983; Fink and Trinkaus 1988). Such cells form a large bleb, which usually engages immediately in circus movement. As cytoplasm of the cell body pours into this self-propagating bleb, the cell undergoes translocation. Then, a new bleb forms, etc.

Individual cells move as described, but in fact most cells in the GR do not move as individuals. The overwhelming majority of cells are more or less in contact with other deep cells most of the time and move in concert with them in dynamic, ever-changing groups or clusters (Figure 2). Indeed, many cells remain surrounded by the very same cells for long periods and move constantly with them, hardly changing their contacts for up to 30 min or more. This movement in clusters is hardly surprising considering the density of the cell population, particularly near the ES.

These cell clusters vary constantly in form and in size, from 2 to 10 cells or so, with individual cells or other clusters joining and leaving frequently and the cluster often splitting into 2-3 smaller clusters. Thus, there are few discrete clusters with much of a life of their own. They are mainly temporary adhesive cell associations.

When clusters split, it generally occurs quite rapidly with quick retraction of both parts, suggesting that they are under tension, as is generally true of cell clusters, whether *in vitro* (Kolega 1981) or *in vivo* (Trinkaus 1988a). Not surprisingly, retraction fibers are often observed between receding sub-clusters and when these rupture retraction accelerates and a wide intercellular space is created. Consistent with their directional movement toward the ES, the subclusters closest to the ES retract more and faster. In time-lapse sequences this will show up as an acceleration of that cell's directional movement. Of course, the subcluster that retracts in the opposite direction moves away from the shield. But soon after, when its retraction has ceased, it resumes its migration toward the ES.

The free surfaces of the marginal cells of these clusters frequently exhibit protrusive activity, both proximally toward the ES and distally away from the ES, and sideways as well. In spite of this multidirectional activity, however, each cluster shows net movement toward the shield. We have not yet been able to determine whether this is related to increased protrusive activity at the proximal or front side of the cluster, or decreased protrusive activity at the distal or rear side of the cluster, since some protrusive activity is ill-defined or invisible because of the imperfect resolution of our video tapes and the crowding of some of the cells. Also, protrusive activity frequently appears at newly free surfaces at the edge of spaces that appear within a cluster and this also could modify the direction of translocation of the whole cluster. Incidentally, this multidirectional protrusive activity at the free edges of the clusters is no doubt responsible for the high degree of tension within each cluster.

Since the only cluster cells observed to show protrusive activity are those with free margins, it would seem that cells within a large cluster that are completely, or almost completely, surrounded by other cells must move passively, dependent for their movement on the motile activity of their neighbors with free surfaces. Also, since cells at the distal periphery of a cluster show net movement toward the shield even when they exhibit protrusive activity at their free surfaces, their directional movement must likewise be passive, due to contact with other cells that are moving directionally; *i.e.*, they are dragged forward in spite of their own backward motile activity.

Contact Inhibition

Inasmuch as protrusive activity of cells within cell clusters is confined to their free surfaces, it is not surprising that this protrusive activity ceases whenever a motile cell surface makes contact with the surface of another cell or cell cluster. This applies particularly to cells moving by means of filolamellipodia. In consequence, cells that are in close contact with other cells within a cluster show very little movement relative to one another. These observations lead to an important conclusion. The vast majority of deep cells moving in the GR toward the ES during *Fundulus* gastrulation exhibit contact paralysis or contact inhibition of cell movement. This is doubtless the underlying cause of their largely monolayered arrangement. Interestingly, and in contrast, the surfaces of cells moving by blebbing locomotion frequently continue blebbing when in contact with other cells, as previously observed by Fink and Trinkaus (1988). Apparently they are non-contact inhibiting.

Important though this contact paralysis is for the detailed movement of individual cells, as during intercalation or invasive activity (see below), it does not significantly impede their net movement toward the ES. The reason for this is that most of the time they are members of directionally moving cell clusters. This is a significant point, because it is widely believed that if cells are contact inhibiting they would not be able to move within a dense cell population *in vivo* unless they become less contact inhibiting or are separated from other cells by ECM. This investigation shows, however, that these restrictions do not apply to cells moving principally as members of cell clusters. If a cluster moves, its constituent cells necessarily also move, but mainly passively as members of the cluster. The directionality of this movement in the GR depends, we suspect, on more net motile activity of the free surface of cells on the proximal side of each cluster, which in turn would be due to extrinsic directional forces (see above).

Intercalation or Invasiveness

We use the terms intercalation or invasiveness interchangeably, indicating movement of cells between other cells. Since most individual cells and all cell clusters of the GR are contact inhibiting, they should not show intercalation unless spaces appear. And, they do not. But this does not eliminate intercalation. In a dynamic, ever-changing cell population like the teleost germ ring, spaces appear constantly between cells and cell clusters; and individual cells, both filolamellipodial and blebbing cells, and cell clusters are often observed moving into such spaces. In a word, they show invasiveness or intercalation. Further, since cell movement and cluster movement in the GR are overwhelmingly in one direction, spaces will invariably appear behind more frequently than in front, providing openings for more distal cells and clusters to penetrate. Thus, once directional movement begins, it will be reinforced by the contact inhibiting properties of the cells themselves.

Since the surfaces of cells moving by blebbing locomotion frequently continue blebbing when in contact with other cells, we were not surprised to observe blebbing cells moving among other cells in the apparent absence of intercellular spaces.

Possible Movement of the I-YSL Surface

Long (1980) has reported that there is a dorsad convergent movement of YSL cytoplasm during gastrulation, based on the movement of injected chalk particles. If such a convergent movement involves the surface of the YSL, it could be a crucial extrinsic factor helping guide deep cells of the GR dorsad, since the surface of the YSL serves as a substratum for their movements. We, therefore, repeated Long's experiment in *Fundulus*, except that instead of using chalk particles we used carbon particles and placed them on the surface of an I-YSL exposed by removal of the blastoderm. Such an exposed I-YSL undergoes epiboly in the absence of the blastoderm (Trinkaus 1951, 1984b). Carbon particles adhere readily to its sticky surface. The fate of the particles was followed in camera lucida tracings. No convergent movement was observed. It seems, therefore, that the dorsal surface of the I-YSL of *Fundulus* does not undergo dorsad convergent movements during gastrulation. Accordingly, it could not by this means serve as an extrinsic directional force guiding cells of the GR toward the ES.

Cytokinesis

In contrast to certain morphogenetic cell movements, like the epibolic spreading of the EVL of *Fundulus*, during which there is virtually no cell division (Betchaku and Trinkaus 1978), deep cells frequently divide during convergence in the GR. We cannot give an accurate figure for the overall rate of cell division because the duration of each of our periods of observation was necessarily limited. But mitoses are evidently frequent. During one 80 minute sequence between stages 16½-17 (mid gastrula) 12% of the 70 cells underwent mitosis.

One can easily distinguish premitotic cells. They round up to spheroids and immediately begin vigorous blebbing. Huge blebs form one after the other, two or three at a time, over the entire cell surface and, as they rapidly retract, new ones form. The duration of this period of frantic blebbing is highly variable, ranging from 6-17 min, and then the cell resumes its spheroid form and is quiescent for several minutes. Again, the duration is highly variable, ranging from 5-16 min. We presume this to be metaphase of mitosis. The next event is a quick elongation of the cell leading to furrow formation, lasting <2 min. This no doubt corresponds to anaphase and telophase. Appearance of the furrow signals the beginning of cytokinesis and the formation of daughter cells. This is usually very brief, occurring in less than a minute. Daughter cells are spheroid immediately after cytokinesis and in some cases their surfaces are quiescent for a short time. Others show some blebbing. In either case, the daughter cells soon begin to move, usually in 2 min or less. Thus, the inexorable migration of these cells and their progeny toward the ES is interrupted for only a short time. We present these details of the ever-fascinating process of cell division for two reasons. 1) Its duration is not long; it constitutes only a short interruption in the onward convergent movement of each deep cell in its dorsad progress toward the ES, which takes several hours. The cells take time out only briefly for the joys of reproduction, divide quickly to form their daughters, then the daughters almost immediately enter the movement of the mass and move themselves, quite as their mother was doing. 2) The rearrangement of each cell's mechanokinetic system from the demands of cytokinesis to those of translocation is exceedingly brief, an important matter that incredibly has not yet received the molecular attention that it deserves.

DISCUSSION

By taking advantage of the favorable optical properties of the *Fundulus* egg, we have been able to answer a number of questions concerning the motile and contact behavior of cells moving directionally in the germ ring.

Although it seems possible that an efficiency of considerably less than 100% would suffice for normal formation of the embryonic shield and its subsequent development, this study shows that <u>all</u> of the cells in the region lateral to the ES move toward and into the ES. Significantly, in another unrelated directional vertebrate cell movement, where individual cells can also be followed–migration of melanocytes of the yolk sac into the pectoral fin bud of the teleost *Blennius pholis*–the efficiency is also 100% (Trinkaus 1988b). Of course, we do not know the reason(s) for this extraordinary efficiency in two unrelated cases, but it suggests that constant directional forces are at work. In *Fundulus*, where time-lapse filming has been applied, we can say more. These forces do not govern every motile act of every cell; they merely impose a directional bias on what would otherwise be random cell movement. Deep cells in the *Fundulus* GR never move in a straight line toward the ES; they always meander considerably. It would certainly be of great interest to know if cells that engage in directional movements in other cell streams, such as the neural crest, likewise show the same overall (net) efficiency, but, nevertheless, also meander considerably.

We have also demonstrated that cells quite close to the ES show faster average net movement toward the ES than those farther away and that it is due to less meandering. This, of course, suggests that the extrinsic force(s) directing their movement dorsad operates in some kind of a gradient. And, this in turn suggests that the high point in this gradient is the ES, a fascinating possibility that could possibly be tested by transplanting the dorsal lip to other regions of the blastoderm, such as the ventral GR and the YS.

Although it is not surprising, it is important to note that cells moving in the GR utilize the same modes of movement that they developed during the late blastula - filolamellipodial and blebbing locomotion. But the former predominates. This holds as well for the cells in the prospective yolk sac that move toward the YS. But in fact those in the GR move quite little by themselves. They are mainly in contact with other cells in dynamic, ever-changing cell clusters. This has developmental significance for three reasons. 1) Although it is only one of two instances where cells have been shown to move in clusters *in vivo* (Trinkaus 1988a), since the moving clusters have been observed in different organisms (*Fundulus* and *Blennius*) and consist of different cell types and since cells in cell streams are often seen clustered together in histological sections in other morphogenetic systems in other organisms (see Trinkaus 1988a), we now have good reason to suspect that this is a commonplace mode of cell movement during development and in the spread of cancer cells. 2) Since cells embedded in a cluster show no protrusive activity and yet move with the cluster toward the ES, much of their directional movement must be passive. Here then, we have yet another example of cells moving passively (see Trinkaus 1984a). Passive cell movement has clearly emerged as an important mode of morphogenetic cell movement. 3) Inasmuch as cells in clusters show protrusive activity only when located at the free surface of a cluster and then only at their own free surfaces, they appear to be contact inhibiting.

Other evidence for contact inhibition comes from observations of individual moving cells. Whenever the motile leading edge of a cell engaging in filolamellipodial movement contacts another cell, its protrusive activity ceases. It shows contact paralysis. It must be concluded, therefore that these deep cells moving *in vivo* in *Fundulus* show contact inhibition of cell movement (see also, Lesseps *et al.* 1979; Van Haarlem 1979). Nevertheless, they move and do so directionally. This is an important observation because it has been thought that contact-inhibiting cells could not move much *in vivo* because they would quickly contact other cells and often be surrounded by them. To counter this, it has been postulated that cells do not actually contact other cells because of intervening ECM. Our work now shows that this postulate may be unnecessary. They need only be members of a directionally moving cell cluster, which moves by virtue of the protrusive activity of those of its constituent cells that possess a free margin. Perhaps this is true of many contact-inhibiting cells that move directionally *in vivo*, like, for example the neural crest (Erickson 1985). Still, however, we must explain how clusters keep moving when the situation becomes crowded, as when cells in the GR have moved closer to the ES.

Fundulus deep cells also engage in intercalation or invasive activity as they converge toward the ES; they move between other cells. But, consistent with their contact inhibition properties, they do so only when a space opens up between those cells. Then, and then only, do they form a protrusion on the side facing the space and move between the previously closely adhering cells. This, we think, might well apply to many systems where the moving cells are contact inhibiting. But it apparently does not apply universally, as, for example, to the intercalating activity of blebbing of cells of the *Fundulus* GR or to the prospective mesodermal cells of *Xenopus*. These move over and between each other quite readily in the apparent absence of intercellular spaces (Keller 1986). Like *Fundulus* blebbing cells, these *Xenopus* cells appear not to be contact inhibiting.

Much has been written about the antagonism between cell movement and cytokinesis. And, it has been suggested that this is why cell movements do not begin in early embryogenesis until cleavage has ceased (Trinkaus 1980). We were, therefore, mildly surprised to observe division of these very actively motile cells of the *Fundulus* GR, for, as expected, they do indeed cease translocation when they enter mitosis. But this takes them out of the action at hand for only a brief period. Then, at the termination of cytokinesis, the daughter cells rapidly join the throng and begin moving with them with net directionality. In consequence, it must be concluded that cell division and mass cell movements are not incompatible. As long as all the cells are not dividing with high frequency, as is the case in the *Fundulus* GR, the effect on the directional movement of the large cell population is minimal.

The reader may have noticed that we have not mentioned involution as another means of cells contributing to the embryonic shield. This should not be taken to imply that involution does not occur in *Fundulus*, as it does in the rosy barb (Wood and Timmermans 1988) and the zebrafish (Warga and Kimmel 1990). It means simply that thus far we have not examined the possibility of involution. Accordingly, we have nothing to say. "Wovon man nicht sprechen kann, darüber muss man schweigen." (Ludwig Wittgenstein).

ACKNOWLEDGEMENTS

This investigation has been supported by a Merit Award from the NCI of the NIH to JPT and a Presidential Young Investigator Award from the NSF and a William and Flora Hewlett Foundation Grant of Research Corporation to RDF. We are indebted to Ray Stevens for use of his Sigma Scan Program.

REFERENCES

Armstrong, P.B. and J.S. Child. 1965. Stages in the normal development of *Fundulus heteroclitus*. *Biol. Bull.* 128:143-168.

Ballard, W.W. 1966. Origin of the hypoblast in *Salmo*. I. Does the blastodisc edge turn inward? *J. Exp. Zool.* 161:201-210.

Ballard, W.W. 1973. Morphogenetic movements in *Salmo gairdneri* Richardson. *J. Exp. Zool.* 184:381-426.

Betchaku, T. and J.P. Trinkaus. 1978. Contact relations, surface activity and cortical microfilaments of marginal cells of the enveloping layer and of the yolk syncytial and yolk cytoplasmic layers of *Fundulus* before and during epiboly. *J. Exp. Zool.* 206:381-426.

Erickson, C.A. 1985. Morphogenesis of the neural crest. p. 481-543. *In: Developmental Biology: A Comprehensive Synthesis. Vol. 2, The Cellular Basis of Morphogenesis.* L.W. Browder (Ed.). Plenum Press, New York.

Fink, R.D. and J.P. Trinkaus. 1988. *Fundulus* deep cells: Directional migration in response to epithelial wounding. *Dev. Biol.* 129:179-190.

Keller, R.E. 1986. The cellular basis of amphibian gastrulation. p. 241-327. *In: Developmental Biology: A Comprehensive Synthesis*. Vol. 2. The Cellular Basis of Morphogenesis. L.W. Browder (Ed.). Plenum Press, New York.

Keller, R.E., J. Shih, and P. Wilson. 1991. Cell motility and regional interactions controlling amphibian gastrulation. p.101-120. *In: Gastrulation: Movements, Patterns, and Molecules*. R. Keller, W.H. Clark, Jr., F. Griffin (Eds.). Plenum Press, New York.

Keller, R.E. and J.P. Trinkaus. 1987. Rearrangement of enveloping layer cells without disruption of the epithelial permeability barrier as a factor in *Fundulus* epiboly. *Dev. Biol.* 120:12-24.

Kolega, J. 1981.The movement of cell clusters *in vitro*: Morphology and directionality. *J. Cell Sci.* 49:15-22.

Lesseps, R., F. Hall, and M.B. Murnane. 1979. Contact inhibition of cell movement in living embryos of an annual fish, *Nothobranchius korthausae*: Its role in the switch from persistent to random cell movement. *J. Exp. Zool.* 207:459-470.

Long, W. 1980. Analysis of yolk syncytium behavior in *Salmo* and *Catostomus*. *J. Exp. Zool.* 214:323-331.

Oppenheimer, J.M. 1936. Processes of localization in developing *Fundulus*. *J. Exp. Zool.* 73:405-444.

Pasteels, J. 1936. Etudes sur la gastrulation des vertébrés méroblastiques. I. Téléostéens. *Arch. Biol.* 47:205-308.

Trinkaus, J.P. 1951. A study of the mechanism of epiboly in the egg of *Fundulus heteroclitus*. *J. Exp. Zool.* 118:269-319.

Trinkaus, J.P. 1967. Procurement, maintenance and use of *Fundulus* eggs. p.113-122. *In: Methods in Developmental Biology*. F.H. Wilt and N.K. Wessells (Eds.). Crowell, New York.

Trinkaus, J.P. 1973. Surface activity and locomotion of *Fundulus* deep cells during blastula and gastrula stages. *Dev. Biol.* 30:68-103.

Trinkaus, J.P. 1980. Formation of protrusions of the cell surface during tissue cell movement. p. 887-906. *In: Tumor Cell Surfaces and Malignancy*. R.O. Hynes and C.F. Fox (Eds.). Alan R. Liss, New York.

Trinkaus, J.P. 1984a. *Cells into Organs, The Forces that Shape the Embryo*. p. 543. Prentice-Hall, Englewood Cliffs, N.J.

Trinkaus, J.P. 1984b. Mechanism of *Fundulus* epiboly–A current view. *Am. Zool.* 24:673-688.

Trinkaus, J.P. 1988a. Directional cell movement during early development of the teleost *Blennius pholis*. I. Formation of epithelial cell clusters and their pattern and mechanism of movement. *J. Exp. Zool.* 245:157-186.

Trinkaus, J.P. 1988b. Directional cell movement during early development of the teleost *Blennius pholis*. II. Transformation of the cells of epithelial clusters into dendritic melanocytes, their dissociation from each other, and their migration to and invasion of the pectoral fin buds. *J. Exp. Zool.* 248:55-72.

Trinkaus, J.P. and C.A. Erickson. 1983. Protrusive activity, mode and rate of locomotion, and pattern of adhesion of *Fundulus* deep cells during gastrulation. *J. Exp. Zool.* 228:41-70.

Trinkaus, J.P., M. Trinkaus, and R.D. Fink. 1992. On the convergent cell movements of gastrulation in *Fundulus*. *J. Exp. Zool*, In press.

Van Haarlem, R. 1979. Contact inhibition of overlapping: One of the factors involved in deep cell epiboly of *Northobranchius korthausae*. *Dev. Biol.* 70:171-179.

Warga, R.M. and C.B. Kimmel. 1990. Cell movements during epiboly and gastrulation in zebrafish. *Development* 108:569-580.

Wood, A. and L.P.M. Timmermans. 1988. Teleost epiboly: a reassessment of deep cell movement in the germ ring. *Development* 102:575-585.

DYNAMICAL MODELS FOR CELL REARRANGEMENT DURING MORPHOGENESIS

Michael Weliky[1*] and George Oster[2]

[1] Group in Neurobiology
* address for correspondence
c/o George Oster
201 Welman Hall
University of California
Berkeley, CA 94720
[2] Departments of Molecular and Cell Biology, and Entomology
201 Wellman Hall
University of California
Berkeley, CA 94720

ABSTRACT

The spatial form of cells and tissues reflect their underlying physical and mechanical properties. These properties are regulated, primarily by the modification of cell cytoskeletal components, to produce a large repertoire of cell behaviors,

Gastrulation, Edited by R. Keller *et al.*
Plenum Press, New York, 1991

including cell shape changes and directed motility. In a number of instances, these behaviors drive tissue morphogenesis during embryological development. Here we propose a mechanical model for studying tissue morphogenesis by cell rearrangement and cell shape change. Our model describes these processes by accounting for the balance of forces between neighboring cells that are junctionally coupled within a tissue. The model is applied to two embryological settings: epiboly in the teleost fish *Fundulus*, and notochord extension in *Xenopus laevis*.

INTRODUCTION

During morphogenesis, the collective behavior of large numbers of cells is reflected in the overall shape changes of a tissue. In a number of cases, it has been demonstrated that these shape changes are the result of cell rearrangements (Keller and Trinkaus 1987; Keller *et al.* 1989). This paper proposes a model for studying the cellular basis of tissue morphogenesis by cell rearrangement in various embryonic settings. Previous models for cell rearrangement have used boundary shortening (Honda *et al.* 1982) or differential adhesion (Sulsky 1982) to generate cell rearrangements. We will describe a mechanical cell model capable of generating contractile and protrusive forces, which incorporates the features of biological cells believed to be responsible for the generation of these forces. By coupling a large number of model cells together we can simulate cell rearrangement within a tissue.

Computer graphics are used to generate an autimated movie showing the dynamical behavior of the model simulations. This can be compared with photomicrographs and time-lapse films of cell rearrangements and shape changes within living embryonic tissue. All forces generated by each simulated cell can be visually monitored and subsequently analyzed throughout all phases of the rearrangement process. This gives us the opportunity to relate the observed phenomenon of cell rearrangement directly to the underlying forces generated by each cell. We attempt to reproduce the experimental observations by modifying the mechanical properties of the model cells. For instance, this allows us to hypothesize various scenarios for how cells may generate different forces dependent upon their positions within the tissue. In so doing, we not only validate the model but gain insight into the pattern of cellular force generation that drives the cell rearrangements and cell shape changes responsible for tissue morphogenesis. By making the appropriate modifications to the model (*e.g.* tissue boundary conditions and cellular mechanical properties), we can study and simulate the mechanical conditions in different embryonic settings. In this paper, we will describe two such efforts.

THE MECHANICAL MODEL

We will first discuss the biophysics of mechanical force generation within a cell. This will be followed by a description of the cell rearrangement model which incorporates these principles. See Weliky and Oster (1990) for a more detailed description of the model.

Sources of Mechanical Force in the Cell Cortex

In vitro cell movement is accomplished by the coordination of protrusive and contractile forces generated within the cell and exerted upon a suitably adherent substratum (Oster 1984; Oster and Perelson 1987). Cell movement and subsequent cell rearrangement within a tissue appears also to depend upon the ability of a cell to adhere to and exert forces upon neighboring cells and extracellular matrix. The cellular cytoskeleton is responsible for providing the mechanical stability of cell shape and for generating the forces that drive cell protrusion and contraction. With regard to cell motility, the most important cytoskeletal components are the actomyosin filamentous structures. Here, we will consider the cortical actomyosin cytogel as the major cytoskeletal component responsible for cellular contractile and protrusive mechanical force generation.

The forces in the cell cortex are: 1) Osmotic and hydrostatic pressure within the cortical cytogel. 2) Elastic forces which include both passive elastic forces of the actin fibers composing the cytogel and active contractile forces generated by actin-myosin interactions.

The osmotic/hydrostatic forces tend to swell the cytogel and cell cortex while the elastic forces tend to contract the cytogel and cell cortex. By altering the balance between the pressure and tension forces the cortical cytogel can expand and contract. This is accomplished by various mechanisms. For example, ionic regulation of solating factors can affect the gel in a number of ways including severing network fibers and breaking crosslinks between fibers as well as depolymerizing fibers. Solation has the general effect of breaking up the network, which reduces the elastic modulus of the gel.

Modeling a Cell

Our mechanical model for a cell must be capable of both contractile and protrusive behavior consistent with the cell mechanisms described in the previous section. This is accomplished by mechanically describing the local balance of elastic and pressure forces within different regions of a cell. The central features of the model are illustrated in (Figure 1). We begin by representing each cell by a two dimensional polygon (*i.e.* as a collection of nodes connected by line segments) with a variable number of sides. At each node we calculate the balance of elastic tension forces and pressure forces.

Internal Pressure. The contributions of both cell hydrostatic and osmotic pressure are combined into a single pressure vector, **P**, which is applied at each cell node (Figure 1a). The pressure vector, **P**, is directed outward from the cell boundary at each cell node.

Because the swelling pressure of a gel is a local effect, the osmotic pressure may be different at different points on the cell periphery. For simplicity, we have assumed in our simulations that the net pressure is the same at each node, and varies inversely with the cell area. Thus if the cell area increases, the pressure will fall, which will tend to restore the cell back to its original surface area. In this way, the cell area will tend to remain almost constant.

Elastic tension forces. Circumferential tension forces encircle the cell perimeter. This represents both the contraction of circumferential actomyosin microfilament bundles and the elastic forces of the cortical cytogel fibers. Two elastic tension vectors

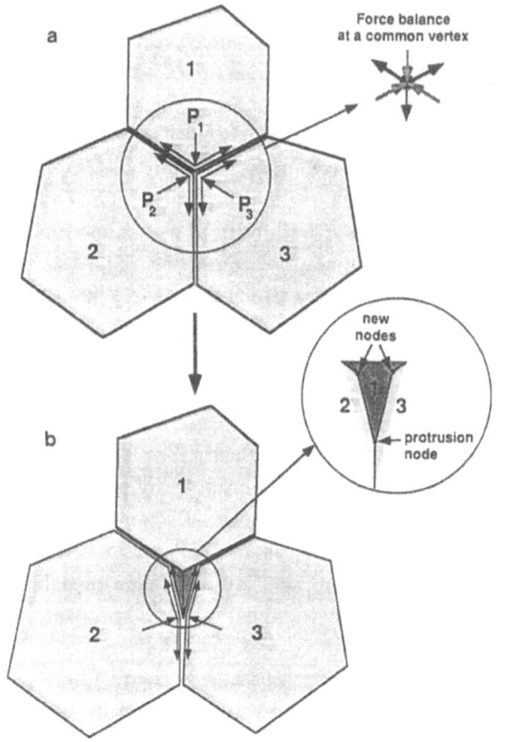

Figure 1. (**a**) Adjacent cells share a common junctional vertex node. The node is in mechanical equilibrium when the pressure and elastic forces from all cells are balanced. (**b**) Cell 1 protrudes when the elastic modulus of its cortex decreases, resulting in an imbalance of forces at the common vertex node. Since protrusion is a local phenomenon we insert two new nodes close to, and on either side of, the protrusion node; the node then slides in the direction of the net force imbalance, which in this case is in the direction of the common boundary of cells 2 and 3.

are applied to each node and are directed along the polygonal cell edges (Figure 1a). The magnitude of these vectors represents the local elastic tension at each cell node. The tension at each node can be independently altered by reducing or increasing the magnitude of the two applied elastic vectors. This represents the local regulation of contractile elements within different regions of the cell.

The net elastic force at each node is the vector sum of the two elastic tension vectors at that node. Note that the net elastic force vector varies with the angle subtended by the edge tensions. This angle represents the local polygonal approximation to the cell surface curvature. The elastic forces at a node where the surface curvature is concave will tend to push the cell surface outwards and straighten while convex regions will contract and be pulled inwards.

The balance of nodal forces. The net force at each cell node is the sum of all the pressure and elastic forces. Thus the shape of the cell boundary reflects the relative balance of elastic and pressure forces at each node. If the elastic tension vectors at all nodes are equal (*i.e.* the cell is contracting equally all around its periphery), the cell will relax to a regular polyhedral form. For a polygonal cell with a large number of faces, the cell will approach a circular shape. The cell shape will remain stable so long as all elastic and pressure forces acting on and within the cell are in mechanical equilibrium. Upsetting the balance of these mutually antagonistic forces at a node can lead either to local cell protrusion or contraction.

Modeling a Tissue

Mechanically coupling cells together. In order to model and simulate a tissue, we mechanically couple a large number of polygonal cells together. This is accomplished by allowing adjacent cells to share common boundary nodes (Figure 1, 2). In this way, mechanical force is transmitted from cell to cell representing junctional coupling. These junctional nodes can slide, and nodes may appear or disappear as necessary, so that the number of polygonal sides is variable. Cell neighbor change occurs when two nodes slide together and change their cellular allegiances (Figure 2). External forces may be applied to the boundaries of the tissue in certain cases. The total force acting at a tissue node is the sum of all net forces from all cells sharing this node plus any external forces.

Active cell rearrangement. Figure 1 shows one mechanism by which cells can move and rearrange within the model tissue. When all pressure and elastic forces at a node are in mechanical equilibrium, the node position will not change (Figure 1a). A cell can intercalate between its neighbors by "actively" regulating the balance of pressure and elastic forces within the cell at a particular node. Unbalanced forces within the cell result in cell protrusion and subsequent cell rearrangement (Figure 1b). Reducing the elastic forces within cell 1 at the central shared node results in internally unbalanced forces for cell 1 at that node. Pressure forces now dominate within cell 1 and consequently the forces between cell 1 and the other two cells are not in mechanical equilibrium, resulting in a net force in the direction of the outward pressure vector of cell 1. Therefore, cell 1 protrudes forward and the shared junctional node moves in the direction of the net force. If cell 1 continues to protrude forward, it will eventually completely separate the other two cells. Note that at protrusion sites we introduce a pair of new nodes so as to model the local nature of cell protrusion.

Passive cell rearrangement. A second mechanism which can also drive cell rearrangements is shown in Figure 2. As an epithelial sheet is stretched by external forces applied to the epithelial margin, cells will "passively" elongate in the direction of the applied force. Cells 1, 2, and 3 share the junctional vertex node circled in Figure 2. Elongation of cells 1 and 3 increases their surface curvature at this node, while elongation of cell 2 decreases its curvature at this shared node. Since the net inward elastic force at a node is large for high cell surface curvature, cells 1 and 3 generate a larger net inward elastic force at the circled node than cell 2. The net force at the node is represented by the vector resultant pointed in the direction of cells 1 and 3. Therefore, the circled node moves to the left. Cell rearrangement occurs in order to restore each cell shape to a more isodiametric polygonal form. When cells subtend an equal angle at a shared node, the net elastic forces generated by each cell are equal and the node is in mechanical equilibrium

Numerical Solution of the Equations of Motion

The mechanical force balance equations are iteratively solved using numerical techniques. The simulation proceeds by repeatedly solving the equations of motions for each cell node during each iteration. During each iteration a node moves a distance proportional to the total force applied to that node. The positions of all nodes are simultaneously updated each iteration. The arrangement resulting from the node displacements in one iteration becomes the starting point for the next iteration. Using computer graphics, we can produce an animated movie showing the successive positions of cells within the model tissue.

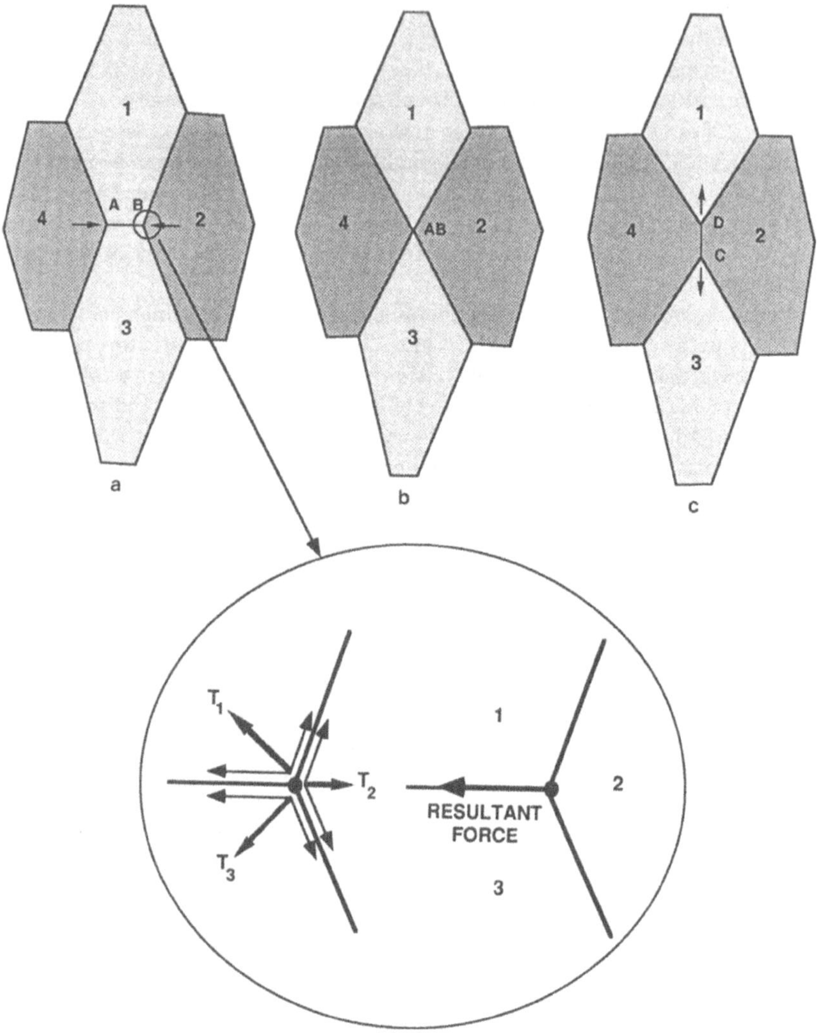

Figure 2. Cell rearrangement occurs as cell nodes move according to the applied mechanical forces. The pair of cells (cells 2 and 4) which are separated in the initial configuration (**a**), establish contact with one another. The remaining pair (cells 1 and 3) which are initially in contact with one another separate in (**b**). Cell rearrangement occurs when two nodes meet, and the junctions "change allegiance". Here, nodes A and B move toward one another in (**a**). During junctional rearrangement, new nodes C and D are created which subsequently separate in (**c**). In the process edge AB has shortened and vanished, to be replace by edge CD. The insert shows how the net force imbalance at node B causes the node to slide to the left as a result of cell stretching and subsequent elongation.

APPLICATIONS OF THE MODEL

In this section, two examples are discussed which demonstrate how the model can be used to gain insight into the mechanisms of cell rearrangement responsible for embryonic tissue morphogenesis. See Oster and Weliky (1990) for additional applications of the model.

Epiboly in *Fundulus*

During early embryonic development of the teleost fish *Fundulus*, the egg is covered by a "cap" of epithelial cells called the enveloping layer (EVL) (Figure 3). During epiboly, this cap is pulled down over the egg by a wave of contraction that sweeps from the animal to the vegetal pole just ahead of the EVL (Keller and Trinkaus 1987). During this process the cells of the EVL rearrange their neighbor relations (Figure 4). Our simulations show that when interior cells are isotropically contractile along all boundaries, simple stress relaxation is sufficient to account for cell rearrangements among interior submarginal cells of the EVL, by the mechanism described in Figure 2.

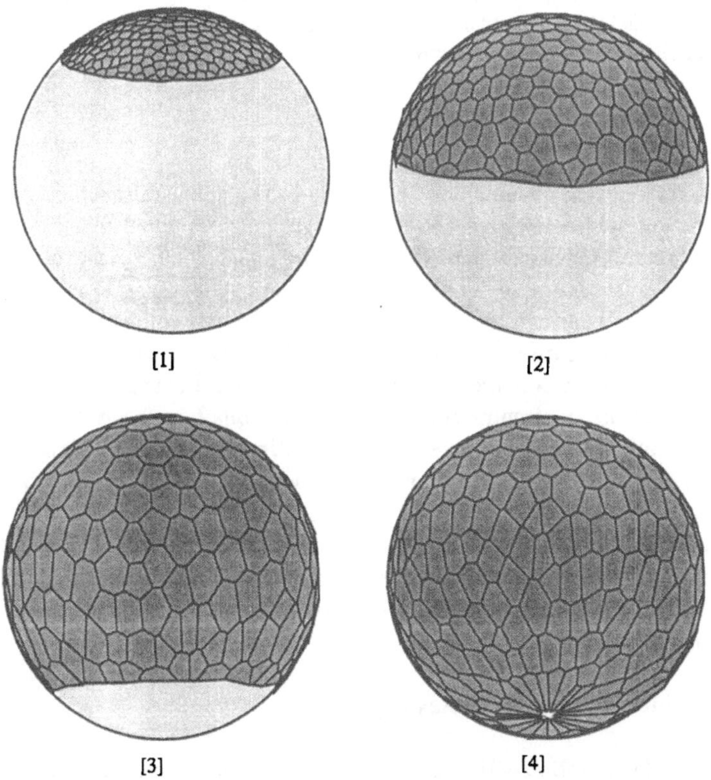

Figure 3. The above sequence show four frames from a computer movie of the simulated epiboly model. The model reproduces the essential features of *Fundulus* epiboly including marginal cell elongation parallel to the EVL boundary [**2**], and later interior cell elongation orthogonal to the boundary [**4**].

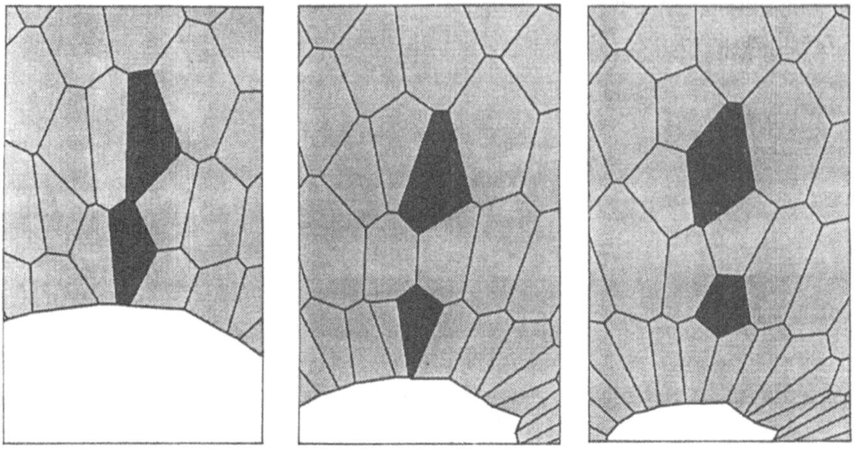

Figure 4. The detailed pattern of marginal and interior cell rearrangements during the later stages of epiboly is reproduced by the model. The lower darkened marginal cell retracts away from the EVL boundary in panel [3]. This occurs after progressive narrowing of its margin along the EVL boundary in panels [1]-[2]. The two darkened cells elongate and subsequently separate in panel [2] by the mechanism described in Figure 2.

As the EVL spreads vegetally, it conforms to the spherical geometry of the egg. Thus, in the animal hemisphere, the EVL margin must lengthen to accommodate the increasing circumference as it spreads toward the equator. Once past the equator, the margin now must decrease as it spreads toward the vegetal pole (Figure 3c). Surprisingly, the experimental observations show that the number of marginal cells continually decreases throughout all stages of epiboly (Figure 5). It is puzzling that the number of marginal cells decreases even in the animal hemisphere where the EVL margin is increasing in circumference. Our simulations have been able to reproduce this behavior only when certain mechanical conditions are satisfied. This has led us to propose a mechanical explanation for this phenomenon, which is described in detail elsewhere (Weliky and Oster 1990). In short, it is required that at the EVL boundary, the leading margin of the cells is always protrusive as described in Figure 2. The simulation shows what happens when net tension forces, as opposed to net protrusive forces, exist at EVL boundary nodes. Under these conditions, the shapes of marginal cells do not appear normal (Figure 6).

Notochord Morphogenesis in *Xenopus laevis*

The second application of the model involves the mechanical factors responsible for elongation and narrowing of the notochord in *Xenopus laevis* from the late gastrula to early neurula stages (Weliky and Oster 1991). Time-lapse cinemicroscopy in living embryos reveals that protrusive activity appears to be inhibited in cell margins that lie along the notochord tissue boundaries. In addition, cells elongate transversely across the width of the notochord as they rearrange (Keller *et al.* 1989). Figure 7 shows two

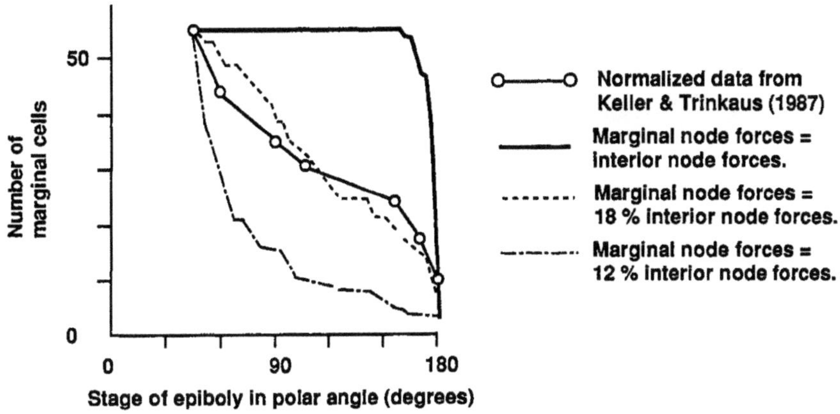

Figure 5. Changing the circumferential elastic force in EVL marginal cells affects the rate at which cells leave the EVL/YSL boundary. When the marginal circumferential force is equal to that of interior cells, marginal cells are always in tension around their entire perimeter, including their leading edges, and the number of marginal cells does not change until near the very end of epiboly. As the elastic force is reduced at nodes along the EVL/YSL boundary, the marginal cells generate a net "protrusive" mechanical force. This condition of reduced cortical tension is necessary to cause marginal cell retraction away from the EVL/YSL boundary. When the circumferential force is quite low (18% of the interior node forces), the simulation results match the experimentally observed data. However, when the circumferential force is too low (12% of the interior node forces), the marginal cell protrusive force is so strong that the rate of marginal cell reduction is much faster than is experimentally observed.

examples of simulations using different rules controlling the model cell behavior. A small region of the notochord is represented by a two dimensional rectangular sheet of cells. The left and right boundaries of the sheet represent the notochord boundaries with the adjacent somitic mesoderm. The actual notochord would extend vertically above and below the simulated cell sheet. As the simulation progresses, the width of the tissue decreases while the length increases. This is exclusively accomplished by cell rearrangements. By the last frames, the number of cells along the horizontal axis has decreased while the number along the vertical axis has increased. In these simulations, "active" cell protrusion drives all cell rearrangements. This is in contrast to epiboly, where "passive" stress relaxation was responsible for all non-marginal cell rearrangements.

In both simulations shown in 7a,b, cell nodes that lie along the right or left boundaries of the sheet are inhibited from protruding. When a cell migrates into the boundary by intercalating between adjacent boundary cells, it is inhibited from continuing to protrude at the boundary nodes and must now protrude only from interior nodes. Figure 7a, shows the results of incorporating unidirectional cell protrusion with random cell movements. Interior cells do not transversely elongate, but rather remain isodiametric in shape. Figure 7b, shows the results of a bidirectional protrusion model where cell-cell interactions, in the form of contact inhibition of protrusive activity, are incorporated in order to guide cell motility. In this case, cells transversely elongate as experimentally observed.

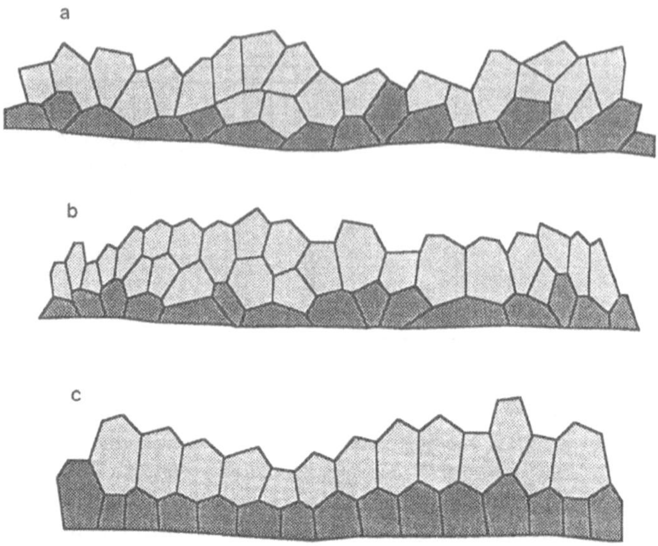

Figure 6. A comparison of EVL marginal cell boundary shapes during the middle stage of epiboly. Marginal cells are darkly shaded, interior cells are more lightly shaded. (**a**) Tracing of marginal cell boundaries from the living *Fundulus* embryo (Keller and Trinkaus 1987). Marginal cells make random angles with the EVL boundary. Some marginal cells elongate parallel to the boundary. (**b**) Result from model simulation. When "protrusive" force exists at cell nodes along EVL margin, cells make random angles with the EVL margin. This is similar to the actual distribution of cell boundary shapes in the real embryo. Some marginal cells also elongate parallel to the boundary as seen in the real embryo. (**c**) Result from model simulation. When marginal cells are exclusively in tension, including nodes along the EVL margin, cells only make 90 degree angles with the EVL margin. This does not match the random angles seen in the real embryo. In addition, none of the marginal cells elongate parallel to the boundary.

The simulations (Figure 7a,b) show that the above rules are sufficient to drive cell rearrangements in such a way that the cell sheet will elongate and narrow as experimentally observed. Inhibiting protrusive activity at the boundary of the sheet effectively biases all marginal cell protrusion towards the interior of the sheet. Marginal cells located at opposite boundaries of the cell sheet converge towards one another, pulling the opposite boundaries closer together, by intercalating between submarginal cells. When a submarginal cell migrates into the boundary, it appears to "stick" and does not usually pull away. This behavior will tend to increase the number of marginal cells along the right and left boundaries of the sheet while reducing the number of interior cells. Consequently, the cell sheet will simultaneously elongate and narrow as cells migrate and crowd into the boundaries from the sheet's interior. In order to reproduce the experimentally observed transverse cell elongation, interior cells must be intrinsically capable of elongation and need some form of cell-cell interactions as guidance cues for directed motility (Figure 7b). We have shown how some simple rules can generate these behaviors. Therefore, the simulation can be used to test hypotheses concerning morphogenetic rules that may constrain embryonic tissue to change shape in specific ways.

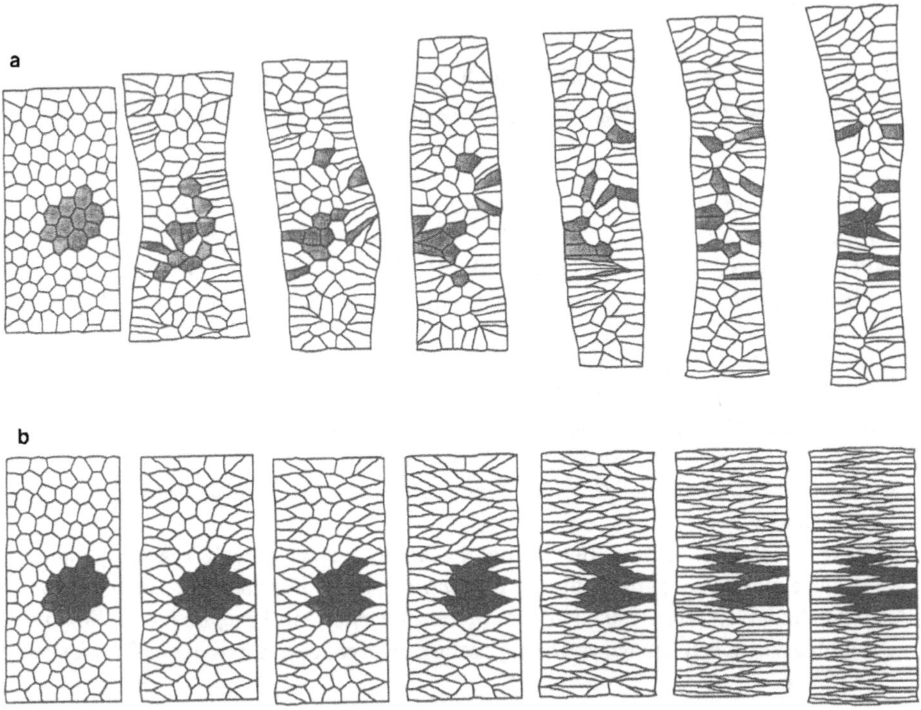

Figure 7. Simulation of cell behavior driving notochord "convergent extension". The time evolution of the sequence of frames is from left to right. (**a**) Unidirectional cell protrusion model. The tissue elongates and narrows while interior cells remain isodiammetric in shape. (**b**) Bidirectional protrusion cell model including cell-cell contact inhibition of protrusive activity. Interior cells transversely elongate into a parallel aligned array. In both (**a**) and (**b**), the patch of darkened labeled cells breaks up and many of these cells are found on the tissue boundaries in the last panel of the sequences. This indicates that large scale cell rearrangements occur as nonlabled and labeled cells intercalate between one another.

CONCLUSION

The model described in this paper provides a general framework for studying the mechanical properties of cells and cell aggregates, as well as for exploring the cellular basis of morphogenetic rules which drive cell rearrangements and cell shape changes. It allows for the testing of assumptions and theories in a precise and rigorous manner incorporating known biological and physical principles. Changing model parameters such as the boundary conditions or the mechanical properties of a motile or adhesive cell subpopulation allows the model to be applied in a wide range of settings in developmental and cell biology.

ACKNOWLEDGEMENTS

This research was supported by NSF Grant No. MCS-8110557 to GFO and a NIH Training Grant No. GM07048 to MW.

REFERENCES

Honda, H., Y. Ogita, S. Higuchi, and K. Kani. 1982. Cell movements in a living mammalian tissue: Long term observation of individual cells in wounded corneal endothelia of cats. *J. Morphol.* 174:25-39.

Keller, R.E. and J.P. Trinkaus. 1987. Rearrangement of enveloping layer cells without disruption of the epithelial permeability barrier as a factor in *Fundulus* epiboly. *Dev. Biol.* 120:12-24.

Keller, R., M.S. Cooper, M. Danilchik, P. Tibbetts, and P. Wilson. 1989. Cell intercalation during notochord development in *Xenopus laevis*. *J. Exp. Zool.* 251:134-154.

Oster, G. 1984. On the crawling of cells. *J. Embryol. Exp. Morphol.* 83:327-364.

Oster, G. and A. Perelson. 1987. The physics of cell motility. *J. Cell. Sci. Suppl.* 8:35-54.

Oster, G. and M. Weliky. 1990. Morphogenesis by cell rearrangement: A computer simulation approach. *Sem. Cell Biol.* 1:313-323.

Sulsky, D. 1982. *Models of Cell and Tissue Movement.* Ph.D. dissertation, New York University.

Weliky, M. and G. Oster. 1990. The mechanical basis of cell rearrangement. I. Epithelial morphogenesis during *Fundulus* epiboly. *Development* 109:373-386.

Weliky, M. and G. Oster. 1991. Notochord morphogenesis in *Xenopus laevis*: Computer simulation of cell behaviors driving tissue convergence and extension. *Development*, In press.

MESODERM CELL MIGRATION IN THE *XENOPUS* GASTRULA

Rudolf Winklbauer, Andreas Selchow, Martina Nagel, Cornelia Stoltz, and Brigitte Angres

Max Planck Institut für Entwicklungsbiologie
Spemannstrasse 35
D-7400 Tübingen
Federal Republic of Germany

ABSTRACT

We analyze the migration of the prospective head mesoderm (HM) across the blastocoel roof (BCR) during gastrulation of the *Xenopus* embryo. Cell spreading and the concomitant appearance of cytoplasmic lamellae depend on the interaction of HM cells with fibronectin (FN) fibrils, which cover the inner surface of the BCR. Isolated HM cells only extend short-lived filiform protrusions on non-adheasive substrates, but form lamelliform protrusions (usually two lamellae appear simultaneously at opposite ends of a cell) on a FN substrate. Isolated bipolar HM cells move in a step-wise mode of translocation, with in-built changes of the direction of migration. This ineffective mode of migration is altered when HM cells move as part of a larger aggregate, as occurs in the embryo, where the HM forms a coherent cell mass. A Ca^{++}-dependent cell-cell adhesion molecule, U-cadherin, mediates aggregate formation, which is one

Gastrulation, Edited by R. Keller *et al.*
Plenum Press, New York, 1991

prerequisite for highly persistent migration. The second requirement is that the mesoderm aggregate moves on a proper substrate. The extracellular matrix of the inner BCR surface can be deposited on a plastic substrate. This conditioned substrate contains directional cues which guide the mesoderm to its target region. The migrating mesoderm becomes visibly polarized on conditioned substrate. Cells appear unipolar and extend protrusions in the direction of migration only, thus underlapping neighboring cells anteriorly. This shingle arrangement of HM cells is also observed in the embryo. We conclude that both cadherin-mediated cell-cell contact and aggregate formation, and a substrate containing guiding cues are required for the unipolar extension of locomotory protrusions, oriented underlapping of neighboring cells, and efficient, persistent, and directional migration of HM cells.

INTRODUCTION

The most active component of the *Xenopus* gastrula is the prospective mesoderm, which drives much of the gastrulation process (Keller 1986, for review). Initially part of the blastocoel roof (BCR), the mesoderm involutes at the blastopore lip and becomes apposed to the inner surface of the BCR. Here it moves directionally towards the animal pole (AP) of the embryo (Figure 1). The cells of the leading edge of the mesodermal mantle, *i.e.* the dorsal prospective head mesoderm, the more lateral heart mesoderm, and the ventral mesoderm, migrate actively by crawling on the inner surface of the BCR (Nakatsuji 1975; Keller and Schoenwolf 1977).

This migratory movement of the mesoderm can be characterized by three basic features. First, in *Xenopus*, the mesoderm moves as a coherent cell mass, and not as a loose stream of individually migrating cells. Second, these aggregated cells move on a plane substrate, the inner cell layer of the BCR, which is covered by a sparse network of ECM fibrils (Nakatsuji and Johnson 1983a; Nakatsuji *et al.* 1985). As in other amphibian embryos, fibronectin (FN) is a major component of these fibrils (Nakatsuji *et al.* 1985) and plays a role in mesoderm cell migration (Winklbauer 1990). Third, the mesoderm moves directionally from the site of involution at the blastopore lip to the AP region of the gastrula.

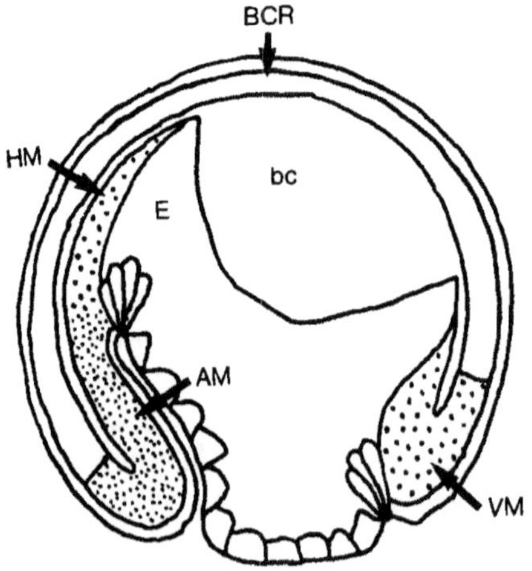

Figure 1. Schematic drawing of a sagittal section through a stage 11 *Xenopus* gastrula. Future dorsal side to the left, animal pole to the top. bc, blastocoel; BCR, blastocoel roof; HM, prospective head mesoderm; AM, prospective axial mesoderm; VM, prospective ventral mesoderm; E, prospective endoderm apposed to the migrating mesoderm.

In the present article, we analyze different aspects of this process. These include analyses of: (1) the motile behavior of mesoderm cells *in vitro* and on their substrate *in situ*, the BCR; (2) details of mesoderm cell interaction with the BCR; (3) mesoderm cell-cell interactions, and how these interactions modify the migratory behavior of these cells; and (4) guiding cues which direct the migrating cells to the AP region on the BCR. In our study, we concentrate on the dorsal, anterior mesoderm, which represents the prospective head mesoderm (HM).

MATERIALS AND METHODS

Embryos and Explants

Embryos, obtained from induced spawnings and artificial fertilization as described (Winklbauer 1986), were staged according to Nieuwkoop and Faber (1967). The jelly coat was removed by treatment with 2% cysteine in Modified Barth's Solution (MBS, Winklbauer 1988) at pH 8, and the vitelline membrane removed manually with forceps. Explants were disected from stage 11 gastrulae with eye-brow hairs and hair-loops (mounted in disposable pipets) in Danilchik's Solution (DS, Keller *et al.* 1985a,b) on a base of plasticine in plastic petri dishes. Explants were secured under a coverslip glass bridge supported by silicone grease at the edges, and cultivated in DS at 21°-24°C.

Preparation of Head Mesoderm Cells

Prospective HM was excised from stage 10.5 gastrulae and dissociated into single cells by incubation in Ca^{++}- and Mg^{++}-free MBS containing 0.1 mM EDTA.

Substrate Coating and Conditioning

Non-adhesive substrate was prepared by incubating the bottom of a plastic dish with 50 mg/ml of bovine serum albumin (BSA; Sigma) for 30 min at room temperature. Preparation of FN substrates was performed by preincubating a marked area of the bottom of a Greiner TC plastic dish with a solution of bovine plasma fibronectin (100 μg/ml) in MBS for 2 hours at room temperature. After washing with MBS, the substrate was incubated with 50 mg/ml of BSA for 30 min to inhibit non-specific attachment of cells. Fibronectin was purified from bovine plasma by affinity chromatography on gelatin sepharose as described (Winklbauer 1988).

To prepare conditioned substrates, the procedure of Nakatsuji and Johnson (1983a) was modified. A strip of BCR extending from the dorsal blastopore lip over the AP to the ventral lip from a stage 10 embryo was excised, placed with its inner surface down on the bottom of a Greiner plastic tissue culture dish, and secured under a coverslip supported by silicone grease. After 2-3 hours in DS (control embryos at stage 11), the outline of the explant, the positions of the dorsal and ventral blastopore lips, and the AP were marked on the bottom of the dish. The explant was then removed by aspiration with buffer and the conditioned substrate saturated with 50 mg/ml of BSA in MBS for 0.5 hours. After each experiment, the conditioned substrate was stained for the presence of FN fibrils, to validate the success of the conditioning procedure.

Recording and Evaluation of Mesoderm Cell Movement

An externally triggered U matic video recorder (Sony), and a Hamamatsu DVS-3000 image processor were used to generate time-lapse recordings (20 sec intervals). Migration of single cells or mesoderm explants *in vitro* was observed on an inverted microscope with phase-contrast optics. Cell migration on the opaque BCR was followed under indirect illumination (Keller and Hardin 1987; Winklbauer 1990). Explant migration on conditioned substrate was quantified by measuring the advance of the center of the mesoderm explant (position of the anterior edge plus position of the posterior edge divided by two, to compensate for the effects of explant spreading) along the dorsal lip-AP axis from time-lapse recordings or from photographs taken at intervals. Movement away from the dorsal lip was assigned positive values (of velocity, or distance), and movement towards it negative values.

Scanning Electron Microscopy

Embryos were fixed in 2.5% glutaraldehyde in MBS at 4° C for 5 hours and washed extensively in phosphate-buffered saline (PBS). Specimens were mounted on 1 × 1 cm plastic squares with tissue glue (Histoacryl Blau, Braun Melsungen AG) or polychloroprene glue (UHU Kontact 2000, UHU GmbH, Brühl) (König 1988, 1990) dissected with tungsten needles as required, and post-fixed with 1% OsO4 in PBS for 1-2 hours at room temperature. The specimens were dehydrated in a graded ethanol series, critical point dried, mounted on stubs, and sputter-coated with gold-palladium. Photographs were taken on Agfapan 100 on a Siemens Autoscan scanning electron microscope.

Antibodies

A rabbit antiserum against *Xenopus* plasma FN was obtained by immunizing with *Xenopus* plasma FN that had been purified by affinity chromatography on gelatin sepharose. Monoclonal antibody (Mab) 6D5 against U-cadherin was prepared by immunizing mice with an electrophoretically purified 90 kD protein fraction obtained from trypsin treated whole *Xenopus* A6 cells (American Type Culture Collection, Bethesda, Maryland, USA). Inert control IgG P3 was produced by the myeloma cell line P3K (Horibata and Harris 1970).

Antibody Staining

Isolated stage 10½ blastopore lip regions were fixed in 20% DMSO and 80% methanol overnight at -20° C. The specimens were incubated in Mab 6D5 to U-cadherin (ascites fluid 1:500) followed by FITC-conjugated goat-anti-mouse F(ab)2 as the secondary antibody (Dianova) and a FITC-conjugated rabbit-anti-goat F(ab)2 antibody specific for the F(ab)2 fragment (Dianova) as a tertiary antibody, each diluted in 20% rabbit serum in PBS, overnight at 4° C. Specimens were dehydrated in DMP (2,3 dimethoxypropane) and embedded in glycerolmethacrylate. Four μm sections were mounted in Mowiol (Hoechst, FRG). BCR or conditioned substrates were stained for the presence of FN fibrils with a rabbit antiserum against affinity purified *Xenopus* plasma FN. Substrate was fixed in 2% formaldehyde in PBS at 4° C, incubated with antiserum (1:200) for 1 hour, washed, and stained with FITC-conjugated goat-anti-rabbit antibody (Dianova, 1:1000).

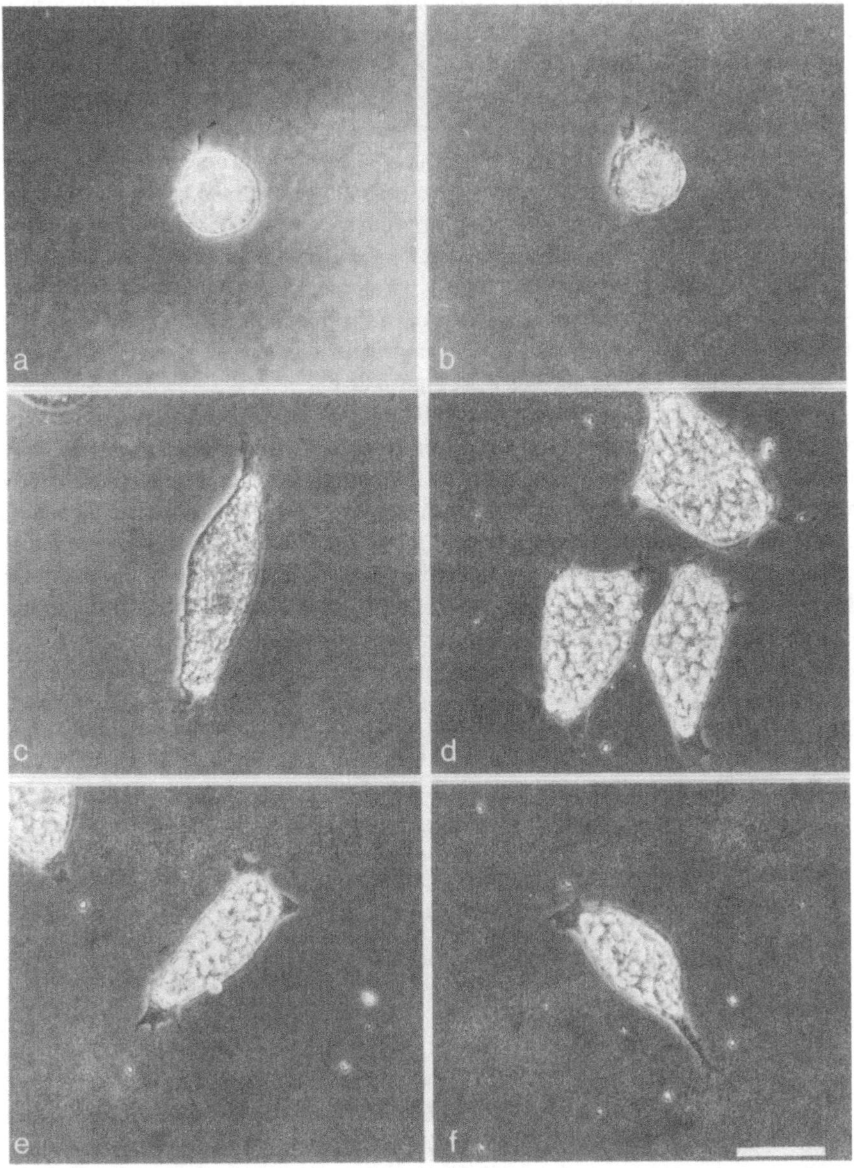

Figure 2. Cytoplasmic protrusions of HM cells *in vitro*. Phase-contrast micrographs of isolated HM cells on tissue culture plastic coated with 50 mg/ml of BSA (**a,b**) or with 100 μg/ml of bovine plasma FN (**c-f**). Filiform protrusions on non-adhesive substrate (**a,b**) or on FN (**c,e**) are labelled by arrow heads. Bar = 50 μm.

RESULTS

Formation of Protrusions by Head Mesoderm Cells *In vitro*

The formation of cytoplasmic processes is essential to the locomotion of crawling cells, and the pattern of protrusive activity determines to a significant extent the rate and mode of movement of a cell. On a non-adhesive substrate, isolated HM cells remain globular and do not spread. Filiform cytoplasmic protrusions extend spontaneously from the yolk-filled cell body (Figure 2a,b), and retract again after one to several minutes. New processes usually appear close to the site of previous ones, and when several protrusions are present at a time, they are often located in close proximity, giving the otherwise radially symmetrical cell an overt polarity (Figure 2a,b).

The protrusive activity of HM cells is altered through interaction with a substrate. On a substrate coated with FN, mesoderm cells adhere, spread, and typically assume a bipolar morphology with lamelliform protrusions extending from opposite ends of the cell (Figure 2c,e,f). Occasionally, cells assume a multipolar shape, with several equally prominent lamellae (Figure 2d). Thus, whereas non-attached cells extend only short-lived filiform protrusions, cells on adhesive substrata form more stable lamelliform protrusions. Additional filiform protrusions, however, may occasionally be present on spread cells (arrow heads in Figure 2c,e).

A time-lapse analysis of the process of substrate attachment shows that HM cells are already bipolar during initial spreading (Figure 3). Several minutes after contact with the FN substrate, two lamellae appear suddenly and simultaneously at opposite ends of the cell (Figure 3b), with one lamella always at the site where filopodia had formed previously on the unspread cell (Figure 3a). This rapid extension of lamellae, and their simultaneous appearance at distinct regions of the cell, are not exhibited by HM cells spreading on lentil lectin or tissue culture plastic (Selchow, unpublished results). This suggests that spreading and lamella formation on FN is a substrate-induced event. In addition, the polarity of a spread HM cell is related to the polarity of a non-attached cell, as defined by the site of filiform protrusion formation.

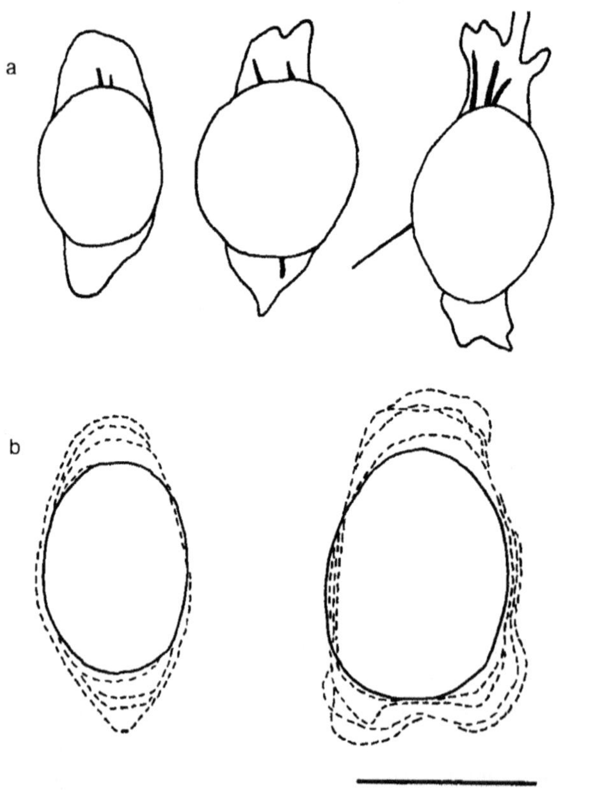

Figure 3. Spreading of HM cells on FN. **(a)** Relationship between a cell's filiform protrusions before spreading, and the appearance of cytoplasmic lamellae during spreading. The last few filopodia formed before spreading, and the later appearing lamellae of a cell are all integrated into a single drawing of the cell, to indicate the relative positions of these protrusions. **(b)** Time course of appearance of lamellae during spreading. Dashed lines indicate outline of cells at 1 min intervals. Bar = 50 μm.

Regulation of Head Mesoderm Cell Protrusive Activity by the Blastocoel Roof

In *Xenopus*, the mesoderm and adjacent endoderm move as a single coherent cell mass (Figure 1). When the upper side of this cell mass, which faces the blastocoel cavity, is compared in the SEM to its lower surface, from which the BCR substrate has been removed after fixation, a clear difference in the extent of protrusion formation is observed. On the upper surface, cells appear closely packed, with extensive lateral contacts, and only a few tiny cytoplasmic processes extend between neighboring cells. On the lower surface originally apposed to the BCR substrate, cells are less densely packed, and numerous filiform and lamelliform protrusions extend from these migratory cells (Figure 4).

Mesoderm also moves as a multi-layered, coherent cell mass in explants consisting of the BCR substrate, the blastopore lip, and the involuted mesoderm (Winklbauer 1990). In cases where the leading edge of the mesodermal cell mass has risen above the substrate, cells at the leading edge not in contact with the substrate had smooth surfaces (Figure 5a), whereas cells which were apposed to the BCR before fixation were less densely packed and with numerous cytoplasmic processes (Figure 5b). This suggests that the inner, matrix-covered surface of the BCR regulates the protrusive activity of the moving mesodermal cells.

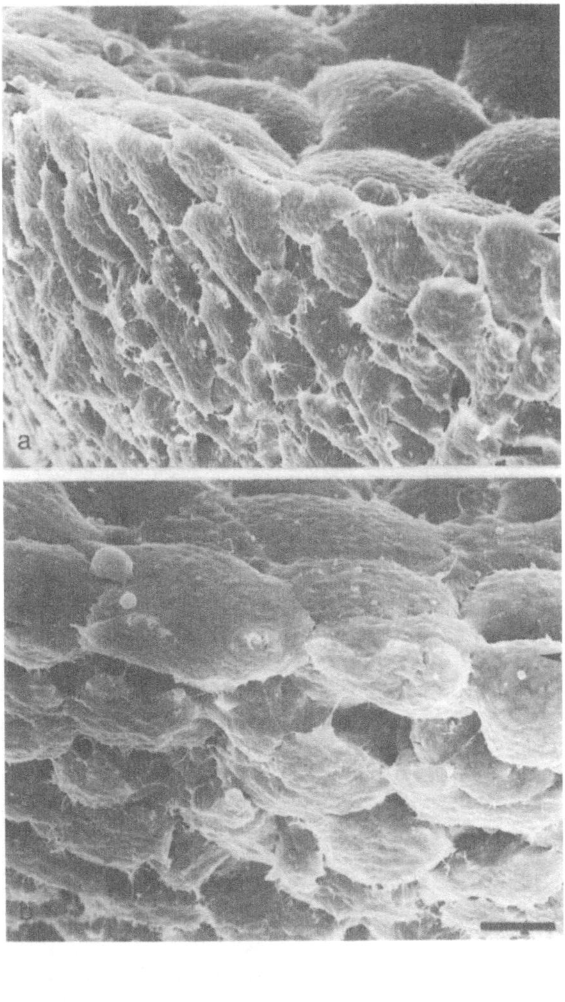

Figure 4. SEM micrographs of the leading edge of the dorsal migratory mesoderm and apposed endoderm. Arrow heads mark the leading edge. Below leading edge mesodermal surface which was in contact with the BCR substrate. BCR has been removed after fixation. Above leading edge blastocoelic surface of mesodermal-endodermal cell mass. Bars = 20 μm.

It may be asked which component of the BCR substrate affects protrusive activity. *In vitro* contact with a FN substrate induces the formation of lamelliform protrusions in HM cells. To test whether FN fibrils of the BCR have a similar effect, the morphology of single HM cells on the BCR was compared under conditions where interaction with FN was permitted or inhibited (Figure 6). Under permissive conditions, HM cells spread moderately on the BCR and extend cytoplasmic processes (Figure 6a,b). On the other hand, in the presence of a GRGDSP peptide, which prevents adhesion of cells to the RGD cell-binding site of FN, cells remain globular (Figure 6c). Also, on a BCR surface devoid of FN-containing fibrils (Winklbauer 1990), cells do not spread (Figure 6d). However, these globular, non-spread cells are nevertheless attached to the BCR (Winklbauer, unpublished results). Thus, interaction with the RGD cell-binding site of FN, or perhaps some other extracellular matrix component with a comparable cell-binding site, is required for cell spreading and the formation of lamelliform protrusions, whereas simple attachment to the BCR is RGD-independent. This suggests that the BCR substrate regulates the protrusive activity of the migratory mesoderm cells through the FN-containing extracellular matrix. Apparently, the effect of FN contained in fibrils on the BCR is comparable to that of artificial FN substrates.

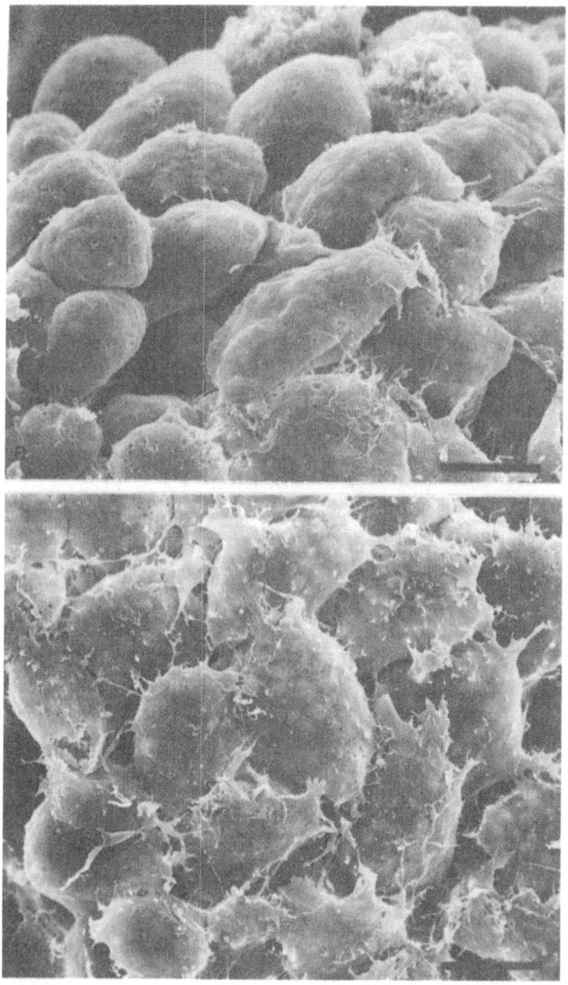

Figure 5. SEM micrograph of the leading edge of mesoderm migrating on the BCR *in vitro*. **a)** Part of the leading edge rising above, and not in contact with the substrate. **b)** Part of the mesoderm near the leading edge which had been in contact with the BCR before fixation. Bars = 20 μm.

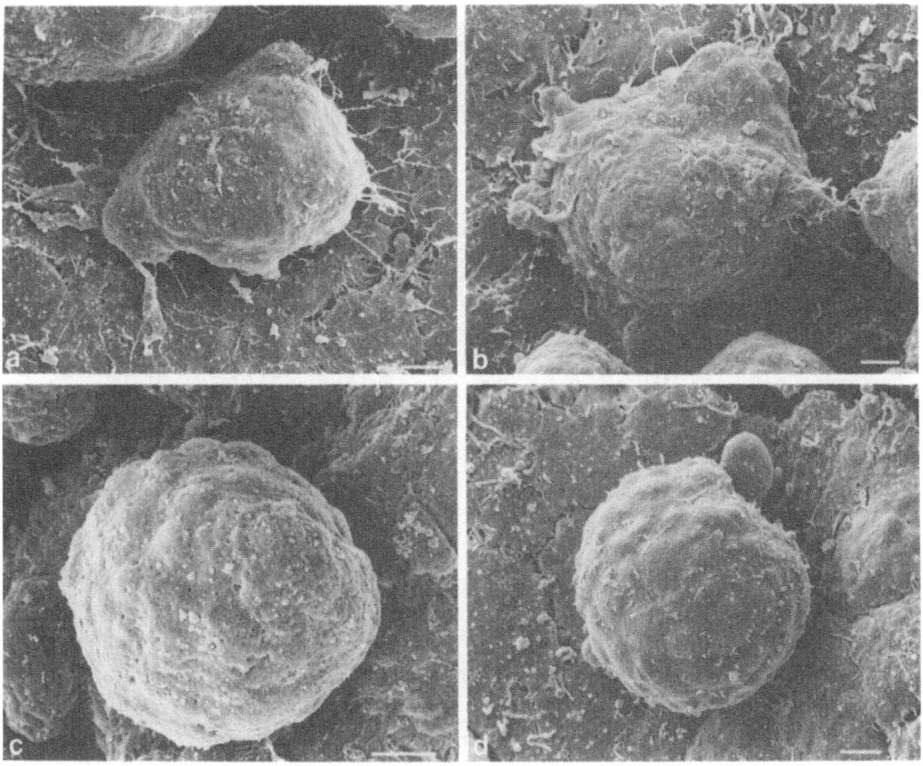

Figure 6. Morphology of isolated HM cells on the BCR, as seen in the SEM. Cells on the inner surface of the BCR, after one hour in GRGESP control peptide solution, 4 mg/ml (**a,b**), or GRGDSP peptide solution, 4 mg/ml (**c**). (**d**) Cell on outer matrix-free surface of inner BCR layer, after one hour. Bars = 5 μm.

Migratory Behavior of Isolated Head Mesoderm Cells

To link the pattern of protrusive activity of HM cells to their migratory behavior, the mode of locomotion of bipolar HM cells was investigated. From time-lapse video sequences of single cells moving on an *in vitro* FN substrate, an inventory of the behavioral repertoire of these cells was extracted (Figure 7). Two fundamentally different types of movement were observed.

First, cells may move steadily, without a change in the number or distribution of lamellae. Thus, bipolar cells may sometimes glide laterally over a short distance (Figure 7; 1). In other instances, cells which have become unipolar for some period of time, will move steadily and efficiently on a straight pathway, with the single lamella leading its movement (Figure 7; 2).

The second type of movement is characterized by a step-wise advance of the cells. The basic events that underlie this migratory behavior are the formation and the retraction of lamellae, the splitting of a single lamella in two, and the lateral movement of lamellae along the cell circumference. Often, the cell moves abruptly forward when a lamella detaches from the substrate and retracts (Figure 7; 3). In rare cases, the other lamella of the bipolar cell is retained, and the cell continues to move in a unipolar fashion, as described above (Figure 7; 2). Typically, however, the remaining lamella is

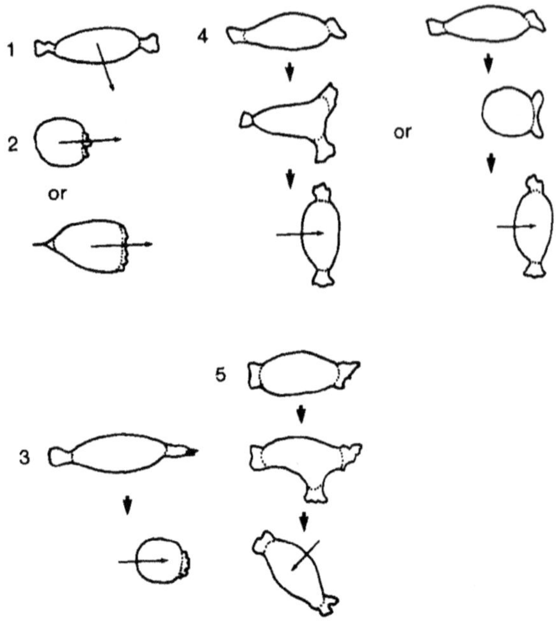

Figure 7. Behavioral repertoire of HM cells. Cell bodies with attached lamelliform protrusions are outlined. Types of movement are numbered as referred to in the text. Horizontal or oblique arrows indicate the direction of cell movement; vertical arrows indicate transitions between different stages in the translocation process.

split, and the two lamellae derived from it move apart and lead to the spreading of the cell in a direction which is perpendicular to the original direction of movement (Figure 7; 4). This process may be repeated several times. In other cases, a third lamella appears, before an original one retracts. Again, this leads to a step-wise translocation of the cell for a short distance, with a concomitant change in direction (Figure 7; 5). Occasionally, cells are multipolar, and their movement is determined in a more complex way, which nevertheless involves the same basic features just described, *i.e.* the appearance, lateral movement, and retraction of lamellae (not shown). Figure 8 shows an example of a single mesoderm cell moving on FN *in vitro*. After the initial bipolar spreading, the cell repeatedly retracts one of its lamellae, and snaps forward to the position of the remaining one, which is then split in two, leading to the re-spreading of the cell in a new orientation. In this way, the cell moves around randomly and performs characteristic turns. After one hour, the cell is found close to the original starting position. A short episode of efficient unipolar migration brings the cell to a new, distant location, where it becomes bipolar again and continues to move step-wise, as before. This mode of cell migration, which seems linked to the bipolar structure of these cells, is not well suited for the persistent, directional migration towards a defined target region.

The step-wise, intermittent mode of migration, with built-in changes in direction, is not altered when isolated mesoderm cells move on their *in vivo* substrate. In the SEM, cells seeded onto an explanted piece of BCR often appear bipolar, as on FN *in vitro* (Figure 9). From time-lapse films, cell trails can be reconstructed which show the characteristic, abrupt turns observed for cells moving on FN *in vitro*, and cells move persistently only over short distances (Figure 10a). However, even this limited persistence is lost when cell interaction with FN is inhibited by an RGD-containing peptide, and cells move on convoluted, random pathways (Figure 10b). As shown above, under these conditions cells attach to the BCR, but remain globular and do not form stable, lamelliform cytoplasmic protrusions. Apparently, FN-induced lamella

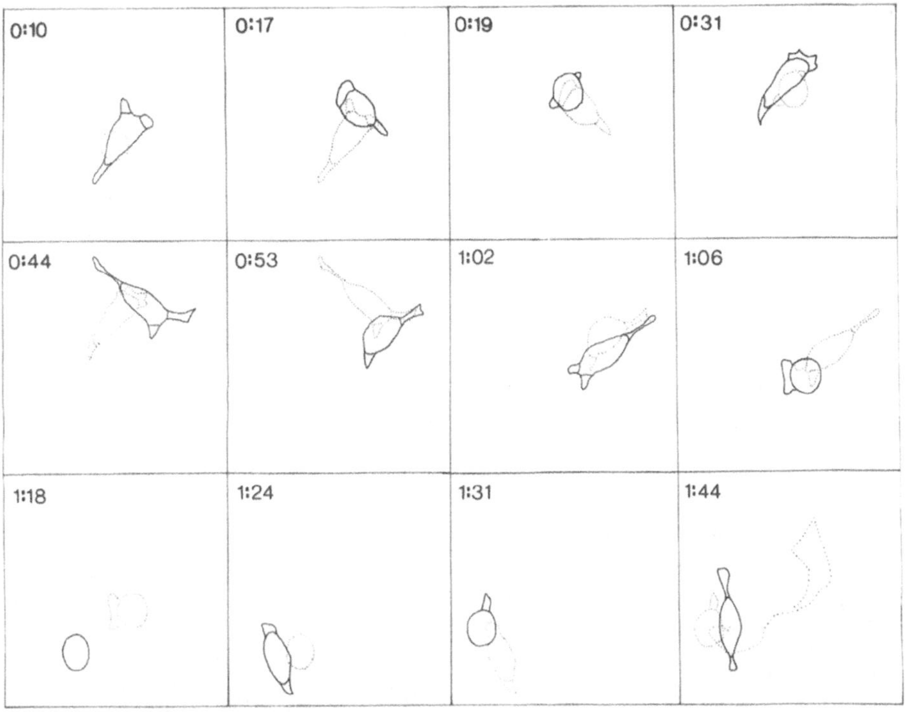

Figure 8. *A HM cell migrating on FN in vitro*. The outline and position of the cell was drawn at characteristic stages from a time-lapse video recording (20 sec intervals). In each frame, the eventual outline of the cell in the previous frame (dotted lines), and the outline at the time indicated in the frame (solid lines) is drawn. Time is in hours and minutes. Bottom right frame also shows the pathway of the cell during 1:44 hours of migration.

formation stabilizes the movement of inherently motile cells such that some limited degree of persistence is attained. However, the bipolar nature of the cells prevents them from a long-range persistent translocation.

Cell-Cell Adhesion in the Migrating Head Mesoderm

When the BCR is removed from a living *Xenopus* gastrula, the mesoderm remains as a coherent cell mass which can be excised and moved around as a whole. In Ca^{++}-free buffers, however, the mesoderm can be dissociated into single cells, indicating that a Ca^{++}-dependent cell-cell adhesion mechanism preserves the integrity of the mesodermal layer.

Cadherins are Ca^{++}-dependent cell-cell adhesion molecules (Takeichi 1988). A novel *Xenopus* cadherin (U-cadherin) was found to be present in the early *Xenopus* embryo (Angres *et al.* 1991). Immunofluorescent staining with an antibody against U-cadherin reveals its presence both on the cells of the BCR, and on the surface of the migrating mesoderm cells (Figure 11).

When a piece of HM is explanted in the presence of control antibody onto FN *in vitro*, it spreads and behaves like an epithelial cell cluster cultured *in vitro*. Cells of the explant glide past each other, but only very rarely manage to break loose and move

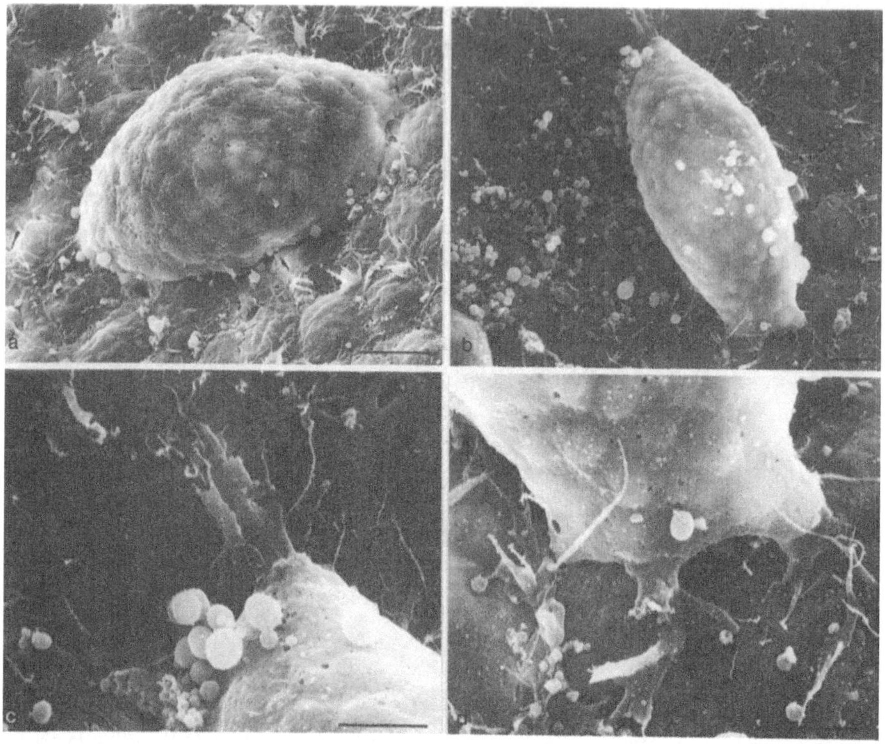

Figure 9. Bipolar spreading of HM cells on the inner surface of the BCR. Single HM cells after one hour on explanted BCR (**a,b**). The upper and lower end of the spindle-shaped cell in (**b**), with cytoplasmic processes, are shown in (**c**) and (**d**), respectively. Bars = 10 μm (**a,b**), or 5 μm (**c,d**).

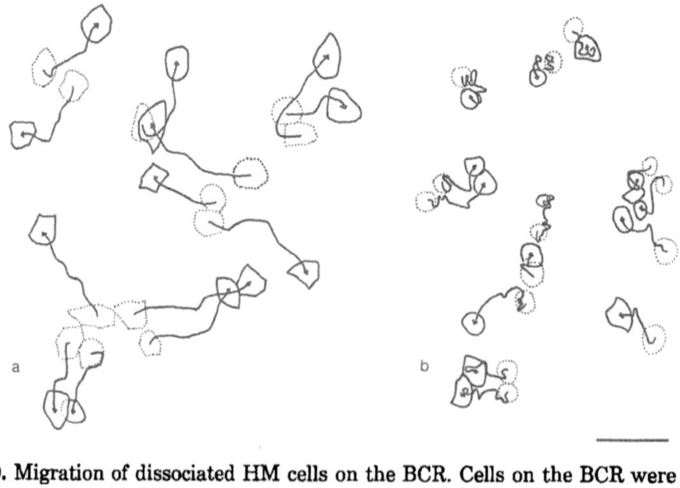

Figure 10. Migration of dissociated HM cells on the BCR. Cells on the BCR were visualized by indirect illumination and filmed. Cell paths were traced from projections of film frames (1-2 min intervals). The initial (dotted outline) and the final (solid outline) positions of cells, and the cells paths during a 1 hour interval are indicated for a control explant (**a**), and for cells migrating in the presence of 4 mg/ml of GRGDSP peptide (**b**). Bar = 100 μm. From Winklbauer (1990), with permission from *Academic Press*.

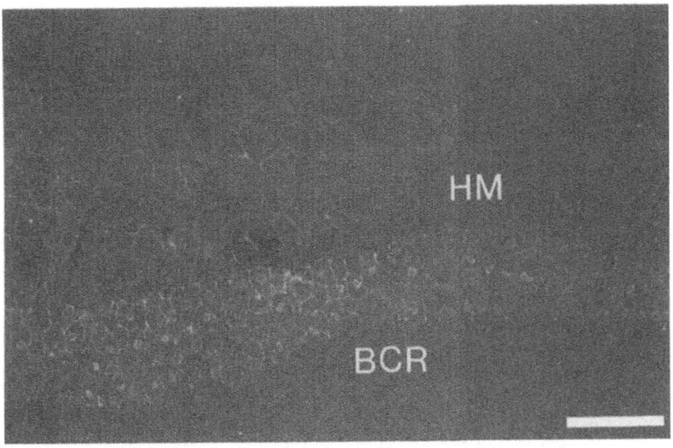

Figure 11. Immunofluorescence staining of migratory HM and BCR substrate with U-cadherin antibody. Bar = 100 μm.

away from the explant (Figure 12a). In contrast, in the presence of 10 μg/ml of a monoclonal antibody against U-cadherin, the cells disperse over the substrate, and the explant completely disintegrates (Figure 12b). This provides strong evidence that the Ca^{++} dependent cohesion of the migratory mesoderm is mediated by U-cadherin.

Migratory Behavior of Head Mesoderm Cells in Aggregates

To see how the migratory behavior of mesoderm cells is modified through mutual contact in a cell aggregate, the trails of isolated cells were compared with the pathways of cells moving as part of a larger mesoderm explant on BCR conditioned substrates (Figure 13). On such a substrate, single HM cells are bipolar and move as on FN *in vitro*, and the cell trails show frequent abrupt turns (Figure 13a). In contrast, cells that move as part of a larger mesoderm explant migrate with high persistency along straight pathways (Figure 13b). Thus, the inclusion of cells in an aggregate through cadherin-mediated mutual cell adhesion stabilizes the migratory behavior of HM cells.

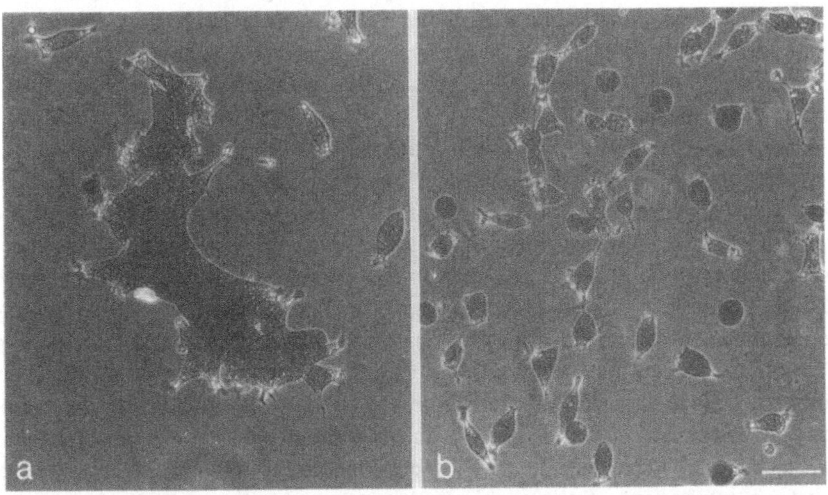

Figure 12. Behavior of HM explants on FN *in vitro*. Explants were cultured in the presence of an inert P3 control antibody (10 μg/ml) (**a**), or of 10 μg/ml of an antibody against U-cadherin (**b**). Photographs were taken after 3 hours in culture. Bar = 100 μm.

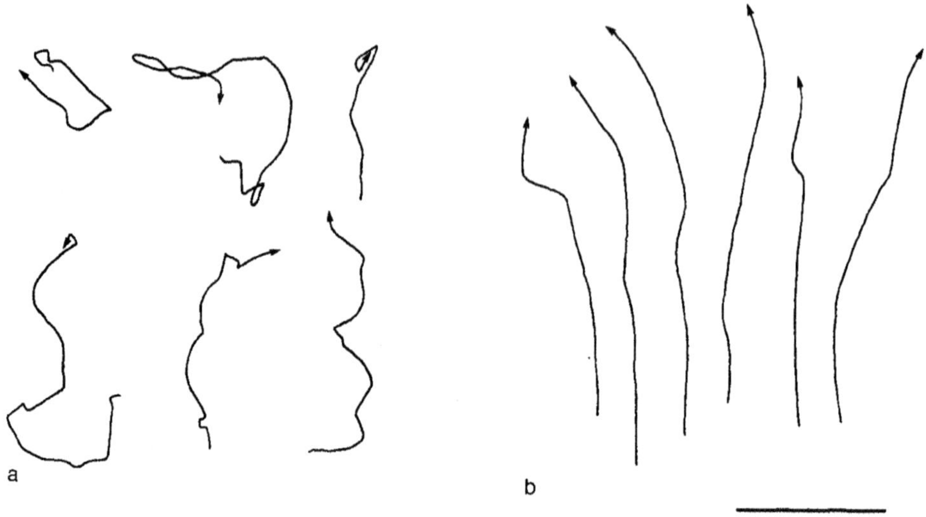

Figure 13. Persistence of HM cell migration. Cell trails of isolated HM cells moving on conditioned substrate (**a**), or of HM cells moving as part of a larger HM explant on conditioned substrate (**b**) were reconstructed from video time-lapse recordings (20 sec intervals). Cells were followed for 2 hours. Bar = 100 μm.

Unfortunately, the activity of cytoplasmic protrusions in HM cell aggregates can only be observed at the margin of explants. To obtain an impression of substrate-related protrusive activity of the cells in the central part of an explant, pieces of HM migrating on conditioned substrates, and explants simply spreading on FN, were fixed and processed for SEM. On FN, cells apposed to the substrate are in close mutual contact. Cytoplasmic protrusions are found mainly on marginal cells, and not much underlapping of sub-marginal cells is observed (Figure 14a). In contrast, underlapping of neighboring cells by extended lamelliform protrusions is common on conditioned substrate. Moreover, these underlapping lamellae are preferentially oriented in the direction of migration (Figure 14b).

Such oriented underlapping of migratory mesoderm cells is also observed in the embryo (Figure 15). When the dorsal BCR is removed from a fixed mid-gastrula, the surface of the mesoderm normally in contact with the substrate is exposed. It shows a pronounced shingle arrangement of cells of the anterior mesoderm, where a cell typically underlaps its anterior neighbor, in the direction of migration. Thus, both on conditioned substrate and on the BCR, migrating mesodermal cells in an aggregate appear unipolar, with lamelliform and filiform protrusions preferentially extending in the direction of migration. This could explain the high persistence of migration of mesodermal cells in larger cell aggregates, as opposed to single cells. Since isolated cells are bipolar and do not move persistently on conditioned substrate or on the BCR, and since, on the other hand, cells do not show oriented underlapping in cell aggregates on FN, both a component of the *in vivo* substrate, and cell aggregation seem to be required for the expression of a unipolar cell structure and persistent locomotion.

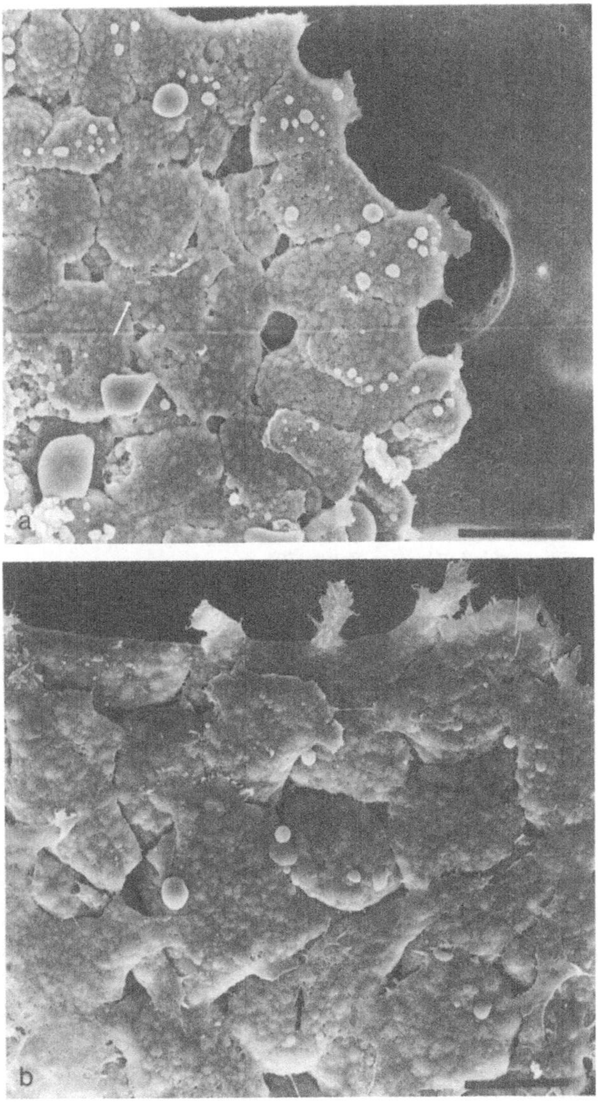

Figure 14. HM explants viewed in the SEM from the substrate side. The lower surface of HM explants on FN **(a)** or on conditioned substrate **(b)** is shown. Arrow, direction of migration on conditioned substrate. Bars = 50 μm.

Directional Movement of the Mesoderm

The extracellular matrix of the BCR consists of a network of FN-containing fibrils which can be visualized by immunofluorescence staining with antibodies against FN (Figure 16a). When a strip of BCR, extending from the dorsal lip to the AP, is cultured with its matrix side down on a plastic substrate for 2 hours, this fibrillar network is transferred to the substrate, as revealed by antibody staining (Figure 16b). FN fibril formation can be inhibited in two independent ways. When substrate conditioning is performed in the presence of a GRGDSP peptide (not shown) or of cytochalasin B (Figure 16c), FN is deposited in a diffuse, punctate pattern, and no extended network of FN fibrils is formed.

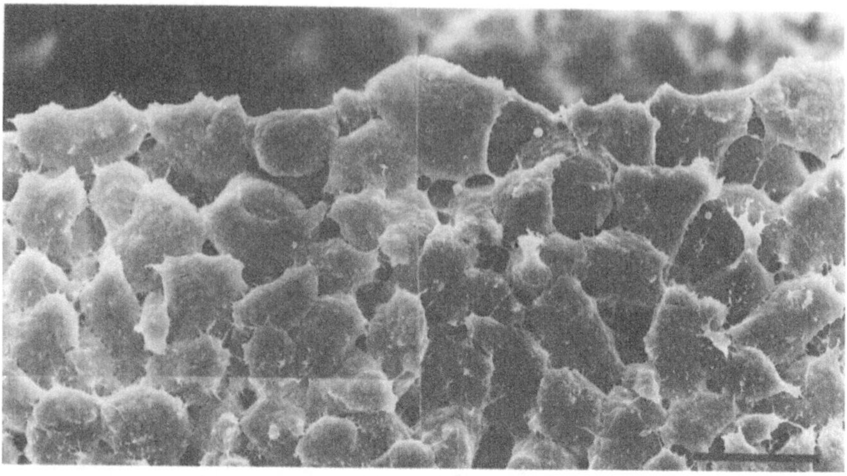

Figure 15. Arrangement of migrating mesoderm cells in the embryo. Shingle arrangement of dorsal mesoderm cells near the leading edge, as viewed from the substrate side. Direction of migration is to the top. Bar = 50 μm.

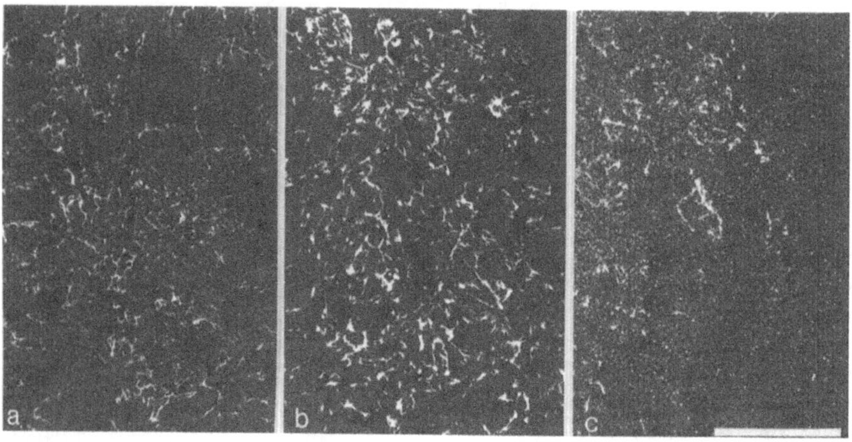

Figure 16. FN fibrils of the BCR extracellular matrix. Inner surface of the BCR (**a**), conditioned substrate (**b**), and substrate conditioned in the presence of 20 μg/ml of cytochalasin B (**c**) was immunostained with rabbit antiserum against *Xenopus* FN and FITC-conjugated second antibody. Bar = 50 μm.

On normal conditioned substrate, a mesoderm explant always moves towards the AP position of the conditioned substrate when placed near the dorsal lip region of the substrate, regardless of the explants orientation, (Figure 17a,b). This shows that the extracellular matrix of the inner surface of the BCR contains cues which are sufficient to guide the migrating mesoderm to its target region.

On conditioned substrate where FN fibril formation has been inhibited, the directionality of mesoderm movement is lost. Both on substrate conditioned in the presence of RGD peptide (Figure 17c) or cytochalasin B (Figure 17d), mesoderm explants are able to spread and move, demonstrating that the substrate is able to support cell adhesion and migration. However, movement seems random, being sometimes towards and sometimes away from the blastopore lip, and sometimes only spreading of the explant in both directions is observed. These experiments demonstrate a correlation between the presence of FN fibrils, and directional mesoderm cell migration.

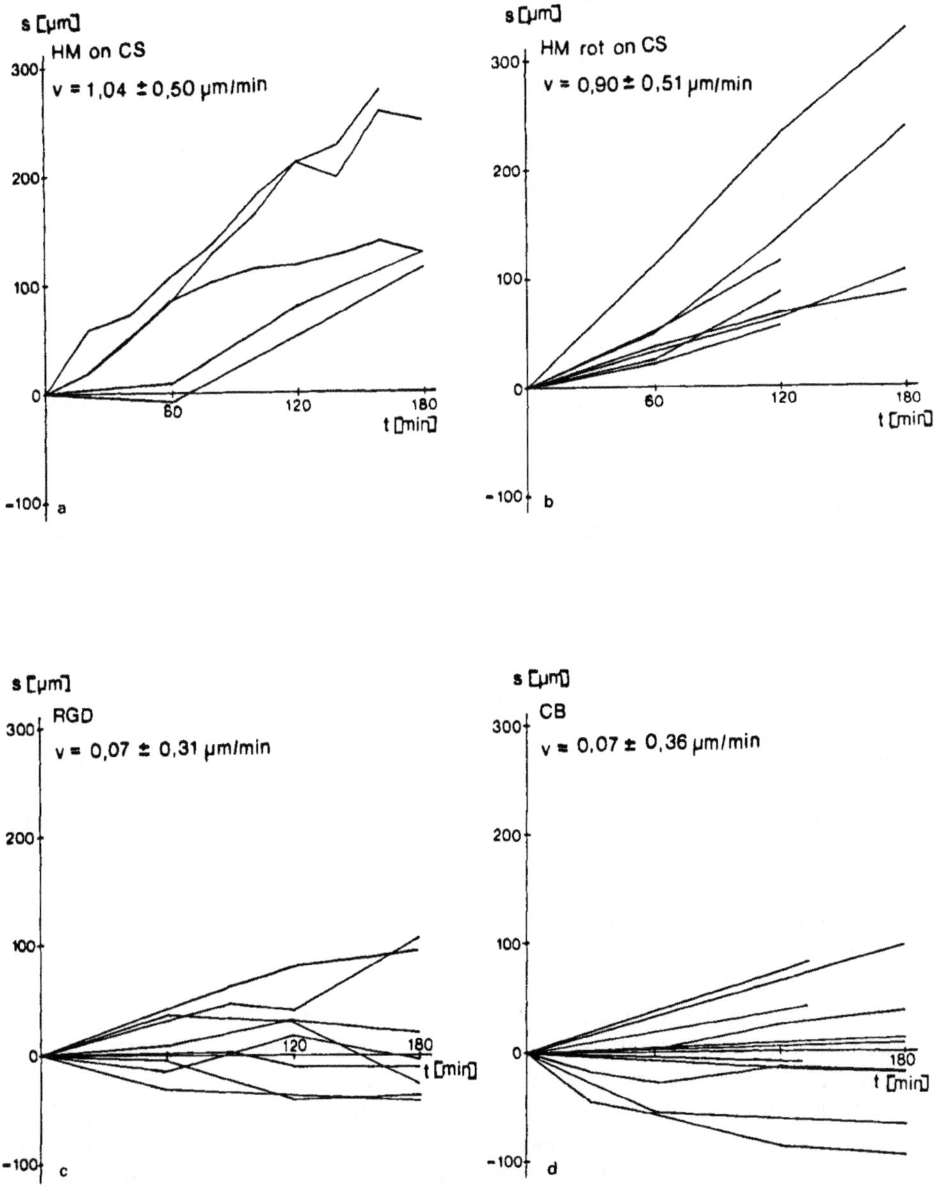

Figure 17. HM explant movement on conditioned substrate. HM was placed either in normal orientation (**a**) or rotated 180° (**b**), on conditioned substrate close to the position of the dorsal lip. Similarly, HM was placed between the dorsal lip and the AP on substrate conditioned in the presence of 4 mg/ml of GRGDSP peptide (**c**), or of 20 μg/ml of cytochalasin B (**d**). Horizontal axis represents time in minutes, vertical axis indicates the distance of the explant from the starting position.

DISCUSSION

The Role of Fibronectin in Mesoderm Cell Migration

In *Xenopus*, the involuted mesoderm moves as a coherent cell mass across the BCR. The mesodermal surface in contact with the BCR substrate is characterized by numerous cytoplasmic protrusions, as compared to the free surface exposed to the blastocoel. This indicates that the BCR might regulate the protrusive activity of the overlying mesoderm. We propose that the extracellular matrix molecule FN induces the formation of cytoplasmic processes in mesodermal cells.

Migratory HM cells spontaneously extend filiform protrusions when placed on a non-adhesive substrate. However, these protrusions are short-lived, and apparently not stabilized by contact with FN (Selchow, unpublished results). In contrast, when the globular cell body comes into contact with a FN-coated substrate, the cell spreads, and stable lamelliform processes are formed. These lamellae appear to be formed de novo, and not to be derived from pre-existing filiform protrusions. Thus, FN is capable of inducing lamelliform protrusions in HM cells *in vitro*.

FN is present in the amphibian gastrula as a network of fibrils on the inner surface of the BCR (Boucaut and Darribère 1983a,b; Boucaut *et al.* 1985; Nakatsuji *et al.* 1985; Darribère *et al.* 1985). In *Xenopus*, this network is very sparse, covering 1-4% of the BCR surface (Nakatsuji and Johnson 1983a; Keller and Winklbauer 1990), and mesoderm cells attach to the BCR in the absence of cell-FN interaction (Winklbauer, unpublished results). However, mesoderm cell spreading on the BCR (Winklbauer 1990; this article) and the concomitant appearance of cytoplasmic protrusions apposed to the substrate is a FN-dependent process. This suggests that the role of FN in the *Xenopus* gastrula is not to promote the simple adhesion of migratory mesoderm cells to their BCR substrate, but to induce the formation of cytoplasmic protrusions which serve as locomotory organelles for moving cells. This function of FN seems especially appropriate under conditions where FN is present in a discrete fibrillar distribution, and not as a homogeneous layer.

Induction of protrusion formation ascribes FN an essential role in mesoderm cell migration. Non-spread, globular HM cells are intrinsically motile, and move around on the BCR, but at very low persistence, on convoluted, random pathways. The effect of cell spreading and protrusion formation is that cell locomotion is stabilized and becomes more persistent, leading to more efficient translocation (Winklbauer 1990; this article). In agreement with this, inhibition of cell-FN interaction in the intact embryo with antibodies against FN (Boucaut *et al.* 1984a), or against a putative FN receptor (Darribère *et al.* 1988, 1990), with an RGD-containing peptide which competitively blocks the cellular FN receptor (Boucaut *et al.* 1984b), or with exogenous tenascin (Riou *et al.* 1990), always results in an arrest of gastrulation in the *Pleurodeles* embryo (see article by Boucaut, this volume). This is consistent with the idea that mesodermal cells interact with FN *in vivo*, and that this interaction is essential for mesodermal cell migration.

The Role of Substrate-dependent Guiding Cues in Mesodermal Cell Migration

Two basic types of mechanisms may be discerned which could direct moving cells to their target. For one, diffusible substances may control the direction of cell migration by positive or negative chemotaxis. Second, substrate-dependent mechanisms can guide migrating cells, *e.g.* by haptotaxis (Carter 1965), or contact guidance (Weiss 1961). Both mechanisms have been proposed to explain the directional migration of the mesoderm in the amphibian embryo. Kubota and Durston (1978) suggested negative chemotaxis to account for directional mesoderm cell movement in the Axolotl, whereas Shi *et al.* (1989) provided strong evidence for a substrate-dependent guiding mechanism in *Pleurodeles*.

However, it is not obvious why any of these mechanisms should actually be required, since it seems that simple population pressure could lead to directional mesodermal movement in *Xenopus*. The involuted mesoderm forms a coherent cell mass around the circumference of the embryo, and free BCR substrate to be invaded by mesoderm cells is only available in the direction of the animal pole. Moreover, cells are constantly added to the migrating mesoderm posteriorly by the process of involution. This could be sufficient to lead to the observed directional advance of the leading edge of the mesodermal mantle. Biassing random cell movement in the animal-vegetal direction could facilitate this process. Indeed, such a mechanism has been proposed to explain the directional migration of mesoderm cells in the amphibian gastrula (Nakatsuji and Johnson 1983b). But the results of Shi *et al.* (1989), and our experiments with conditioned substrates conclusively show that the extracellular matrix of the BCR contains guiding cues which are sufficient to direct the migrating mesoderm to its target region.

A gradient of some ECM component could lead to directional mesoderm movement, *e.g.* by the mechanism of haptotaxis (Carter 1965). Alternatively, the substrate could be oriented and have vector-like properties. Our experiments where inhibition of FN fibril formation is correlated with the loss of directional migration are consistent with the latter possibility. Thus, we favor the hypothesis that some structure of the BCR extracellular matrix, probably the FN-containing fibrils, is polarized, as was originally proposed by Nakatsuji and Johnson (1983b), and gives directionality to migrating mesoderm cells.

The migrating mesoderm is visibly polarized both in the embryo, and in explants moving on conditioned substrate, showing a characteristic shingle arrangement of cells, with oriented underlapping of cell processes in the direction of migration. Such a polarized structure of the mesoderm is not observed in explants on a homogeneous FN substrate, and it is therefore probably the result of an interaction with the polarized structure of the extracellular matrix substrate. This suggests that the movement of the mesodermal cell aggregate is not driven by the activity of marginal cells of the leading edge, as it is the case in epithelial clusters moving *in vitro* (Dipasquale 1975; Kolega 1981). Instead, all cells in contact with the substrate, at least in a zone several cell rows deep behind the leading edge, are actively migrating and contribute to the movement of the mesoderm. This may be particularly important when the moving cell aggregate does not consist of a single cell layer, but of a solid mass. In this case, most of the cells are not in contact with the substrate, but are passively transported by the undermost, crawling cell population. In this situation, extension of the edge of the aggregate by the motile activity of marginal cells only would lead to the progressive thinning of the mesodermal and apposed endodermal cell mass, and its spreading over the BCR surface, instead of driving the inward movement of the whole vegetal cell mass.

The Role of Cell-Cell Adhesion in Mesoderm Cell Migration

In the intact embryo, as well as in explants moving directionally on conditioned substrate, HM cells appear unipolar, with one leading cytoplasmic process each, when viewed from the substrate side. Also, on a non-adherent substrate, isolated HM cells have a unipolar symmetry, defined by the site of spontaneous protrusion formation. However, on FN, single HM cells are typically bipolar or even multipolar, with two or more prominent lamellae, respectively, but only very rarely unipolar (as a consequence of the temporary retraction of a lamella). The axis of a bipolar, spread cell is related to that of an unspread cell, since, during spreading, one lamella extends at the former site of spontaneous protrusion formation.

The bipolar structure of HM cells, together with their pattern of protrusive activity, leads to a peculiar mode of migration of these cells, which is characterized by a step-wise advance, and frequent turns, so that long-range persistence is low (Nakatsuji and Johnson 1982; this article). This mode of translocation is obviously not very efficient, and not well suited for directional migration. The bipolar structure of HM cells and the ensuing migratory behavior is not an artifact caused by the *in vitro* FN substrate. Isolated HM cells are also bipolar on conditioned substrate, and on their *in vivo* BCR substrate. Moreover, migration shows the same characteristics on conditioned substrate and on the BCR as on FN *in vitro*. This means that HM cell movement on the BCR is certainly more persistent when cells are allowed to interact with FN fibrils, than when this interaction is inhibited, but cell trails still lack the persistence of the same cells when they are part of the mesodermal aggregate.

The possibility remains that the bipolar spreading and locomotion of HM cells is caused by the artificial isolation of cells. In the embryo, cadherin-mediated cell-cell adhesion gives the mesoderm the property of a tightly coherent cell aggregate. Moreover, small explants of HM behave like epithelial clusters *in vitro*. On FN, a HM explant spreads, and its marginal cells project lamelliform protrusions, whereas submarginal cells show little protrusive activity and underlapping of neighboring cells. This behavior is typical for patches of epithelial cells in culture (*e.g.* Dipasquale 1975; Kolega 1981). In contrast, directional underlapping and a unipolar cell morphology is observed only in mesoderm explants on conditioned substrate, and *in situ*. Thus, simply being part of a mesodermal aggregate is not sufficient for the development of a unipolar cell structure, but neither is the oriented *in vivo* substrate alone capable of imposing a unipolar morphology on isolated HM cells. Instead, both conditions have to be met to generate a unipolar cell morphology: cells have to be part of a larger aggregate, and this aggregate has to move on a substrate which contains directional cues.

We can only speculate about the mechanisms by which cell-cell contact allows for the efficient polarization of migratory mesoderm cells. Cadherin-mediated cell-cell contact could extinguish the pre-existing polarity as expressed in unspread cells, thus allowing the oriented *in vivo* substrate to impose its own inherent polarity on the migrating mesoderm cells. This in turn would lead to the unipolar extension of locomotory protrusions, oriented underlapping, and efficient, persistent, and directional migration.

ACKNOWLEDGEMENTS

We thank Andrea Belkacemi for technical assistance, Roswitha Grömke-Lutz for help with the photographical work, and Gerd König, Peter Hausen, and Ray Keller for helpful comments on the manuscript. Ray Keller drew our attention to the shingle arrangement of migrating mesoderm cells. Part of this work was supported by an EMBO long-term fellowship to R. Winklbauer and NIH Grant No. HD 18979 to Ray Keller.

REFERENCES

Angres, B., A. Müller, and P. Hausen. 1991. Differential expression of two cadherins in *Xenopus laevis*. *Development*. 111:829-844.

Boucaut, J.-C. and T. Darribère. 1983a. Fibronectin in early amphibian embryos: Migrating mesodermal cells contact fibronectin established prior to gastrulation. *Cell Tissue Res*. 234:135-145.

Boucaut, J.-C. and T. Darribère. 1983b. Presence of fibronectin during early embryogenesis in the amphibian *Pleurodeles waltlii*. *Cell Differ*. 12:77-83.

Boucaut, J.-C., T. Darribère, H. Boulekbache, and J.-P. Thiery. 1984a. Prevention of gastrulation but not neurulation by antibodies to fibronectin in amphibian embryos. *Nature* 307:364-367.

Boucaut, J.-C., T. Darribère, T.J. Poole, H. Aoyama, K.M. Yamada, and J.-P. Thiery. 1984b. Biologically active synthetic peptides as probes of embryonic development: A competitive peptide inhibitor of fibronectin function inhibits gastrulation in amphibian embryos and neural crest cell migration in avian embryos. *J. Cell Biol.* 99:1822-1830.

Boucaut, J.-C., T. Darribère, D.-L. Shi, H. Boulekbache, K.M. Yamada, and J.-P. Thiery. 1985. Evidence for the role of fibronectin in amphibian gastrulation. *J. Embryol. Exp. Morphol.* 89 (Suppl.):211-227.

Boucaut, J.-C., T. Darribère, D. Shi, J.-F. Riou, K.E. Johnosn, and M. Delarue. 1991. Amphibian Gastrulation: The Molecular Bases of Mesodermal Cell Migration in Urodele Embryos. p. 169-184. *In: Gastrulation: Movements, Patterns, and Molecules*. R. Keller, W.H. Clark, Jr., F. Griffin (Eds.). Plenum Press, New York.

Carter, S.B. 1965. Principles of cell motility: The direction of cell movement and cancer invasion. *Nature* 208:1183-1187.

Darribère, T., H. Boulekbache, D.-L. Shi, and J.-C. Boucaut. 1985. Immunoelectron microscopic study of fibronectin in gastrulating amphibian embryos. *Cell Tissue Res*. 239:75-80.

Darribère, T., K. Guida, H. Larjava, K.E. Johnson, K.M. Yamada, J.-P. Thiery, and J.-C. Boucaut. 1990. *In vivo* analyses of integrin ß1 subunit function in fibronectin matrix assembly. *J. Cell Biol*. 110:1813-1823.

Darribère, T., K.M. Yamada, K.E. Johnson, and J.-C. Boucaut. 1988. The 140-kDa fibronectin receptor complex is required for mesodermal cell adhesion during gastrulation in the amphibian *Pleurodeles waltlii*. *Dev. Biol.* 126:182-194.

Dipasquale, A. 1975. Locomotory activity of epithelial cells in culture. *Exp. Cell Res.* 94:191-215.

Horibata, K. and A.W. Harris. 1970. Mouse myelomas and lymphomas in culture. *Exp. Cell Res.* 60:61-77.

Keller, R.E. 1986. The cellular basis of amphibian gastrulation. p. 241-327. *In: Developmental Biology: A Comprehensive Synthesis. Vol.2. The Cellular Basis of Morphogenesis*. L.W. Browder (Ed.). Plenum Press, New York.

Keller, R.E., M. Danilchik, R. Gimlich, and J. Shih. 1985. The function and mechanism of convergent extension during gastrulation of *Xenopus laevis*. *J. Embryol. Exp. Morphol.* 89(Suppl.):185-209.

Keller, R.E. and G.C. Schoenwolf. 1977. An SEM study of cellular morphology, contact and arrangement as related to gastrulation in *Xenopus laevis*. *Wilhelm Roux's Arch. Dev. Biol.* 182:165-186.

Keller, R.E. and J. Hardin. 1987. Cell behaviour during active cell rearrangement: Evidence and speculations. *J. Cell Sci. Suppl.* 8:369-393.

Keller, R. and R. Winklbauer. 1990. The role of the extracellular matrix in amphibian gastrulation. *Sem. Dev. Bio.* 1:25:33.

Kolega, J. 1981. The movement of cell clusters *in vitro*: Morphology and directionality. *J. Cell Sci.* 49:15-32.

König, G. 1988. A method for mounting specimens for scanning electron microscopy. *Trends Genet.* 4:270.

König, G. 1990. *Untersuchungen zur Determination der planaren Zellpolarität in den epidermalen Cilienzellen von Embryonen des südafrikanischen Krallenfrosches Xenopus laevis.* Thesis, Universität Tübingen.

Kubota, H.Y. and A.J. Durston. 1978. Cinematographical study of cell migration in the opened gastrula of *Ambystoma mexicanum*. *J. Embryol. Exp. Morphol.* 44:71-80.

Nakatsuji, N. 1975. Studies on the gastrulation of amphibian embryos: Cell movement during gastrulation in *Xenopus laevis* embryos. *Wilhelm Roux's Arch. Dev. Biol.* 178:1-14.

Nakatsuji, N. and K.E. Johnson. 1982. Cell locomotion *in vitro* by *Xenopus laevis* gastrula mesoderm cells. *Cell Motil.* 2:149-161.

Nakatsuji, N. and K.E. Johnson. 1983a. Comparative study of extracellular fibrils on the ectodermal layer in gastrulae of five amphibian species. *J. Cell Sci.* 59:61-70.

Nakatsuji, N. and K.E. Johnson. 1983b. Conditioning of a culture substratum by the ectodermal layer promotes attachment and oriented locomotion by amphibian gastrula mesodermal cells. *J. Cell Sci.* 59:43-60.

Nakatsuji, N., M.A. Smolira, and C.C. Wylie. 1985. Fibronectin visualized by scanning electron microscopy immunocytochemistry on the substratum for cell migration in *Xenopus laevis*. *Dev. Biol.* 107:264-268.

Nieuwkoop, P.D. and J. Faber. 1967. *Normal Table of Xenopus laevis (Daudin)*. 2nd edition. North-Holland, Amsterdam.

Riou, J.-F., D.-L. Shi, M. Chiquet, and J.-C. Boucaut. 1990. Exogenous tenascin inhibits mesodermal cell migration during amphibian gastrulation. *Dev. Biol.* 137:305-317.

Shi, D.-L., T. Darribère, K.E. Johnson, and J.-C. Boucaut. 1989. Initiation of mesodermal cell migration and spreading relative to gastrulation in the urodele amphibian *Pleurodeles waltl*. *Development* 105:351-363.

Takeichi, M. 1988. The cadherins: Cell-cell adhesion molecules controlling animal morphogenesis. *Development* 102:639-655.

Weiss, P. 1961. Guiding principles in cell locomotion and cell aggregation. *Exp. Cell Res.* 8 (Suppl.):260-281.

Winklbauer, R. 1986. Cell proliferation in the ectoderm of the *Xenopus* embryo: Development of substratum requirements for cytokinesis. *Dev. Biol.* 118:70-81.

Winklbauer, R. 1988. Differential interaction of *Xenopus* embryonic cells with fibronectin *in vitro*. *Dev. Biol.* 130:175-183.

Winklbauer, R. 1990. Mesoderm cell migration during *Xenopus* gastrulation. *Dev. Biol.* 142:155-168.

AMPHIBIAN GASTRULATION: THE MOLECULAR BASES OF MESODERMAL CELL MIGRATION IN URODELE EMBRYOS

Jean-Claude Boucaut*, Thierry Darribère*, De Li Shi*, Jean-Francois Riou*, Kurt E. Johnson#, and Michel Delarue*

* Centre National de la Recherche Scientifique
Universite Pierre et Marie Curie
Laboratorie de Biologie Experimentale URA-CNRS-1135
9, Quai Saint-Bernard, 75005
Paris, FRANCE

Department of Anatomy
The George Washington University Medical Center
2300 I Street N.W.
Washington, DC 20037 USA

Gastrulation, Edited by R. Keller *et al.*
Plenum Press, New York, 1991

INTRODUCTION

During the early developmental period of the vertebrate embryo, called gastrulation, changes in cell shape, cell number, and cell-cell associations produce fundamental changes in embryonic morphology. Selected populations of cells are designated to perform particular ensembles of cell movements. Typically, morphogenetic cell movements are regulated in a repeatable pattern from embryo to embryo. These morphogenetic cell movements lead to the organization of an embryo with three primary germ layers: ectoderm, mesoderm, and endoderm. It is difficult enough to understand how cells move from one location to another inside the embryo but even more mysterious why they choose one particular pathway for this locomotion from among the large number of pathways theoretically available to them.

The late amphibian blastula consists of embryonic cells arranged around a central cavity known as the *blastocoel*. The blastula possesses a darkly pigmented animal cap composed of small blastomeres with relatively little yolk and a vegetal region composed of large, pale, yolk-laden blastomeres. In between the animal and vegetal regions, one finds an intermediate marginal zone composed of blastomeres of intermediate size, pigmentation, and yolkiness. At the beginning of gastrulation, *bottle cells* form in the dorsal marginal zone (DMZ) *via* the contraction of their apical surfaces (Baker 1965). This initial step in gastrulation results in the formation of the dorsal lip of the *blastopore* (Figure 1A). The blastopore grows laterally and ventrally from the dorsal lip, eventually forming a crescent shaped slit. A broad sheet of cells above the dorsal lip of the blastopore (the primordium of the chordamesoderm) moves toward, and then over, the dorsal lip in a movement known as *involution*. During this process cells within this sheet become rearranged by intercalation. During intercalation, cells change neighbors and regroup so that they are converted from a short broad group into a long narrow group. This movement has been called *convergent extension* by Keller (1984) and is thought to be the main engine producing the driving force for gastrulation in *Xenopus*. Convergent extension is an autonomous movement which occurs spontaneously in isolated tissue fragments *in vitro* (Keller 1984, 1986; Keller *et al.* 1985).

In the urodele embryo, it appears that migration of the mesodermal cells is more important in gastrulation than their convergent extension (Shi *et al.* 1987) and thus we will focus on this process, with special attention to the factors that *guide* the mesodermal cells in their migration.

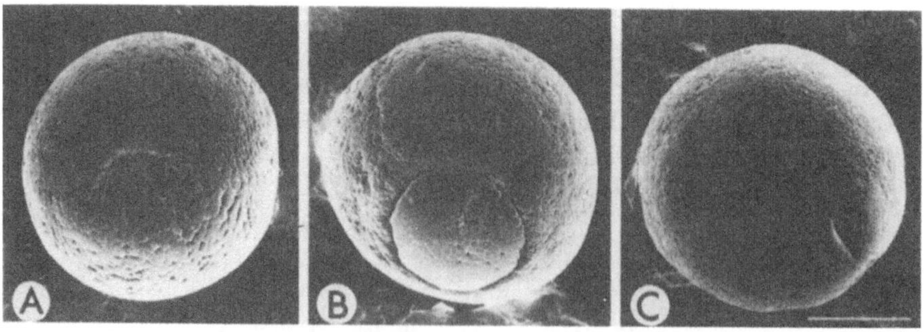

Figure 1. Gastrulation of *Pleurodeles waltl.*, vegetal view. **A**, blastopore is a depressed area just beginning to invaginate. **B**, blastopore is a complete circle. **C**, yolk plug disappears from the surface and blastopore closes. Bar, 500 μm. From Boucaut *et al.* 1990 used with permission.

It should be pointed out that significant differences exist between different amphibian species *vis a vis* the origin and initial arrangement of mesodermal cells. In *Xenopus* gastrulae, the DMZ consists of a superficial epithelium and a deep layer of four to six cells thick. *Xenopus* mesodermal cells arise entirely from deep cells in the DMZ (Keller 1975, 1976; Smith and Malacinski 1983). In contrast, in urodele gastrulae such as *Ambystoma* and *Pleurodeles*, studies with vital dyes, cell surface labelling, and lineage tracers have shown that superficial cells of the DMZ contribute to mesodermal structures (Vogt 1929; Smith and Malacinski 1983; Lundmark 1986). In addition, evidence has been provided that urodele mesodermal cell ingression and migration may play a significant role in convergent extension (Lundmark 1986).

As chordamesodermal cells involute during gastrulation, their space on the outer surface becomes occupied by spreading presumptive ectodermal cells in a process termed *epiboly*, closely coordinated with convergent extension. Blastopore growth continues as the crescent becomes a circle (Figure 1B) and then the diameter of the circular blastopore shrinks by constriction. By the end of gastrulation, the presumptive endoderm is drawn completely inside the embryo (Figure 1C).

A new cavity, the *archenteron*, is formed as the invagination site at the blastopore deepens and expands toward the animal pole inside the embryo. The leading edge of the archenteron is a population of cells that adhere to and spread on the inner surface of the blastocoel roof. The cells at the leading edge of the growing archenteron form broad lamellipodia with numerous filopodia on the basal surface of cells of the blastocoel roof. These cells attach preferentially to this inner surface and are stimulated to move away from the blastopore by several factors which we will discuss later.

It is now well established that the extracellular adhesive glycoprotein fibronectin (FN) and its major cell surface receptor $\alpha_5\beta_1$ integrin (INT) play a key role in the initiation and maintenance of the many diverse changes in adhesion and spreading that accompany cell migration (Yamada 1983; Thiery *et al.* 1985). One major goal of our current research is to understand the molecular mechanisms governing mesodermal cell migration in gastrulating amphibian embryos. Progress has been made in the last decade in understanding the origin, structure and function of the FN-rich fibrillar network on the basal surface of the blastocoel roof in amphibian embryos. The particular emphasis of this review is on a cellular and molecular analysis of the morphogenetic functions of both FN and INT using the amphibian urodele gastrula as a model system. We will discuss the experimental inhibition of mesodermal cell interaction with FN and an analysis of FN assembly. We will also point out areas that are not well understood and therefore warrant further investigation.

INTERACTION OF MIGRATING MESODERMAL CELLS WITH EXTRACELLULAR FIBRONECTIN FIBRILS

Fibronectin Fibrils in Amphibian Gastrulae

Nakatsuji *et al.* (1982) and Boucaut and Darribère (1983a,b) discovered a dense network of extracellular fibrils lining the basal surface of epithelial cells of the blastocoel roof (Figure 2). These fibrils are sparse prior to gastrulation, increase in abundance at the beginning of gastrulation, and continue to accumulate during

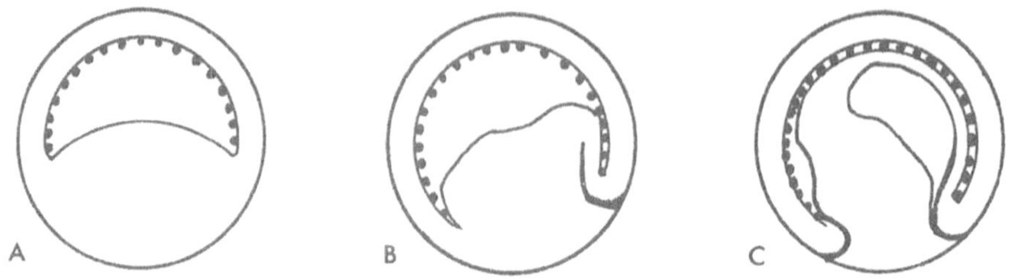

Figure 2. Diagrams showing the distribution of extracellular fibrils (heavy dots) in early amphibian embryos. **A**, Blastula stage; **B**, early gastrula stage; **C**, mid-gastrula stage.

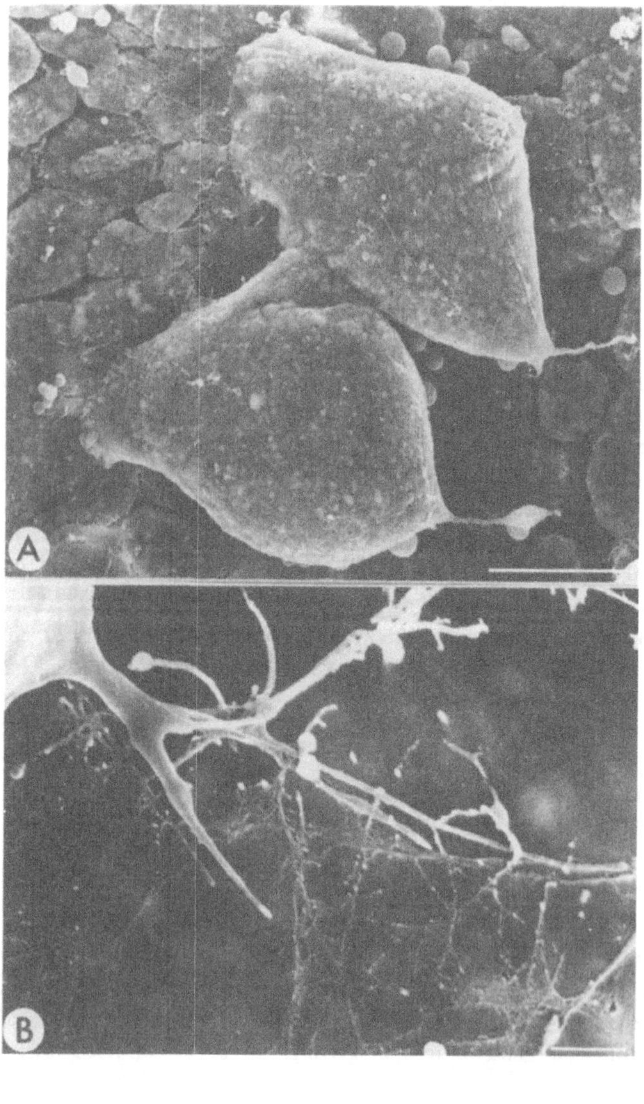

Figure 3. Scanning electron micrographs of migrating mesodermal cells on the inner surface of the blastocoel roof in a urodele (*Ambystoma maculatum*) gastrula. **A**, Two migrating mesodermal cells. The animal pole is to the left and the dorsal lip of the blastopore is to the right. The cells have large lamellipodia on the leading edge directed toward the animal pole and a uropod-like attachment to the blastocoel roof on the trailing edge. Bar, 50 μm. **B**, Filopodia on the leading edge of a lamellipodium associated with the fibrillar ECM on the basal surface of the blastocoel roof. Bar, 1 μm. From Boucaut *et al.* 1990 used with permission.

gastrulation. Migrating mesodermal cells on the basal surface of the blastocoel roof (Figure 3A) possess lamellipodia and filopodia (Figure 3B) in close association with extracellular fibrils. These fibrils have been observed in all 9 species of amphibian embryos examined to date including the urodeles *Ambystoma maculatum*, *A. mexicanum*, *Pleurodeles waltl.*, and *Cynops phyrrhoghaster* and the anurans *Xenopus laevis*, *Rana pipiens*, *Rana sylvatica*, *Bufo bufo*, and *Bufo calamita* (Boucaut and Darribère 1983a,b; Johnson 1984; Nakatsuji 1984; Nakatsuji and Johnson 1983a,b; 1984a,b; Nakatsuji *et al.* 1985a,b; Delarue *et al.* 1985; Boucaut *et al.* 1990; Johnson *et al.* 1990). These fibrils contain FN (Boucaut and Darribère 1983a,b; Boucaut *et al.* 1984 a 1985; Darribère *et al.* 1984; 1985; Johnson *et al.* 1987; Lee *et al.* 1984; Nakatsuji *et al.* 1985b) (Figure 3) and laminin (Darribère *et al.* 1986; Nakatsuji *et al.* 1985a). Synthesis of FN occurs at low levels prior to gastrulation but in quite active amounts during gastrulation. As synthesis proceeds, FN-containing fibrils accumulate preferentially on the inner surface of the roof of the blastocoel (Boucaut and Darribère 1983a,b; Lee *et al.* 1984) presumably because receptors for FN, possibly integrins, are localized there (see Darribère *et al.* 1988).

Organization of Fibronectin Fibrils

In early embryos, of *Pleurodeles waltl.*, the time course and pattern of normal fibril formation *in vivo* was established by indirect immunofluorescence using an antibody directed against FN. In eight-cell embryos, there was no FN on the basal surface of blastomeres at the animal pole. In morulae, no FN fibrils were yet observed. However, non-fibrillar fluorescent labelling was detected either as a ring around blastomeres near cell-cell contacts or as minute specks on the surface of each blastomere (Figure 4A). The first FN fibrils were observed two hours later in early blastulae. They were located primarily at the edges of the basal surfaces of the smallest blastomeres of the blastocoel roof (Figure 4B). In mid-blastulae, fibrils were apparent around the cell peripheries and across adjacent cell boundaries (Figure 4C). They were smaller than those observed in early blastulae and were arranged perpendicular to the plane of contact between blastomeres. Within six hours, fibrils elongated from these peripheral sites toward the central portion of the cell surface over the nucleus. By the late blastulae stage, fluorescent staining for FN was well developed over most of the basal cell surface (Figure 4D). FN fibrils were still more abundant at the periphery of each cell. At this stage, two fibril morphologies were distinguished; short and thin fibrils and longer, thicker fibrils. Twenty-four hours after fertilization, in early gastrulae, FN fibrils formed a complex anastomosing fibrillar matrix which covered the entire basal surface of the blastocoel roof. During gastrulation, the extracellular meshwork of FN fibrils showed further increases in density and complexity (Figure 4E).

Extracellular Fibrils Guide Mesodermal Cell Locomotion

Migrating mesodermal cells in urodele gastrulae use an anastomosing network of extracellular matrix fibrils on the inner surface of the ectoderm layer as their substratum for cell migration (Nakatsuji *et al.* 1982; Nakatsuji and Johnson 1983a; Nakatsuji and Johnson 1984b). Boucaut *et al.* (1984a) have shown that when the animal cap was inverted 180° (so that the surface originally facing the perivitelline space was facing the blastocoel), grafted explants did not provide a suitable substratum for

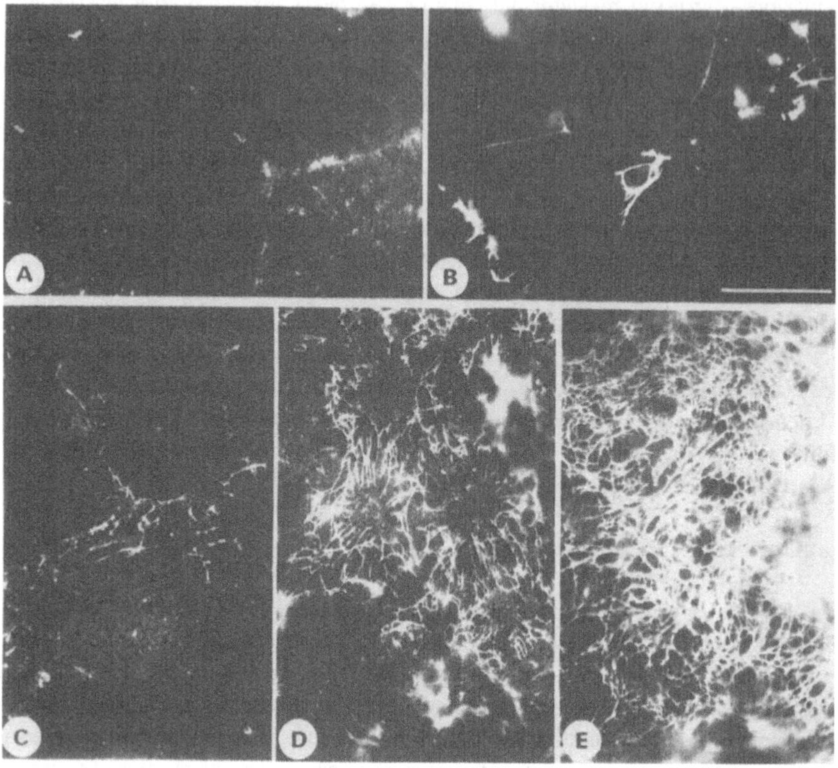

Figure 4. Analysis of FN-fibril formation. Whole mounts of the blastocoel roof from *Pleurodeles waltl.* embryos were analyzed by indirect immunofluorescence with purified anti-FN IgG. **A**, Morula. Fluorescent staining is observed in areas of cell-cell contact and more weakly on the cell surface. No fibrils are detected at this stage. **B**. Early blastula. A faint fluorescent ring surrounds blastomeres. The first FN-fibrils appear at the edge of some blastomeres. **C**, Mid-blastula. Fibrillar arrays of FN staining are concentrated at the periphery of blastomeres. **D**, Late blastula. FN fibrils extend from the cell periphery toward a central supranuclear zone. **E**, Mid-gastrula. A complex, extensively anastomosed network of FN covers the entire inner surface of the blastocoel roof. Bar, 10 μm. From Darribère *et al.* 1990 with permission.

mesodermal cell migration. During normal migration filopodia of mesodermal cells often show close association with fibrils, suggesting that the fibrils serve as a preferential adhesion site for filopodia and lamellipodia. Furthermore, the presence of the statistically significant alignment of such fibril networks along the blastopore-animal pole axis (Nakatsuji *et al.* 1982), indicates an interesting possibility for an actual role *in vivo* for guidance by an aligned fibril network (Nakatsuji *et al.* 1982; Nakatsuji 1984), a contact guidance system postulated long ago by Weiss (1945).

To test this hypothesis, Nakatsuji and Johnson (1983a) and Shi *et al.* (1989) examined the behavior of cells on extracellular fibril network conditioned cover slips (Figure 5). Presumptive ectoderm was dissected from the dorsal part of early gastrulae of *Ambystoma maculatum* or *Pleurodeles waltl.* and explanted, blastocoelic surface down, onto plastic cover slips. After incubation to allow transfer of the fibrillar extracellular matrix to coverslips, the direction of the animal poles was noted and the

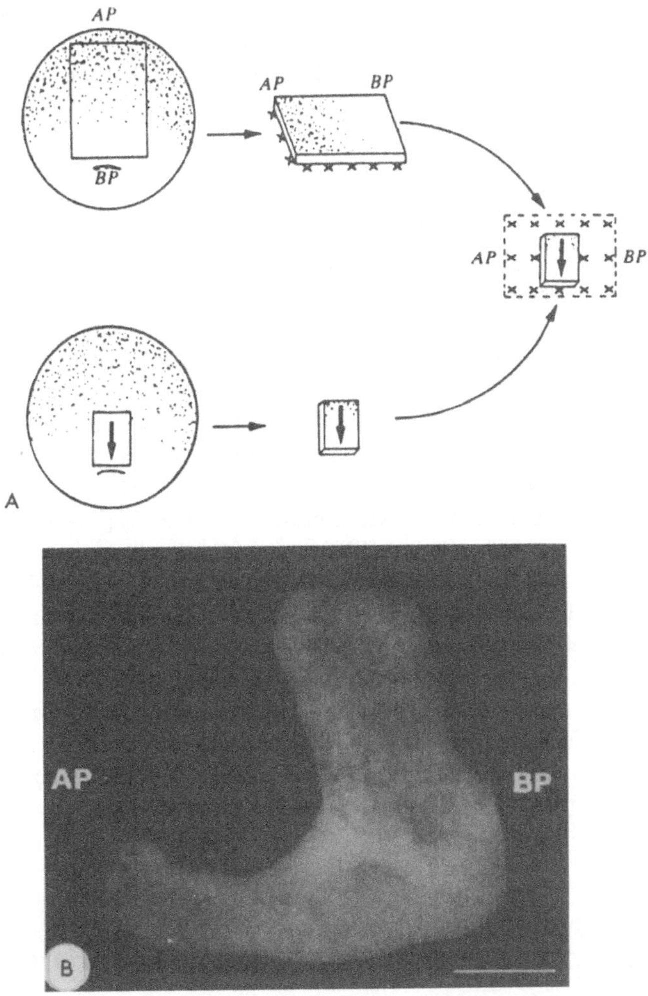

Figure 5. Orientation of DMZ outgrowth toward the animal pole (AP) on an ECM-conditioned substratum. **A.** Diagram of explants containing DMZ and adjacent ectoderm. These explants conditioned the substratum by deposition of ECM from the basal surface of the explant. The original animal pole (AP) to blastopore (BP) axis (thick arrow) was perpendicular to that of the fragment used to condition the substratum. **B.** After 24 hours of culture, a cohesive cell sheet has spread from the DMZ explant toward the animal pole region of the conditioned substratum but stopped once it reached the edge of the conditioned area. From Boucaut *et al.* 1990 used with permission.

explant was removed. Dissociated mesoderm cells isolated from gastrulae or dorsal marginal zone explants were then seeded on the condition surface of the coverslip. Both isolated cells and explants adhered to conditioned surfaces, extended lamellipodia and filopodia, and migrated actively on the conditioned surfaces. Isolated cells moved preferentially toward the animal pole direction of the conditioned substratum. Similarly, the outgrowth of mesodermal cells from DMZ explants was distorted toward the animal pole direction of conditioned substrata (Figure 5). These observations suggest that it is possible to guide mesodermal cell migration by ECM components *in vivo* and *in vitro* and they have lead us to present the hypothesis that the fibrillar ECM guides migrating mesodermal cells from the dorsal lip of the blastopore toward the animal pole by contact guidance.

Behavior of Cells on Fibronectin- or Laminin-Coated Substrata

We have analyzed the attachment and migration of mesodermal cells on ECM substrates in culture. Culture dish surfaces coated with fibronectin were found to be good substrata for attachment and migration by *Pleurodeles* or *Ambystoma* gastrula mesodermal cells or dorsal marginal zone explants (Darribère *et al.* 1988, Shi *et al.* 1989). For example, symmetrical outgrowth of mesodermal cells from dorsal marginal zone explants occurs on FN-coated substrata. This outgrowth does not occur on laminin coated substrata. Nakatsuji (1986) also showed that type IV collagen and heparin sulfate supported neither adhesion nor movement of *Xenopus* gastrula mesodermal cells. Further support for the involvement of FN in *Pleurodeles* cell migration comes from the fact that mesodermal cell outgrowth is inhibited by Fab' fragments of anti-FN IgG, Fab' fragments of anti-INT IgG, and GRGDS peptides. Furthermore, Darribère *et al.* (1988) showed that Fab' fragments of anti-INT caused detachment of *Pleurodeles* cells previously attached to FN-coated substrata. Another important finding was that an anisotropic distribution of fibronectin molecules *in vitro* guided the migration of the mesodermal cells. For example, *Bufo bufo japonicus* gastrula mesodermal cells accumulated inside an area coated with fibronectin (Nakatsuji 1986). *Rana pipiens* cells adhered preferentially and moved along FN strands (Johnson *et al.* 1990). These observations reinforce the hypothesis that it is possible to guide the mesodermal cells by an isotropic distribution of ECM *in vivo* and *in vitro*.

STUDIES WITH PROBES TO DISRUPT THE INTERACTION OF MIGRATING MESODERMAL CELLS WITH EXTRACELLULAR FIBRONECTIN FIBRILS

Studies with Antibodies and Peptides

Based on the observations that FN fibrils localize to the roof of the blastocoel and mesodermal cells migrate on fibrillar matrices *in vivo* and *in vitro*, we proposed that these extracellular fibrils serve as a substratum for mesodermal cells to adhere, spread, and migrate in urodele gastrulae. To test this hypothesis, we injected probes to disrupt cell-FN interaction and examined the effects on gastrulation. First, when living embryos were injected with Fab' fragments of anti-FN IgG at the early gastrula stage, gastrulation was blocked (Figure 6). The antibodies interacted with FN-containing fibrils *in vivo* in such a way as to prevent cell adhesion to the fibrils. Similar injections at the late gastrula stage had no noticeable effect on neurulation (Boucaut *et al.* 1984a). Second, when early gastrulae were injected with synthetic peptides representing the major cell-binding domain of FN, again gastrulation was blocked completely. A peptide representing the collagen-binding domain of FN had no effect on gastrulation (Boucaut *et al.* 1984b; 1985). Third, when Darribère *et al.* (1988) injected Fab' fragments of anti-INT IgG into living early gastrulae, once again, gastrulation was completely inhibited.

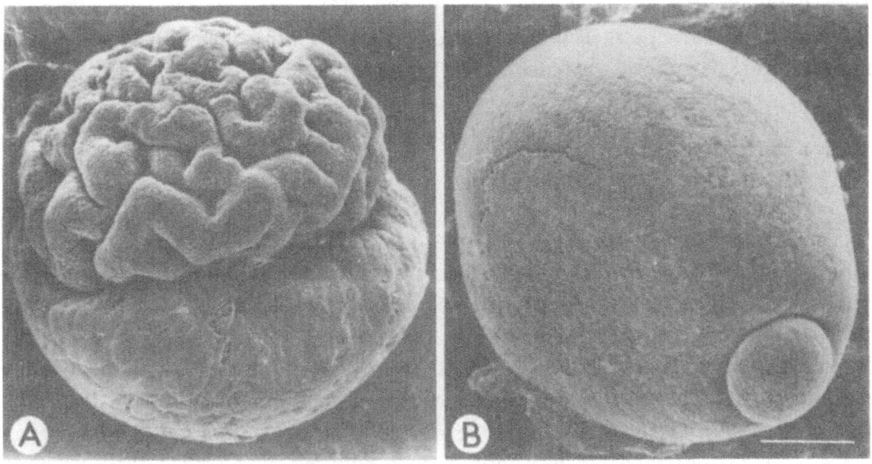

Figure 6. Scanning electron micrograph of the *in vivo* effect of Fab' fragments of anti-FN during gastrulation. *Pleurodeles waltl.* embryos were microinjected at the early gastrula stage and observed 24 hours later. **A**, Injection of Fab' fragments of anti-FN. This result is typical of the more severely arrested embryos. A complete inhibition of gastrulation was observed. Note the highly convoluted roof of the blastocoel, circular blastopore, and smooth exposed endodermal mass. **B**, Control embryo injected with Fab' preimmune IgG. The embryo underwent normal gastrulation. An early neural plate formed in this control embryo indicating that primary embryonic induction has taken place. Bar, 250 μm. From Boucaut *et al.* 1990 used with permission.

All three reagents prohibit the interaction between migrating mesodermal cells and the FN fibrils deposited on the inner surface of the roof of the blastocoel, albeit by different mechanisms. The anti-FN antibodies coat FN-fibrils (Darribère *et al.* 1985), preventing attachment by migrating mesodermal cells. The anti-INT antibodies bind to mesodermal cell surface INT and thereby prevent cell adhesion to FN-containing fibrils. The synthetic peptides bind to cell surface INT and thus prohibit cells from adhering to the FN fibrils. Each reagent has the crucial effect of preventing the interaction between migrating mesodermal cells and ECM deposited on the inner surface of blastocoel roof.

Scanning Electron Microscopy of Probed Embryos

Boucaut *et al.* (1984a,b; 1985) and Darribère *et al.* (1988) performed extensive scanning electron microscopic (SEM) observations on *Pleurodeles waltl.* embryos injected with Fab' anti-FN, Fab' anti-INT, and synthetic peptides to the cell binding domain of FN. All these reagents produce strikingly similar embryos. Injected embryos formed a blastopore in an appropriate location. The blastopore formed a conspicuous circular constriction in the equatorial region which divided the embryo into two hemispheres. The animal hemisphere became extensively convoluted. In contrast, the vegetal hemisphere remained smooth (Figure 6).

When probed embryos were fractured and examined in the SEM the mesodermal cells formed a ring-like collection in the marginal zone but they failed to migrate across the smooth inner surface of the roof of the blastocoel, presumably because they were unable to gain an appropriate foot-hold there. Migrating mesodermal cells projected small filopodia and lamellipodia toward the FN-fibril strewn substratum but examination of stereopairs of images revealed that the mesodermal cells were unable to attach their peripheral protrusions to the extracellular matrix.

Studies with Tenascin

Tenascin (TN) is a noncollagenous glycoprotein found in the extracellular matrix. TN was first identified by Chiquet and Fambrough (1984a,b) and was shown to exhibit temporal and spatial regulation during development. Subsequently, it was demonstrated that TN modifies integrin-mediated cell attachment to FN (Crossin *et al.* 1986; Chiquet-Ehrismann *et al.* 1986, 1988; Mackie *et al.* 1988; Riou *et al.* 1988). Recently, we have shown that TN has striking effects on mesodermal cell migration in *Pleurodeles* (Riou *et al.* 1990). Outgrowth of mesodermal cells in explants of the dorsal marginal zone on FN-coated substrata was inhibited by addition of TN to culture media. In addition, cell outgrowth was inhibited on substrata coated with FN and TN (Figure 7). When TN was injected into the blastocoel of living embryos, gastrulation was significantly inhibited. The morphology of TN-inhibited embryos is reminiscent of embryos injected with the other reagents that disrupt mesodermal cell-extracellular matrix interaction. Finally, a monoclonal antibody which masks the cell binding site of TN reverses the inhibitory effects of TN *in vitro* and *in vivo*. These results provide additional evidence to support the hypothesis that cell binding to the FN-rich extracellular matrix on the basal surface of the blastocoel roof is necessary for normal gastrulation.

PERTURBATION OF FIBRONECTIN FIBRIL FORMATION INHIBITS MESODERMAL CELL MIGRATION

Experimental Disruption of Fibronectin Matrix Assembly

Recently, we have examined the factors governing FN assembly into fibrils on the inner blastocoel roof in *Pleurodeles waltl.* embryos (Darribère *et al.* 1990a). Native FN begins to assemble at the early blastula stage and progressively forms a complex extracellular matrix. We injected FITC-labelled bovine plasma FN into the blastocoel of living embryos and observed that the exogenous labelled-FN was assembled into fibrils in the same spatio-temporal pattern as observed for endogenous FN. Fibrillogenesis of exogenous FITC-FN was inhibited in a dose-dependent manner by both the GRGDS peptide and monospecific antibodies to amphibian integrin β_1 subunit. Injection of antibodies to the cytoplasmic domain of integrin β_1 subunit produce a reversible inhibition of FN-fibril formation that follows early cell lineages (Figure 8) and causes delays in development (Figure 9). Together, these data indicate that *in vivo*, the integrin β_1 subunit and the RGDS recognition sequence are essential for the proper assembly of FN fibrils. Also, they suggest that normal gastrulation requires normal assembly of a FN-rich fibrillar extracellular matrix.

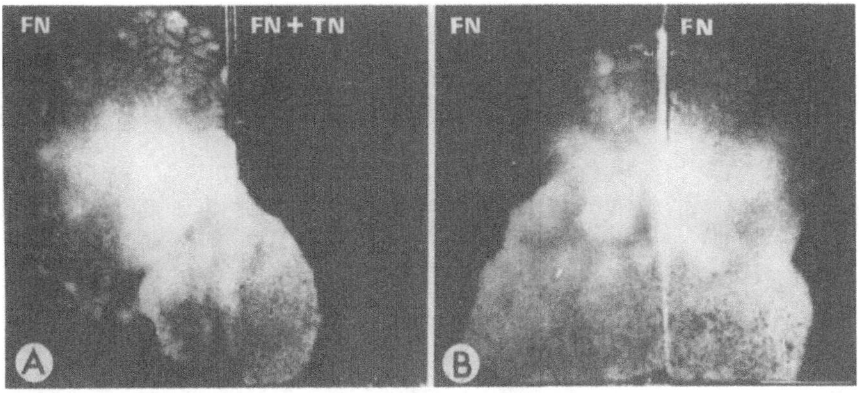

Figure 7. Migratory behavior of mesodermal cells *in vitro* from *Pleurodeles waltl.* gastrulae on a substrate conditioned by FN and TN. **A**, An explant containing the DMZ and adjacent ectoderm was placed at the edge of one track conditioned by FN (left) and another conditioned by FN and TN (right). The migrating mesodermal sheet fails to spread on FN- and TN-coated substratum. The vertical lines through the micrographs represent the boundaries between each type of substratum. **B**, Control where both tracks were conditioned by FN alone. The cell sheet spread equally in both tracks. Bar, 500 μm. From Boucaut *et al.* 1990 used with permission.

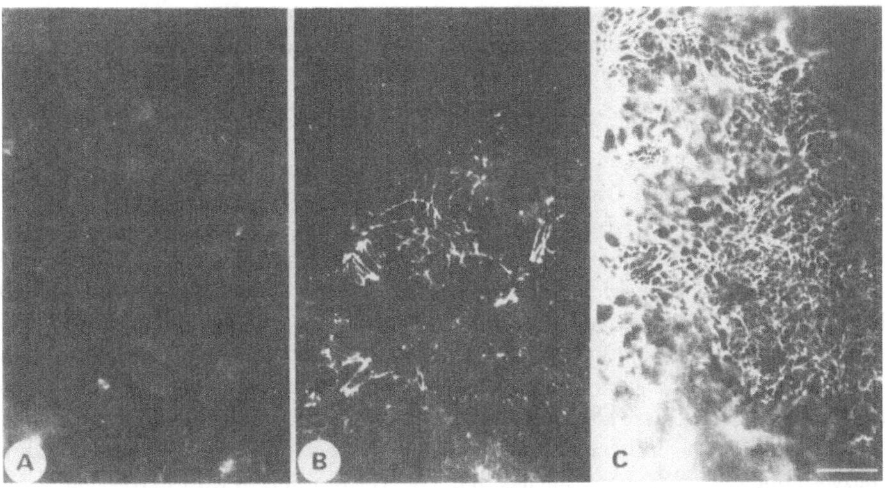

Figure 8. Inhibition of *in vivo* FN-fibril formation. Monovalent antibodies against the cytoplasmic domain of integrin β_1 (50 ng/embryo) were injected into uncleaved *Pleurodeles waltl.* embryo. Embryos were maintained at 18°C, then dissected after the indicated times of incubation. Immunodetection of FN was done on whole mounts of the blastocoel roof. **A**, Twenty four hours after injection, the embryo has reached the late blastula stage, but no staining for FN can be detected. **B**, The first FN fibrils appear at the early gastrula stage. They are distributed around the periphery of cells. Their pattern is comparable to normal embryos observed at the early blastula stage (17 hours earlier). **C**, Control injected with a mixture of anti-β_1 COOH antibodies and purified amphibian integrin β_1. Immunodetection of FN was performed the same as in B. A well-developed FN-meshwork was present. Bar, 5 μm. From Darribère *et al.* 1990 with permission.

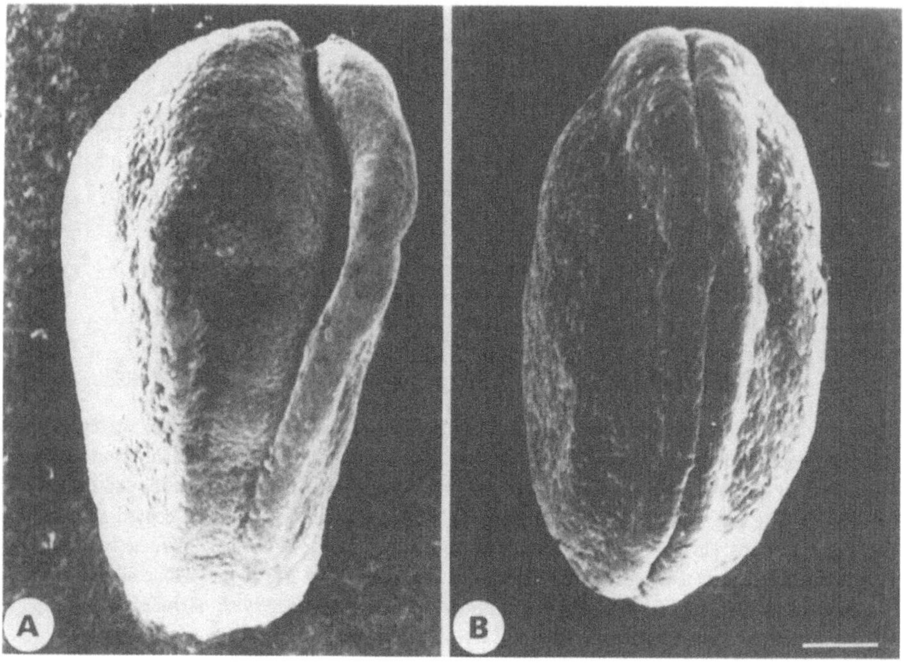

Figure 9. Embryos injected with antibodies to the cytoplasmic domain of integrin β_1 subunit. **A**, Embryo at the two cell stage was injected into the left blastomere with 100 ng of anti-β_1 COOH Fab'. At the time of observation 72 hours later, the left neural fold is defective. **B**, Control experiment. Anti-β_1 Fab' was preincubated with amphibian integrin β_1 and injected into the left blastomere at the two-cell stage. Neurulation occurs normally. Bar, 0.3 mm. From Darribère *et al.* 1990 with permission.

Studies with a Mutant Defective in FN-Fibrillogenesis

Darribère *et al.* (1991) studied FN and INT in the ac/ac ("ascite caudale") maternal effect mutation in *Pleurodeles waltl*. In progeny from homozygous ac/ac females, cleavage is apparently normal but then all embryos eventually exhibit the "ectodermal syndrome" (Beetschen and Fernandez 1979). These progeny have a pitted or furrowed appearance in the animal hemisphere. The morphogenetic movements of epiboly and archenteron formation are also disturbed, sometimes leading to partial exogastrulation. The phenotype of severely arrested embryos is strikingly similar to the appearance of embryos injected with Fab' antibodies to FN or INT or injected with RGDS-containing peptides (Figure 9). The synthesis of FN and $\alpha_5\beta_1$ INT and the ability of mutant mesodermal cells to adhere and migrate on FN-coated substrata were comparable to wild type embryos. In contrast, ac/ac progeny show a conspicuous defect in the assembly of either endogenous FN or exogenous injected FITC-FN into a complex fibrillar ECM. The FN present in ac/ac mutant progeny exists in minute specks (Figure 10A) rather than a complex fibrillar ECM characteristic of wild-type embryos (Figure 10B). Although it is unclear why fibrillogenesis is defective in ac/ac progeny, the morphological similarity between ac/ac progeny and probed embryos once again suggests that a normal fibrillar ECM is required to support mesodermal cell migration and gastrulation in *Pleurodeles waltl*.

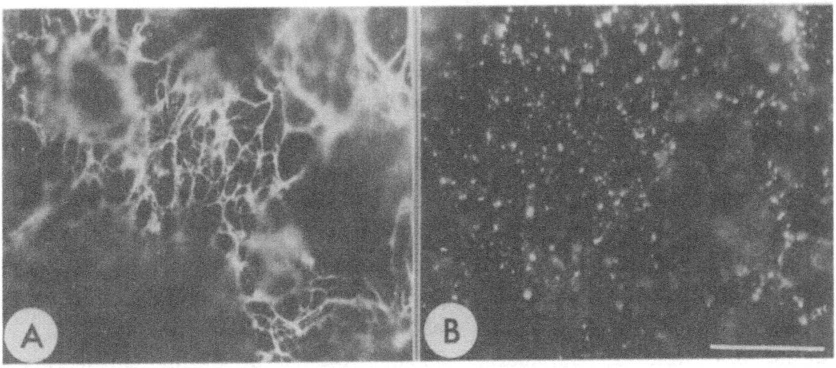

Figure 10. *In situ* distribution of fibronectin in ac/ac mutant embryos. Whole-mount specimens of the inner surface of the blastocoel roof of mid-gastrula stage embryos. **A**, Control immunofluorescent staining performed with anti-FN IgG on the blastocoel roof of normal mid-gastrula. FN-fibrils are clearly apparent. **B**, Immunofluorescent staining for FN in a mutant embryo corresponding to control mid-gastrula. Fluorescent granules are present on the basal surface of cells. Bar, 10 μm.

CONCLUSIONS AND PERSPECTIVES

With our clearer understanding of the role of FN and INT in mesodermal cell migration, we can begin to ask questions concerning the control of FN synthesis and distribution during gastrulation. In the future we hope to learn more about the factors that control assembly of FN on the basal surface of the blastocoel roof. Recombinant DNA technology has allowed cloning of the genes for FN and integrin β_1 subunit in *Xenopus* (DeSimone and Hynes 1988). We have begun to clone these genes in *Pleurodeles*, and we plan to raise antibodies against specific domains of these important molecules. In addition, we plan to create fusion proteins to provide new insights concerning the assembly of FN and its role in cell behavior.

Ultimately, we may be able to introduce and regulate the expression of novel or modified genes for extracellular matrix components. Expression of novel integrin receptors may be used in studying integrin mediated responses of mesodermal cells to changing features of their environment. Using these methods we hope to learn more about the signals that control the initiation and the direction of mesodermal cell migration.

ACKNOWLEDGEMENTS

We are grateful to Dr. J.P. Thiery and Dr. K.M. Yamada for their continual help and encouragement. The worked presented in this paper was supported by grants from CNRS (UA 1135), Ministère de l'Education (ARU), ARC and by Université P.M. Curie (France) to JCB and NSF Grant PCM-8400256 to KEJ.

REFERENCES

Baker, P.C. 1965. Fine structure and morphogenic movements in the gastrula of the tree frog, *Hyla regilla*. *J. Cell Biol.* 24:95-116.

Beetschen, J.C. and M. Fernandez. 1979. Studies on the maternal effect of the semi-lethal factor *ac* in the salamander *Pleurodeles waltlii*. p. 269-286. *In: Maternal Effects in Development*. D.R. Newth and M. Balls (Eds.). Cambridge University Press, Cambridge.

Boucaut, J.-C. and T. Darribère. 1983a. Fibronectin in early amphibian embryos: Migrating mesodermal cells are in contact with a fibronectin-rich fibrillar matrix established prior to gastrulation. *Cell Tissue Res.* 234:135-145.

Boucaut, J.-C. and T. Darribère. 1983b. Presence of fibronectin during early embryogenesis in the amphibian *Pleurodeles waltlii*. *Cell Differ.* 12:77-83.

Boucaut, J.-C., T. Darribère, H. Boulekbache, and J.P. Thiery. 1984a. Prevention of gastrulation but not neurulation by antibody to fibronectin in amphibian embryos. *Nature* 307:364-367.

Boucaut, J.-C., T. Darribère, T.J. Poole, H. Aoyama, K.M. Yamada, and J.P. Thiery. 1984b. Biologically active synthetic peptides as probes of embryonic development: A competitive peptide inhibitor of fibronectin function inhibits gastrulation in amphibian embryos and neural crest cell migration in avian embryos. *J. Cell Biol.* 99:1822-1830.

Boucaut, J.-C., T. Darribère, D.L. Shi, H. Boulekbache, K.M. Yamada, and J.P. Thiery. 1985. Evidence for the role of fibronectin in amphibian gastrulation. *J. Embryol. Exp. Morphol.* 89 (Suppl.):211-227.

Boucaut, J.-C., K.E. Johnson, T. Darribère, D.-L. Shi, K.-F. Riou, H. Boulekbache, and M. Delarue. 1990. Fibronectin-rich fibrillar extracellular matrix controls cell migration during amphibian gastrulation. *Int. J. Dev. Biol.* 34:139-147.

Chiquet, M. and D.M. Fambrough. 1984a. Chick myotendinous antigen. I. A monoclonal antibody as a marker for tendon and muscle morphogenesis. *J. Cell Biol.* 98:1926-1936.

Chiquet, M. and D.M. Fambrough. 1984b Chick myotendinous antigen. II. A novel extracellular glycoprotein complex consisting of large disulfide-linked subunits. *J. Cell Biol.* 98:1937-1947.

Chiquet-Ehrismann, R., P. Kalla, C.A. Pearson, K. Beck, and M. Chiquet. 1988. Tenascin interferes with fibronectin action. *Cell* 53:383-390.

Chiquet-Ehrismann, R., E.J. Makie, C.A. Pearson, and T. Sakakura. 1986. Tenascin: An extracellular matrix protein involved in tissue interactions during fetal development and oncogenesis. *Cell* 47:131-139.

Crossin, K.L., S. Hoffman, M. Grumet, J.P. Thiery, and G.M. Edelman. 1986. Site-restricted expression of cytotactin during development of the chick embryo. *J. Cell Biol.* 102:1917-1930.

Darribère, T., D. Boucher, J.-C. Lacroix, and J.-C. Boucaut. 1984. Fibronectin synthesis during oogenesis and early development of the amphibian *Pleurodeles waltlii*. *Cell Differ.* 14:171-177.

Darribère, T., H. Boulekbache, D.L. Shi, and J.-C. Boucaut. 1985. Immunoelectron microscopic study of fibronectin in gastrulating amphibian embryos. *Cell Tissue Res.* 239:75-80.

Darribère, T., K. Guida, H. Larjava, K.E. Johnson, K.M. Yamada, J.-P. Thiery, and J.-C. Boucaut. 1990. *In vivo* analyses of integrin β_1 subunit function in fibronectin matrix assembly. *J. Cell Biol.* 110:1813-1823.

Darribère, T., J.-F. Riou, K. Guida, A.-M. Duprat, J.-C. Boucaut, and J.-C. Beetschen 1991. A maternal-effect mutation disturbs extracellular matrix organization in the early *Pleurodeles waltl* embryo. *Cell Tissue Res.* 263:507-514.

Darribère, T., J.-F. Riou, D.L. Shi, M. Delarue, and J.-C. Boucaut. 1986. Synthesis and distribution of laminin-related polypeptides in early amphibian embryos. *Cell Tissue Res.* 246:45-51.

Darribère, T., K.M. Yamada, K.E. Johnson, and J.-C. Boucaut. 1988. The 140 kD fibronectin receptor complex is required for mesodermal cell adhesion during gastrulation in the amphibian *Pleurodeles waltlii. Dev. Biol.* 126:182-194.

Delarue, M., T. Darribère, C. Aimar, and J.-C. Boucaut. 1985. Bufonid nucleocytoplasmic hybrids arrested at the early gastrula stage lack a fibronectin-containing fibrillar extracellular matrix. *Wilhelm Roux's Arch. Dev. Biol.* 194:275-280.

DeSimone, D.W. and R.O. Hynes. 1988. *Xenopus laevis* integrins. Structural and evolutionary divergence of integrin ß subunits. *J. Biol. Chem.* 263:5333-5340.

Johnson, K.E. 1984. Glycoconjugate synthesis during gastrulation in *Xenopus laevis. Am. Zool.* 24: 605-624.

Johnson, K.E., J.-C. Boucaut, T. Darribère, and J.-F. Riou. 1987. Fibronectin in normal and gastrula arrested hybrid frog embryos. *Anat. Rec.* 218:68A.

Johnson, K.E., T. Darribère, and J.-C. Boucaut. 1990. Cell adhesion to extracellular matrix in normal *Rana pipiens* gastrulae and in arrested hybrid gastrulae *Rana pipiens* ♀ × *Rana esculenta* ♂. *Dev. Biol.* 137:86-99.

Keller, R.E. 1975. Vital dye mapping of the gastrula and neurula of *Xenopus laevis* I. Prospective areas and morphogenetic movements in the superficial layer. *Dev. Biol.* 42:222-241.

Keller, R.E. 1976. Vital dye mapping of the gastrula and neurula of *Xenopus laevis* II. Prospective areas and morphogenetic movements in the deep region. *Dev. Biol.* 51:118-137.

Keller, R.E. 1984. The cellular basis of gastrulation in *Xenopus laevis*: Postinvolutional convergence and extension. *Am. Zool.* 25:589-602.

Keller, R.E. 1986. The cellular basis of amphibian gastrulation. p. 241-327. *In: Developmental Biology: A Comprehensive Synthesis. Vol. 2. The Cellular Basis of Morphogenesis.* L.W. Browder (Ed.). Plenum Press, New York.

Keller, R.E., M. Danilchik, R. Gimlich, and J. Shin. 1985. Convergent extension by cell intercalation during gastrulation in *Xenopus laevis*. p. 111-141. *In: Molecular Determinants of Animal Form.* G.M. Edelman, (Ed.). Alan R. Liss, New York.

Lee, G., R. Hynes, and M. Kirschner 1984. Temporal and spatial regulation of fibronectin in early *Xenopus* development. *Cell* 36:729-740.

Lundmark, C. 1986. Role of bilateral zones of ingressing superficial cells during gastrulation of *Ambystoma mexicanum*. *J. Embrol. Exp. Morphol.* 97:47-62.

Mackie, E.J., R.P. Tucker, W. Halfter, R. Chiquet-Ehrismann, and Epperlein, H.H. 1988. The distribution of tenascin coincides with pathways of neural crest cell migration. *Development* 102:237-250.

Nakatsuji, N. 1984. Cell locomotion and contact guidance in amphibian gastrulation. *Am. Zool.* 24:615-627.

Nakatsuji, N. 1986. Presumptive mesodermal cells from *Xenopus laevis* gastrulae attach to and migrate on substrata coated with fibronectin or laminin. *J. Cell Sci.* 86:109-118.

Nakatsuji, N., A. Gould, and K.E. Johnson. 1982. Movement and guidance of migrating mesodermal cells in *Ambystoma maculatum* gastrulae. *J. Cell Sci.* 56:207-222.

Nakatsuji, N., K. Hashimoto, and M. Hayashi. 1985a. Laminin fibrils in newt gastrulae visualized by immunofluorescent staining. *Dev. Growth & Differ.* 27:639-643.

Nakatsuji, N. and K.E. Johnson. 1983a. Conditioning of a culture substratum by the ectodermal layer promotes attachment and oriented locomotion by amphibian gastrula mesodermal cells. *J. Cell Sci.* 59:43-60.

Nakatsuji, N. and K.E. Johnson. 1983b. Comparative study of extracellular fibrils on the ectodermal layer in gastrulae of five amphibian species. *J. Cell Sci.* 59:61-70.

Nakatsuji, N. and K.E. Johnson. 1984a. Experimental manipulation of a contact guidance system in amphibian gastrulation by mechanical tension. *Nature* 307:453-455.

Nakatsuji, N. and K.E. Johnson. 1984b. Substratum conditioning experiments using normal and hybrid frog embryos. *J. Cell Sci.* 68:49-67.

Nakatsuji, N., M.A. Smolira, and C.C. Wylie. 1985b. Fibronectin visualized by scanning electron microscopy immunocytochemistry on the substratum for cell migration in *Xenopus laevis* gastrula. *Dev. Biol.* 107:264-268.

Riou, J.-F., D.-L. Shi, M. Chiquet, and J.-C. Boucaut. 1988. Expression of tenascin in response to neural induction in amphibian embryos. *Development* 104:511-524.

Riou, J.-F., D.-L. Shi, M. Chiquet, and J.-C. Boucaut. 1990. Exogenous tenascin inhibits mesodermal cells migration during amphibian gastrulation. *Dev. Biol.* 137:305-317.

Shi, D.-L., T. Darribère, K.E. Johnson, and J.-C. Boucaut. 1989. Initiation of mesodermal cell migration and spreading relative to gastrulation in the urodele amphibian *Pleurodeles waltl* gastrulae. *Development* 105:223-236.

Shi, D.-L., M. Delarue, T. Darribère, J.-F. Riou, J.-C. Boucaut. 1987. Experimental analysis of the extension of the dorsal marginal zone in *Pleurodeles waltl* gastrulae. *Development* 100:147-161.

Smith, J.C. and G.M. Malacinski. 1983. The origin of the mesoderm in the anuran, *Xenopus laevis*, and a urodele, *Ambystoma mexicanum*. *Dev. Biol.* 98:250-254.

Thiery, J.P., J.L. Duband, and A. Delouvée. 1985. The role of cell adhesion in morphogenetic movements during early embryogenesis. p. 169-196. *In: The Cell In Contact.* G.M. Edelman and J.P. Thiery (Eds.). Wiley, New York.

Vogt, W. 1929. Gestaltungsanalyse am Amphibienkeim mit Örtlicher Vitalfärbung. II. Teil. Gastrulation and Mesodermbildung bei Urodelen und Anuren. *Wilhelm Roux' Arch. Entwicklungsmech. Org.* 120:384-706.

Weiss, P. 1945. Experiments on cell and axon orientation *in vitro*: The role of colloidal exudates in tissue organization. *J. Exp. Zool.* 100:353-386.

Yamada, K. M. 1983. Cell surface interactions with extracellular materials. *Annu. Rev. Biochem.* 52:761-799.

THE EXPRESSION OF FIBRONECTINS AND INTEGRINS DURING MESODERMAL INDUCTION AND GASTRULATION IN *XENOPUS*

Douglas W. DeSimone[1], Jim C. Smith[2], James E. Howard[2], David G. Ransom[1], and Karen Symes[2,3]

[1] University of Virginia Health Sciences Center,
Department of Anatomy and Cell Biology and the
Molecular Biology Institute,
Charlottesville, VA 22908, USA
[2] Laboratory of Embryogenesis
National Institute for Medical Research
The Ridgeway, Mill Hill
London NW7 1AA, UK
[3] Present Address: Department of Cell and Molecular Biology
385 LSA, University of California,
Berkeley, CA 94720, USA

ABSTRACT

Fibronectins (FNs) and integrins are first expressed in *Xenopus* embryos during the mid to late blastula stages. FN is synthesized in both animal and vegetal halves of the embryo but becomes localized to the roof of the blastocoel during gastrulation. Integrins are expressed in all regions of the early embryo. Structural heterogeneity of

Gastrulation, Edited by R. Keller *et al.*
Plenum Press, New York, 1991

FN isoforms during embryogenesis occurs by alternative splicing of a common FN transcript, whereas integrin diversity is generated by the expression of several distinct integrin $\alpha\beta$ heterodimers. The timing of expression for these molecules suggests that they may play important roles in supporting and/or controlling morphogenetic events in the early embryo.

We have investigated the roles played by these proteins in supporting the gastrulation-like movements that occur in animal pole tissue in response to mesoderm-inducing factors. *Xenopus* animal pole ectoderm was isolated from stage 8 embryos and exposed to the XTC mesoderm inducing factor (XTC-MIF; a *Xenopus* homologue of activin A). Animal pole ectoderm treated with XTC-MIF, like stage 10 dorsal marginal zone, will adhere and spread on FN coated surfaces. Uninduced animal pole ectoderm adheres poorly and does not spread on FN. The ability to spread on FN in response to XTC-MIF is also retained by single cells derived from dissociated animal pole tissue. This defines one of the few mesoderm-specific responses to induction that has been demonstrated for single cells. FN-mediated cell spreading is inhibited in the presence of the synthetic peptide Gly-Arg-Gly-Asp-Ser-Pro (GRGDSP), which corresponds to one of the active cell binding sites on the FN molecule. However, the gastrulation-like movements associated with elongation of XTC-MIF induced animal pole ectoderm are not inhibited by the GRGDSP peptide. These results indicate that convergent extension does not depend on cell adhesion to FN. Furthermore, scanning electron microscopy and cell marking techniques suggest that although cellular activity is enhanced following induction, no long range cell mixing occurs during elongation of induced explants. We are now investigating whether the changes in cell adhesion observed following induction with XTC-MIF are controlled by the expression of integrin receptors and ECM molecules such as FN.

INTRODUCTION

Fibronectins (FNs) are large, multifunctional adhesive glycoproteins that are widely distributed in vertebrate extracellular matrices (ECMs). In the past decade, substantial evidence has accumulated to suggest that FNs play important roles in supporting the adhesive and migratory properties of cells both *in vivo* and *in vitro* (Hynes 1990). In *Xenopus* embryos, FN is first synthesized from maternal mRNAs around the time of the mid blastula transition (MBT: Newport and Kirschner 1982) whereupon the protein becomes localized to the ECM that forms along the inside of the blastocoel roof (Nakatsuji and Johnson 1983; Lee *et al.* 1984). Because of this interesting spatiotemporal distribution, it has been suggested that FN may be involved in many of the morphogenetic movements associated with gastrulation in amphibians.

The most compelling experimental evidence in support of a direct role for FN during gastrulation comes from work done in urodeles (for review see Johnson *et al.* 1991; Boucaut and Johnson, this volume). Gastrulation can be perturbed in these embryos following injection into the blastocoel of agents known to interfere with FN mediated cell attachment *in vitro*, such as anti-FN antibodies or synthetic peptides containing the Arg-Gly-Asp (RGD) cell binding site of the FN molecule (Boucaut *et al.* 1984a; 1984b).

In order to improve our understanding of the precise roles played by FN during morphogenesis it is of course necessary to define and characterize the cell-surface receptors for FNs that are coincidently expressed in the embryo. The recently identified integrin family of cell adhesion molecules includes several transmembrane receptors that can account for most, if not all, FN mediated cell adhesion (Albelda and Buck 1990; Hynes; 1990). All integrins are expressed at the cell-surface as heterodimers of non-covalently associated α and β subunits. Different $\alpha\beta$ combinations differ in their ligand binding specificities. At least 11 distinct α subunits have thus far been identified in humans, each of which can form heterodimers with one or more of a smaller number of β subunits (at least six) to generate a complex family of receptors with extensive structural and functional diversity.

It is now apparent that much of the diversity of integrin structure evident in mammals is conserved among lower vertebrate and invertebrate species (DeSimone and Hynes 1988; Marcantonio and Hynes 1988). For example, cDNAs encoding integrins β_1, β_2, β_3 and a novel β subunit (β_x) have been obtained for *Xenopus* (DeSimone and Hynes 1988; Ransom and DeSimone 1990). The integrin β_1 sub-family is composed of several receptors for ECM molecules including three with known FN binding activity ($\alpha_3\beta_1$, $\alpha_4\beta_1$ and $\alpha_5\beta_1$). In *Xenopus*, β_1 mRNA levels increase at the mid blastula stage (DeSimone and Hynes 1988) when zygotic mRNA expression begins (Newport and Kirschner 1982). It is unclear how many different integrins are expressed during gastrulation but immunoprecipitation experiments indicate that several $\alpha\beta_1$ heterodimers are synthesized by early gastrula stages (Smith *et al.* 1990). Microinjection of antibodies directed against β_1 integrins into *Pleurodeles* embryos has been shown to interfere with mesodermal cell adhesion and gastrulation (Darribere *et al.* 1988; 1990), consistent with earlier results obtained by perturbing cell adhesion to FN (Boucaut *et al.* 1984a; 1984b).

The timing of expression of FNs and integrins suggests that these molecules may be involved in controlling the initiation of gastrulation in amphibians. One way to investigate this takes advantage of the recent isolation of potent mesoderm-inducing factors such as basic fibroblast growth factor (bFGF) and XTC-MIF, a *Xenopus* homolog of activin A (see review by New and Smith 1990).

It has been known since the work of Nieuwkoop (1969) that the mesoderm of *Xenopus* (Nieuwkoop and Sudarwati 1971) and other amphibian embryos (Nieuwkoop 1969) arises through an inductive interaction in which cells of the vegetal hemisphere act on overlying animal pole cells. The inductive effect of the vegetal hemisphere can be mimicked by bFGF and XTC-MIF, which cause animal caps to form mesodermal rather than ectodermal tissue. It is the cells of the mesoderm that drive much of gastrulation and, consistent with this, one of the earliest responses of animal pole cells to XTC-MIF and bFGF is the onset of gastrulation-like movements (Symes and Smith 1987; Cooke and Smith 1989; Smith *et al.* 1990). This observation makes it possible to manipulate various aspects of gastrulation and then to ask whether there are concomitant changes in expression of FN or integrins. Our recent results, described in this paper, confirm that FN/integrin interactions play an important role in gastrulation. However, the initiation of gastrulation movements is not correlated with changes in the expression of FN, nor of the integrin β_1 chain. Future work, discussed below, will concentrate on other integrin subunits and on possible post-translational modifications that may influence integrin function.

MATERIALS AND METHODS

Preparation and Handling of Oocytes, Embryos and Embryo Fragments

Methods used for artificial fertilization of *Xenopus* eggs and embryo culture were as described previously (Smith *et al.* 1990). Embryos were staged according to Nieuwkoop and Faber (1967). Dissections were performed on manually devitellinized, mid-blastula stage embryos, which were dissected into animal, marginal zone and vegetal pole regions with sharpened tungsten needles. Follicle cell layers were manually dissected from stage VI oocytes with watchmaker's forceps.

XTC-mesoderm-inducing-factor (XTC-MIF)

XTC-MIF was purified from heated XTC-conditioned medium as previously described (Smith *et al.* 1988; 1990). Explants were incubated with 5-20 units/ml of XTC-MIF in 75% normal amphibian medium (NAM: Slack 1984), where 1 unit is defined as the minimum quantity/ml of medium necessary for induction to occur (Cooke *et al.* 1987).

Lineage Labelling

To analyze cell mixing during XTC-MIF-induced convergent extension, embryos received injections of 10 nl rhodamine-lysine-dextran (RLDx: Gimlich and Braun 1985) into one blastomere at the two-cell stage. These "hemi-labelled" embryos were allowed to develop until the mid-blastula stage when animal pole regions were dissected as usual and incubated in NAM with or without XTC-MIF. Induced and uninduced explants were fixed at intervals, embedded in wax, sectioned, and examined to determine the extent of cell mixing.

Scanning Electron Microscopy (SEM)

Mid-blastula stage animal pole regions were dissected and cultured in 20 units/ml partially purified XTC-MIF or a control solution lacking XTC-MIF. When sibling control embryos reached stages 10 and 12.5, ten explants from each culture were fixed and processed for SEM using a modification of the method of Wollweber, Stracke and Gothe (1981). The explants were fixed for 48 h in 2% glutaraldehyde, 0.1% Na cacodylate (pH 7.3) at 4°C, and postfixed for 1.5 h in 1% osmium tetroxide in the same buffer followed by 1 h in 1% tannic acid, 50 mM Na cacodylate, both at room temperature. The explants were then incubated for 1 h in 0.5% uranyl acetate in dH_2O and dehydrated in an ethanol series. After dehydration, the explants were critical point dried with CO_2. The samples were then mounted on SEM stubs, gold coated by sputtering (6 nm), and examined in a JEOL-35 CF scanning electron microscope.

Metabolic Labelling and Immunoprecipitation of Oocytes, Embryos and Cultured Cells

Stage 10 embryos were metabolically labelled by injecting approximately 1.0 uCi/embryo (in 50 nl) of ^{35}S-Trans-label (ICN) into the blastocoel. Oocytes, follicle cell layers and sub-confluent cultures of chick embryo fibroblasts (CEFs) were continuously labelled for 4-12 h in media containing 100 uCi/ml Trans-label.

Immunoprecipitation procedures were identical to those previously described (Smith *et al.* 1990) using an anti-integrin antibody directed against a synthetic peptide, which corresponds to the conserved carboxy-terminus of the β_1 subunit (Marcantonio and Hynes 1988).

Isolation of RNA and RNase Protection Analyses

The cDNA template used to generate the integrin β_1 RNA probe for RNase protections was derived from a 531 bp AvaII/HindIII subclone of cDNA E1 in pGEM-1 (DeSimone and Hynes 1988). The plasmid template was linearized with EcoR1 and a ^{32}P-UTP labelled, anti-sense transcript synthesized with T7 polymerase. The complete transcript length is 580 nt including 49 nt of plasmid sequence, and protects a 531 nt fragment of the β_1 mRNA.

Total RNA was isolated from oocytes and embryos by the proteinase K/LiCl method (Melton and Cortese 1979). RNase protection analyses were carried out with 5 oocyte or embryo equivalents of total RNA per sample according to standard methods (Krieg and Melton 1987).

RESULTS AND DISCUSSION

Behavior of Induced Animal Cap Ectoderm on Fibronectin Substrates

It has been reported by several investigators that prospective mesodermal cells of *Xenopus* tend to spread and migrate on substrates coated with low concentrations of FN, whereas cells from the animal cap require much higher concentrations (Nakatsuji 1986; Komasaki 1988; Winklbauer 1988, 1990). Two techniques have been used to demonstrate this difference in adhesion. In one, intact pieces of prospective mesoderm, or animal cap tissue, are placed on a FN-coated substrate (see Shi *et al.* 1989; Winklbauer 1990), and in the other, dispersed cell populations are used (Nakatsuji 1986; Winklbauer 1990). Both assays show that the animal pole cells adhere poorly whereas the mesodermal cells spread and migrate. We have investigated whether animal pole cells exposed to the mesoderm-inducing factor XTC-MIF acquire the ability to spread on FN. Our results, both with intact caps (Figure 1) and with dispersed cells (see Figure 1 in Smith *et al.* 1990), show that induced cells spread in a manner similar to the behavior of cells of the dorsal marginal zone (prospective mesoderm). As with cells of the marginal zone, the spreading of induced animal pole cells on FN is inhibited by the peptide GRGDSP, which corresponds to the primary FN cell attachment site (Pierschbacher and Ruoslahti 1984). As we discuss below, this ability to adhere to FN must be due to a change in the expression of specific cellular receptors for FN.

Cell Mixing Experiments

The above experiments show that treatment of *Xenopus* animal pole regions with XTC-MIF induces one gastrulation-specific response: the ability of cells to spread and migrate on FN. However, another important type of cell behavior that occurs at

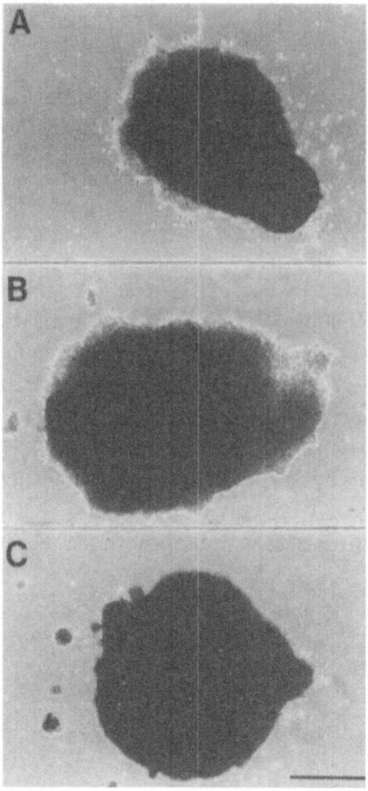

Figure 1. The cells of XTC-MIF-treated *Xenopus* animal pole regions, like those of the dorsal marginal zone, adhere to and migrate on a fibronectin-coated surface; those of uninduced animal pole regions do not. **(A)** The dorsal marginal zone region of an early gastrula embryo was placed on tissue-culture plastic coated with 50 μg/ml fibronectin so that the deep cells were in contact with the substrate. Notice cells migrating away from the explant. **(B)** A *Xenopus* animal pole region exposed to XTC-MIF from the mid-blastula to the early gastrula stage and then placed on a similar fibronectin-coated surface. Notice that many cells have attached to the substrate and are moving away from the explant. **(C)** The cells of a control animal pole explant do not attach to the substrate. Scale bar in (C) is 200 μm.

gastrulation is *convergent extension*, during which cells undergo active rearrangement and intercalation, resulting in a dramatic narrowing and elongation of the prospective mesoderm (see chapter by Keller *et al.* in this volume). Symes and Smith (1987) have shown that XTC-MIF causes intact animal pole regions to undergo convergent extension in a manner very similar to the dorsal marginal zone, and this offers an opportunity to study the molecular and cellular bases of the phenomenon.

Our first experiments indicated that incubation of intact animal caps in the presence of the peptide GRGDSP does not inhibit XTC-MIF induced convergent extension (Smith *et al.* 1990). This might suggest that convergent extension movements do not require RGD-mediated interactions, but as discussed below, other interpretations are possible. More recently, we have investigated to what extent cell mixing provides the driving force for elongation of XTC-MIF induced explants. *Xenopus* embryos were labelled by the injection of RLDx into one blastomere at the two-cell stage. Sections of intact "hemi-labelled" embryos at the mid blastula stage (Figure 2 A,B) show that little mingling of labelled and unlabelled cells occurs during cleavage stages, as has also been reported by Wetts and Fraser (1989). Similarly, when animal pole explants are fixed and sectioned about 30 min after dissection, there is seen to be little mixing of labelled and unlabelled blastomeres, irrespective of whether the explants are cultured in XTC-MIF or a control solution (Figure 2 C-F). We therefore went on to examine cell mixing in explants allowed to develop until the equivalent of stage 12.5, by which time animal pole regions exposed to XTC-MIF had undergone considerable elongation, while control explants remained spherical.

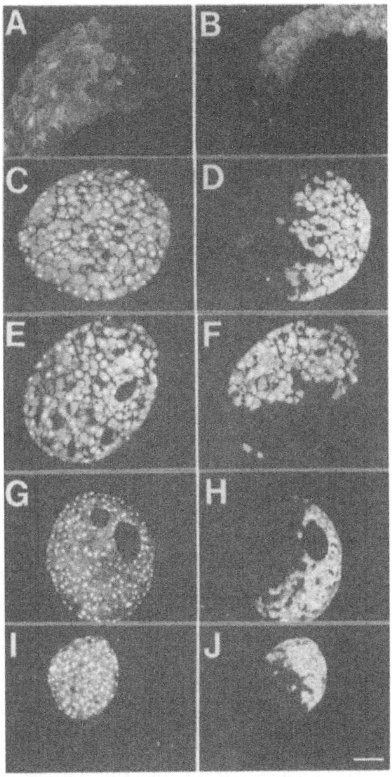

Figure 2. Little cell mixing occurs during XTC-MIF-induced elongation of *Xenopus* animal pole explants. *Xenopus* embryos received injections of 10 nl rhodamine-lysine-dextran into one of their blastomeres at the two cell stage. **(A,B)** Examples of such embryos were fixed and sectioned at the mid- to late blastula stage to confirm that little cell mixing occurs during cleavage stages. **(C-F)** Animal pole explants fixed 30 min after dissection and sectioned in a plane containing the original animal-vegetal axis demonstrate that dissection itself does not cause cell mixing. **(C,D)** Uninduced explants. **(E,F)** Explants treated with XTC-MIF. C and E are stained with 4',6-diamidino-2-phenylindole-dihydrochloride (DAPI), and D and F show rhodamine fluorescence. **(G-J)** Animal pole explants fixed and sectioned at the equivalent of the early neurula stage show that little cell mixing occurs during convergent extension. **(G,H)** Uninduced explants. **(I,J)** Explants treated with XTC-MIF. Notice the smaller cross-sectional area of the elongated induced explant. **G** and **I** are stained with DAPI and **H** and **J** show rhodamine fluorescence. Notice that labelled cells remain in a discrete group in both the induced and the uninduced explant. Scale bar in J is 200 μm, and applies to all frames.

In both induced and uninduced explants, blastomeres mixed with each other to a very limited extent and little difference could be discerned between the two (Figure 2 G-J). In both, some individual cells were found a few cell diameters away from the main group, but the overall impression is that convergent extension does not involve long-range cell migration but rather a short-range and perhaps directed exchange of neighbors. Our ability to determine the extent of short-range cell-mixing is limited due to the size of the labelled population of cells used in these experiments (see also Keller and Tibbetts 1989).

Keller and Hardin (1987) have suggested that such exchange of neighbors might occur through repetitive changes in shape and "jostling" of adjacent cells. This idea is supported by the appearance of induced and uninduced explants in the scanning electron microscope. At the equivalent of the early gastrula stage, when the explants are "rounding-up", induced and uninduced explants appear similar; the apices of the surface cells are generally rounded (Figure 3 A,B). By stage 12.5, however, the external surface of uninduced tissue has become smooth and the cells form a flat epithelial-like sheet (Figure 3 D). Induced explants, by contrast, still consist of cells with rounded apices (Figure 3 C).

These differences in surface relief may reflect any number of differences between induced and uninduced cells including changes in adhesivity, motility and cytoskeletal organization. It is reasonable that the curved apices are more deformable than the flattened ones, and this could mean that mechanical changes have occurred in the cortical cytoskeleton of induced cells, making it easier to stretch. Increased deformability

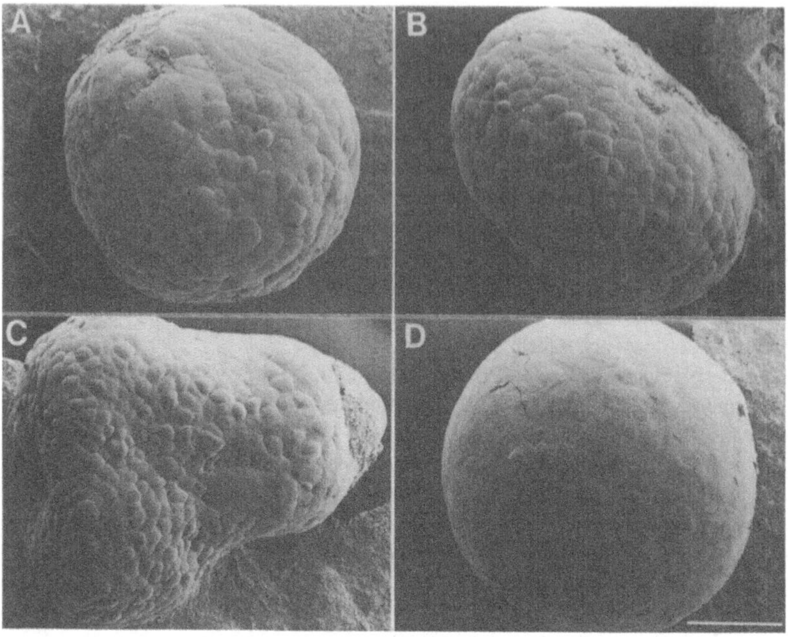

Figure 3. Scanning electron microscopy of induced and uninduced explants shows that induced explants display a higher level of cellular activity. **(A,B)** Induced **(A)** and uninduced (B) explants fixed at the late gastrula stage. **(C,D)** Induced **(C)** and uninduced **(D)** explants fixed at the gastrula stage. Notice that the induced explant has elongated and consists of more rounded cells compared with the control. Scale bar in **(D)** is 100 μm, and applies to all frames.

or extensibility of the cell surface has been correlated with the onset of motility, specifically the formation of lamellipodia and lobopodia in fish gastrulation (see Tickle and Trinkaus 1973). Thus we suggest that one of the changes in the induced cells may be an increased deformability that would allow formation of the lamelliform and filiform protrusions involved in cell intercalation (see article by Keller *et al.*, this volume).

Integrin and FN Expression in Early Embryos

In this and previous studies (Smith *et al.* 1990) we have confirmed that one of the earliest responses to mesoderm induction in *Xenopus* is the ability of induced cells to adhere, spread and migrate on FN coated surfaces. These data are consistent with the idea that FN mediated cell adhesion may play an important part in the process of gastrulation. Precisely how FN adhesion fits into the overall pattern of cellular rearrangements that unfold during gastrulation remains unclear. The ability to adhere and migrate on FN, however, must ultimately depend upon the array of specific cellular receptors for FN that are expressed on a given cell type.

There are at least two major pieces of evidence that implicate a role for specific FN receptors in amphibian gastrulation. The first includes evidence from several laboratories that mesodermal cells are able to migrate on FN substrates, both *in vivo* and *in vitro*, and that this interaction is important for gastrulation to proceed (reviewed by Johnson *et al.* 1991 and see above). The second observation is that FN fibrils are preferentially localized to the roof of the blastocoel. Furthermore, Darribere *et al.* (1990) have recently obtained evidence in *Pleurodeles*, which suggests that FN assembly into fibrils is dependent upon proper integrin receptor function in the cells lining the roof of the blastocoel. Our current working hypothesis is that the localized deposition of FN into the blastocoel roof and the migratory behavior of mesodermal cells on FN are distinct, integrin-dependent processes. Determining the overall pattern of integrin expression in the embryo is, therefore, a crucial first-step to understanding how cell-FN interactions are mediated throughout development.

In an earlier study we reported results of Northern blot experiments, demonstrating that integrin β_1 mRNAs begin to accumulate by the mid blastula stage in *Xenopus*, which suggests that integrins are first expressed as a result of zygotic transcription (DeSimone and Hynes 1988). In this study we have used RNase protection analyses to re-investigate integrin β_1 mRNA levels in stage VI oocytes and early embryos (Figure 4). The increased sensitivity afforded by this procedure (Krieg and Melton 1987) reveals that integrin β_1 mRNAs are indeed present in the oocyte (Figure 4, VI) and early cleavage stage embryos (not shown) as maternal mRNAs. By stage 8 there is a several-fold increase in β_1 mRNA levels, presumably as a result of new transcription at the MBT (Figure 4, st. 8). Integrin mRNA levels steadily rise during gastrulation and continue to increase as development proceeds (Figure 4). Immunoprecipitation and Western blot results, however, reveal that the integrin β_1 protein is not detected until after stage 8 (DeSimone, unpublished observations).

The β_1 family of integrins includes three receptors that are known to bind to FN ($\alpha_3\beta_1$, $\alpha_4\beta_1$ and $\alpha_5\beta_1$; see Hynes 1990). Because one of the earliest responses to induction is the ability to spread and migrate on FN (Figure 1), we were interested in learning whether overall integrin β_1 levels are elevated in induced tissues. We previously demonstrated that little or no difference in the amounts of integrin β_1 or FN is detected in XTC-MIF induced animal cap explants compared to uninduced controls (Smith *et al.* 1990). These results suggest that the initiation of gastrulation is unlikely to be controlled simply by the expression of FNs or β_1 integrins in response to mesoderm induction. In addition, synthetic peptides containing the RGD sequence do not appear to affect the timing or extent of convergent extension in XTC-MIF induced explants, even at concentrations known to inhibit mesoderm adhesion to FN *in vitro* (Smith *et al.* 1990; Winklbauer 1990). There are several possible interpretations for these RGD results in addition to the obvious suggestion that convergent extension is unlikely to involve RGD dependent (*e.g.*, FN-integrin mediated) adhesive mechanisms.

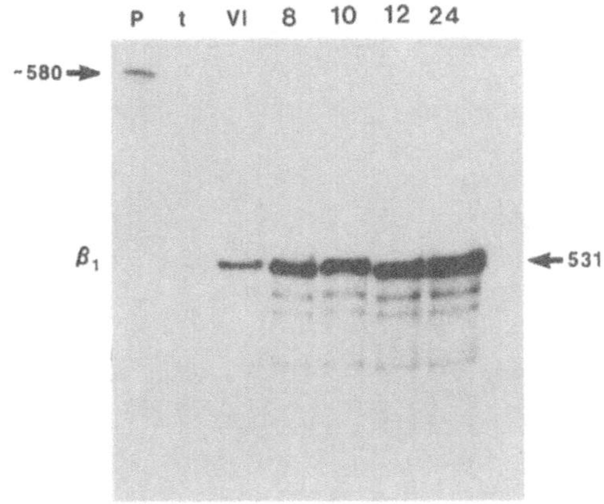

Figure 4. RNase protection analysis reveals that integrin β_1 mRNA is present in the oocyte with new β_1 transcripts appearing around the time of the mid-blastula transition. (P) Probe alone, unhybridized transcript not digested with RNase, (VI) Stage VI oocyte, (8) mid-blastula, (10) early and (12) late gastrula, (24) tailbud stage. Probe protects a 531 nt fragment of the β_1 mRNA. Five oocyte or embryo equivalents of RNA are represented in each lane.

From a technical standpoint, for example, it is difficult to ascertain during the time course of these experiments whether the RGD peptide can effectively resist degradation, infiltrate a multicellular explant, and still remain at high enough local concentrations to affect integrin-ECM interactions.

Alternative explanations for these RGD results emerge from what is currently understood about integrin biology in general and integrin-FN interactions in particular. The "classic", high-affinity FN receptor ($\alpha_5\beta_1$), as originally described, binds to the RGD site located in the central cell binding domain of the FN molecule (Pytela *et al.* 1985). However, Wayner *et al.*, (1989) have demonstrated that the integrin $\alpha_4\beta_1$ receptor binds to a different sequence within the alternatively spliced CS-1 (V-region) segment of FN. This sequence is also conserved in the V-region of *Xenopus* FN (DeSimone *et al.*, submitted). It is possible, therefore, that some of the integrin-FN interactions observed during embryogenesis are not RGD-dependent. It is also important to note that the list of additional integrins (both β_1 and non-β_1) with demonstrated FN binding activity (both RGD and non-RGD dependent) continues to grow (Albelda and Buck 1990; Hynes 1990).

We have recently established that the majority of the β_1 subunit expressed during gastrulation is an immature form of the protein, which is not present at the cell surface. Appreciable amounts of β_1 are not expressed at the cell surface until later in development as β_1 synthesis increases. This pattern of expression is demonstrated in Figure 5. The anti-β_1 antibody immunoprecipitates both mature (Figure 5, arrowhead) and immature (open arrowhead) forms of the protein from metabolically labelled gastrulae (st. 12). The immature form is several fold more abundant and cannot be chased into the mature form until much later in development (not shown). Higher molecular weight bands correspond to associated α subunits that are co-precipitated with the β_1 proteins. This is more clearly demonstrated in immunoprecipitates of *Xenopus* ovary tissue and isolated follicle cell layers (α and β, indicated by brackets) where the majority of β_1 evident after 4-6 hours of continuous labelling is the mature form. Follicle cells actively synthesize integrins whereas no integrin synthesis is detected in "follicle cell-free" oocytes (DeSimone, unpublished observations). A similar pattern of integrins is synthesized by chick embryo fibroblasts after a comparable labelling period and is included as a control in Figure 1 (CEF). It is presently unclear whether the relatively low levels of integrins expressed at the surface during gastrulation are sufficient to account for the adhesive behavior of mesodermal cells on FN.

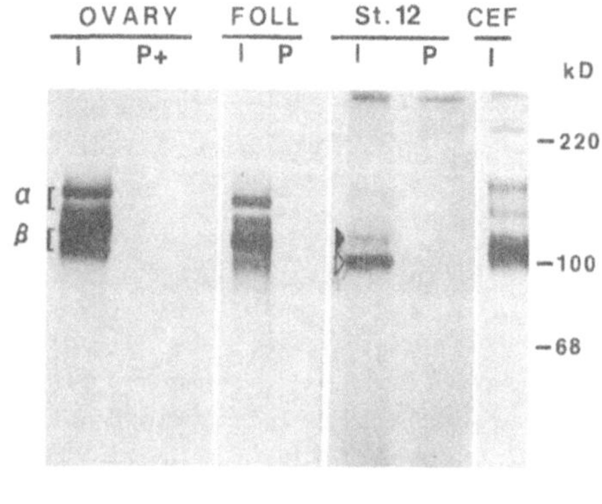

Figure 5. Immunoprecipitation of ^{35}S-methionine labelled gastrulae (st. 12), ovary, follicle cells (FOLL), and chick embryo fibroblasts (CEF) with an integrin β_1 subunit-specific antibody under non-reducing conditions identifies both β and associated α subunits (brackets). Solid arrowhead in the gastrula lane (st. 12) indicates the position of the mature integrin β_1 subunit. (I), anti-β_1 immune; and (P), pre-immune serum; (+) immune serum plus synthetic peptide control.

Integrin Heterogeneity During Development

Clearly, in order to understand FN function during development it is necessary to have a complete picture of all the FN receptors expressed (at the cell-surface) at a given stage. Recently, we devised a rapid strategy for obtaining *Xenopus* cDNAs that encode multiple β integrin subunits (Ransom and DeSimone 1990). Degenerate oligodeoxynucleotide primers were designed, based on regions of highest identity among existing β integrin sequences from human, mouse, avian and *Xenopus*. These primers were then used to amplify integrin cDNAs by polymerase chain reaction (PCR) methods from *Xenopus* tissues, cell lines, and various embryonic stages. We have thus far obtained cDNAs encoding *Xenopus* homologs of the mammalian β_2 and β_3 subunits in addition to a novel integrin designated β_x and our previously isolated β_1 subunit. RNase protection analyses reveal that β_3 mRNA is present in the oocyte and early embryo with increased expression beginning during neurulation. The β_2 mRNA is not expressed in the embryo until the late tailbud stage. These results provide evidence that integrin expression in the embryo is complex. Multiple integrins imply cellular interactions with multiple ligands and, at least in the case of FN, several integrins may interact with a single ligand. We are now undertaking experiments to determine if integrin heterogeneity provides positional information in the early embryo. As pointed out earlier, for example, it is possible that distinct FN-receptors may be localized in the cells lining the roof of the blastocoel and on the surfaces of migrating mesodermal cells.

Conclusions

Our failure to detect any change in the synthesis of FNs or integrin β_1 in XTC-MIF (activin) induced explants (Smith *et al.* 1990) is somewhat surprising in view of what is known about the actions of TGFβ. TGFβs have been shown to affect the synthesis of ECM molecules and integrins in many cells and tissues (reviewed by Massague 1990). For example, integrin levels can go up, down or remain the same in response to TGFβ although a given response is dependent upon the cell type and the specific TGFβ used (Heino *et al.* 1989; Heino and Massague 1989). The fact that animal cap explants rapidly begin to spread and migrate on FN in response to XTC-MIF, however, strongly suggests that integrin function and/or expression are somehow affected by the factor (Figure 1). We plan to investigate whether postranslational modifications of existing integrins might account for this rapid change in FN binding affinity, as has been suggested for other systems (Adams and Watt 1990). Our initial studies have focused on the overall expression of the integrin β_1 subunit in response to XTC-MIF. The availability of antibody and cDNA reagents for additional *Xenopus* integrins will now also make it possible for us to determine if the expression of other integrins is altered in response to mesoderm induction.

ACKNOWLEDGEMENTS

The authors are grateful to Liz Hirst for undertaking the scanning electron microscopy reported in this study. We also thank Ray Keller for helpful discussions and comments on this manuscript. This research was supported by a grant from USPHS

(RO1 HD26402 to DWD) and by the Medical Research Council of Great Britain. DWD is a recipient of a Pew Scholars Award in the Biomedical Sciences.

REFERENCES

Adams, J.C and F.M. Watt. 1990. Changes in keratinocyte adhesion during terminal differentiation: Reduction in fibronectin binding precedes $\alpha 5\beta 1$ integrin loss from the cell surface. *Cell* 63:425-435.

Albelda, S.M. and C.A. Buck. 1990. Integrins and other cell adhesion molecules. *FASEB J.* 4:2868-2880.

Boucaut, J.-C., T. Darribère, H. Boulekbache, and J.-P. Thiery. 1984a. Prevention of gastrulation but not neurulation by antibody to fibronectin in amphibian embryos. *Nature* 307:364-367.

Boucaut, J.-C., T. Darribère, T.J. Poole, H. Aoyama, K.M. Yamada, and J.-P. Thiery. 1984b. Biologically active synthetic peptides as probes of embryonic development: A competitive peptide inhibitor of fibronectin function inhibits gastrulation in amphibian embryos and neural crest cell migration in avian embryos. *J. Cell Biol.* 99:1822-1830.

Boucaut, J.-C., T. Darribère, D. Shi, J.-F. Riou, K.E. Johnson, and M. Delarue. 1991. Amphibian Gastrulation: The Molecular Bases of Mesodermal Cell Migration in Urodele Embryos. p. 169-184. *In: Gastrulation: Movements, Patterns, and Molecules.* R. Keller, W.H. Clark, Jr., F. Griffin (Eds.). Plenum Press, New York.

Cooke, J. and J.C. Smith. 1989. Gastrulation and larval pattern in *Xenopus* after blastocoelic injection of a *Xenopus* inducing factor: Experiments testing models for the normal organization of mesoderm. *Dev. Biol.* 131:383-400.

Cooke, J., J.C. Smith, E.J. Smith, and M. Yaqoob. 1987. The organization of mesodermal pattern in *Xenopus laevis*: Experiments using a *Xenopus* mesoderm-inducing factor. *Development* 101:893-908.

Darribère, T., K. Guida, H. Larjava, K.E. Johnson, K.M. Yamada, J.-P. Thiery, and J.-C. Boucaut. 1990. *In vivo* analyses of integrin β_1 subunit function in fibronectin matrix assembly. *J. Cell Biol.* 110:1813-1823.

Darribère, T., K.M. Yamada, K.E. Johnson, and J.-C. Boucaut. 1988. The 140 kD fibronectin receptor complex is required for mesodermal cell adhesion during gastrulation in the amphibian *Pleurodeles waltlii*. *Dev. Biol.* 126:182-194.

DeSimone, D.W. and R.O. Hynes. 1988. *Xenopus laevis* integrins: Structural conservation and evolutionary divergence of integrin β subunits. *J. Biol. Chem.* 263:5333-5340.

Gimlich, R.L. and J. Braun. 1985. Improved fluorescent compound for tracing cell lineage. *Dev. Biol.* 109:509-514.

Heino, J., R.A. Ignotz, M.E. Hemler, C. Crouse, J. Massague. 1989. Regulation of cell adhesion receptors by transforming growth factor-β. Concomitant regulation of integrins that share a common $\beta 1$ subunit. *J. Biol. Chem.* 264:380-388.

Heino, J. and J. Massague. 1989. Transforming growth factor β switches the pattern of integrins expressed in MG-63 human osteosarcoma cells and causes a selective loss of adhesion to laminin. *J. Biol. Chem.* 264:21806-21811.

Hynes, R.O. 1990. *Fibronectins*. Springer Verlag, New York.

Johnson, K.E., J.C. Boucaut, and D.W. DeSimone. 1991. The role of the extracellular matrix in amphibian gastrulation. *Curr. Top. Dev. Biol.* In press.

Keller, R.E. and J. Hardin. 1987. Cell behavior during active cell rearrangement: Evidence and speculations. *J. Cell Sci. Suppl.* 8:369-393.

Keller, R.E. and P. Tibbetts. 1989. Mediolateral cell intercalation in the dorsal axial mesoderm of *Xenopus laevis*. *Dev. Biol.* 131:539-549.

Keller, R., J. Shih, and P. Wilson. 1991. Cell Motility, Control and Function of Convergence and Extension During Gastrulation in *Xenopus*. p. 101-120. *In: Gastrulation: Movements, Patterns, and Molecules.* R. Keller, W.H. Clark, Jr., F. Griffin (Eds.). Plenum Press, New York.

Komazaki, S. 1988. Factors related to the initiation of cell migration along the inner surface of the blastocoelic wall during amphibian gastrulation. *Cell Differ.* 24:25-32.

Krieg, P.A. and D.A. Melton. 1987. *In vitro* synthesis with SP6 RNA polymerase. *Methods Enzymol.* 155:397-415.

Lee, G., R.O. Hynes, and M. Kirshner. 1984. Temporal and spatial regulation of fibronectin in early *Xenopus* development. *Cell* 36:729-740.

Marcantonio, E.E. and R.O. Hynes. 1988. Antibodies to the conserved cytoplasmic domain of the integrin β1 subunit react with proteins in vertebrates, invertebrates, and fungi. *J. Cell Biol.* 106:1765-1772.

Massague, J. 1990. The transforming growth factor-β family. *Annu. Rev. Cell Biol.* 6:597-641.

Melton, D.A. and R. Cortese. 1979. Transcription of cloned tRNA genes and nuclear partitioning of a tRNA precursor. *Cell* 18:1165-1172.

Nakatsuji, N. 1986. Presumptive mesodermal cells from *Xenopus laevis* gastrulae attach and migrate on substrata coated with fibronectin or laminin. *J. Cell Sci.* 86:109-118.

Nakatsuji, N. and K.E. Johnson. 1983. Comparative study of extracellular fibrils on the ectodermal layer in gastrulae of five amphibian species. *J. Cell Sci.* 59:61-70.

New, H.V. and J.C. Smith. 1990. Inductive interactions in early amphibian development. *Curr. Opin. Cell Biol.* 2:969-974.

Newport, J., and M. Kirschner. 1982. A major developmental transition in early *Xenopus* embryos. 1. Characterization and timing of cellular changes at the midblastula stage. *Cell* 30:675-686.

Nieuwkoop, P.D. 1969. The formation of mesoderm in urodelean amphibians. I. Induction by the endoderm. *Wilhelm Roux' Arch. Entwicklungsmech. Org.* 162:341-347.

Nieuwkoop, P.D. and J. Faber. 1967. *Normal table of Xenopus laevis (Daudin).* 2nd edition. North-Holland, Amsterdam.

Nieuwkoop, P.D. and S. Sudarwati. 1971. Mesoderm formation in the Anuran *Xenopus laevis* (Daudin). *Wilhelm Roux's Arch. Dev. Biol.* 166:189-204.

Pierschbacher, M.D. and E. Ruoslahti. 1984. Cell attachment activity of fibronectin can be duplicated by small synthetic fragments of the molecule. *Nature* 309:30-33.

Pytela, R., M.D. Pierschbacher, and E. Ruoslahti. 1985. Identification and isolation of a 140 kd cell surface glycoprotein with properties expected of a fibronectin receptor. *Cell* 40:191-198.

Ransom, D.G. and D.W. DeSimone. 1990. Cloning and characterization of multiple integrin α and β subunits expressed in *Xenopus* embryos. *J. Cell Biol.* 111:142a.

Shi, D.-L., T. Darribère, K.E. Johnson, and J.-C. Boucaut. 1989. Initiation of mesodermal cell migration and spreading relative to gastrulation in the urodele amphibian *Pleurodeles waltl. Development* 105:351-363.

Slack, J.M. 1984. Regional biosynthetic markers in the early amphibian embryo. *J. Embryol. Exp. Morphol.* 80:289-319.

Smith, J.C., K. Symes, R.O. Hynes, and D.W. DeSimone. 1990. Mesoderm induction and the control of gastrulation in *Xenopus laevis*: The roles of fibronectin and integrins. *Development* 108:229-238.

Smith, J.C., M. Yaqoob, and K. Symes. 1988. Purification, partial characterization and biological properties of the XTC mesoderm inducing factor. *Development* 103:591-600.

Symes, K. and J.C. Smith. 1987. Gastrulation movements provide an early marker of mesoderm induction in *Xenopus laevis*. *Development* 101:339-349.

Tickle, C. and J.P. Trinkaus. 1973. Change in surface extensibility of *Fundulus* deep cells during early development. *J. Cell. Sci.* 13:721-726.

Wayner, E.A., A. Garcia-Pardo, M.J. Humphries, J.A. McDonald, and W.G. Carter. 1989. Identification and characterization of the lymphocyte adhesion receptor for an alternative cell attachment domain (CS-1) in plasma fibronectin. *J. Cell Biol.* 109:1321-1330.

Wetts, R. and S.E. Fraser. 1989. Slow intermixing of cells during *Xenopus* embryogenesis contributes to the consistency of the fate map. *Development* 105:9-15.

Winklbauer, R. 1988. Differential interaction of *Xenopus* embryonic cells with fibronectin *in vitro*. *Dev. Biol.* 130:175-183.

Winklbauer, R. 1990. Mesodermal cell migration during *Xenopus* gastrulation. *Dev. Biol.* 142:155-168.

Wollweber, L., R. Stracke, and U. Gothe. 1981. The use of a simple method to avoid cell shrinkage during SEM preparation. *J. Microscopy* 121:185-189.

MECHANICS AND GENETICS OF CELL SHAPE CHANGES DURING *DROSOPHILA* VENTRAL FURROW FORMATION

Maria Leptin

Max Planck Institut für Entwicklungsbiologie
Spemannstrasse 35
D-7400 Tübingen
Federal Republic of Germany

INTRODUCTION

The folding of epithelia is an important morphogenetic process that turns flat sheets of cells into more complex three-dimensional structures. It occurs at various stages of embryogenesis in the development of many different organs, and during gastrulation. The first morphogenetic movement during *Drosophila* gastrulation is the invagination of the ventral furrow, which is the beginning of the infolding of the mesoderm. Ventral furrow formation is particularly well suited for the analysis of epithelial folding. First, it is simple and quick; the invaginating cell sheet is a homogeneous, single layer epithelium which changes its shape to form a tube in a period of less than half an hour. Second, no cell division or growth occurs during this process. Finally, many of the genes that determine the fates of the cells in and around the ventral furrow are known. This allows us to define cell populations not only by their behavior, but also by the expression patterns of specific genes. Furthermore, mutations in these genes can be used to change the fates of cells in the embryo in order to test the contributions of different populations to the movements of gastrulation. Eventually, we hope to find out how the genes that determine cell fates regulate the behavior of cells, directing them to undergo the morphogenetic changes of gastrulation. This paper describes the events during ventral furrow formation and

Gastrulation, Edited by R. Keller *et al.*
Plenum Press, New York, 1991

shows how we have made use of *Drosophila* genetics to analyze the forces and mechanisms involved in the invagination of the furrow.

Gastrulation Movements

About 3 hours after fertilization, the *Drosophila* embryo consists of a single epithelial layer of approximately 6000 tall columnar cells (the cellular blastoderm) surrounding a central yolk mass (Figure 1a). During gastrulation, over a period of about 3 hours, various epithelial foldings and movements convert this epithelial sheet into the embryo with its three germ layers and organ rudiments (Figures 1b-d). The mesoderm arises from an invagination along the ventral midline, the ventral furrow (Figure 1b). The endoderm is created by invaginations arising near the anterior and posterior ends of the embryo (Figure 1c).

After the ventral furrow has invaginated, the germ band, consisting of the tube of mesoderm and the overlying ectoderm elongates (possibly by convergent extension--see Wieschaus *et al.*, this volume) in a process called germ band extension. Since the embryo is confined by the vitelline membrane and the chorion that surround it, it does not actually become longer, but instead, the germ band is pushed around the posterior pole of the embryo and doubles up on its dorsal side, with the most posterior end coming to lie directly behind the head (Figure 1d). Later, the germ band contracts back to its initial length. The purpose of germ band elongation and retraction is not known.

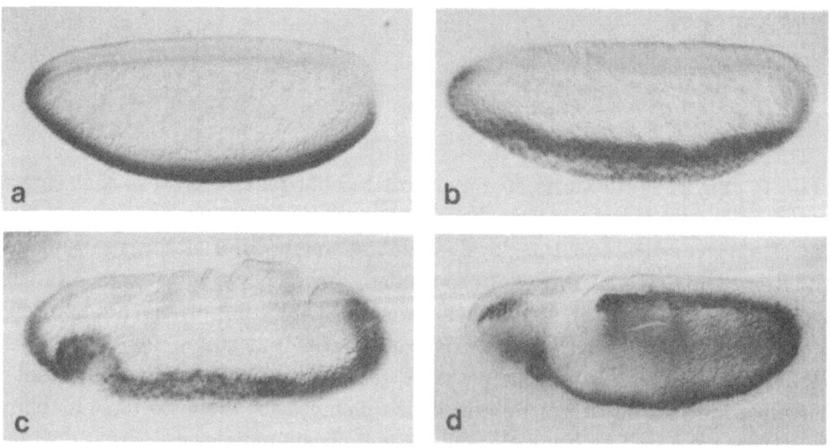

Figure 1. Whole mounts of embryos before and during gastrulation. Dorsal is up, anterior to the left. The embryos were stained with antibodies against the *twist* protein. The secondary antibody was coupled to horsereadish peroxidase, which was visualized in a histochemical reaction giving a dark brown signal. **a.** Late stage of cellularization. The nuclei lined up at the periphery of the egg are being separated by plasma membranes. The *twist* protein is expressed in ventral nuclei. **b.** The cells on the ventral side have begun to invaginate to form the ventral furrow. **c.** The germ band, consisting of the invaginated mesoderm and the overlying ectoderm, is beginning to extend around the posterior end of the embryo onto the dorsal side. Simultaneously, the endoderm is invaginating at the anterior and posterior ends of the ventral furrow, forming the anterior and posterior midgut invaginations. **d.** Fully extended germ band stage. The invaginated mesoderm has spread out as a cell layer underneath the ectoderm.

(These movements are described in more detail in Fullilove *et al.* 1978 and Campos-Ortega and Hartenstein 1985).

Cell Shape Changes During Ventral Furrow Formation

All cells of the cellular blastoderm look morphologically identical until ventral furrow formation begins (Figure 2a). Their basal sides face the interior of the embryo, the apical surfaces towards the outside. The nuclei are lined up very close to the apical cell surfaces and the cell membranes form hemispherical bulges, giving the outer surface of the embryo a cobblestone-like appearance (Figure 2a and Rickoll 1976; Turner and Mahowald 1977). However, already at this stage, gene expression patterns are different between ventral and dorsal cells. The ventral cells express the *twist* gene, whose product is a nuclear protein (Thisse *et al.* 1988), which we use as a marker for prospective mesoderm cells (although some cells at the poles of the embryo which will not become mesoderm also express this gene; Figure 1a).

The first visible difference between prospective mesoderm cells and the rest of the blastoderm is in the appearance of their surface. The cobblestone appearance becomes obliterated and the ventral surface of the embryo appears smoother over a width of about 20 cells (E. Wieschaus, personal communication; Leptin and Grunewald 1990). A few minutes later, two different types of cellular behavior are distinguishable among these smoother cells, defining two subpopulations of the prospective mesoderm, a central and a peripheral population (Figure 2b and Figure 3; Leptin and Grunewald 1990). The central population is the ventralmost band, about 10-12 cells wide, along the ventral midline of the embryo (grey cells in Figure 3). The peripheral population (stippled in Figure 3) consists of the two bands of cells, each 4-5 cells wide, bordering the central population. The cells with weakly stained nuclei (stippled nuclei in Figure 3) at the edge of the prospective mesoderm are not mesoderm precursors, but mesectoderm cells, which will form part of the nervous system (Campos-Ortega and Hartenstein 1985; Crews *et al.* 1988). The nuclei of the central population move away from the apical towards the basal surfaces of the cells, and the cells' apical ends flatten, and later constrict (Figure 2b). Thus, the shapes of the cells change, and probably as a result of this, an indentation–the ventral furrow–is formed (Figure 2c). The nuclei of the peripheral cells remain at the apical surfaces. As the central cells begin to change shape, the apical surfaces of the peripheral cells become stretched (Figure 2c), and the basal ends narrow. The differences in cell behavior between central and peripheral cells are apparently not merely different mechanical effects of subcellular events common to both populations. The two populations most likely differ in their genetic program. One would therefore predict the existence of a gene or genes that determines this difference *via* being expressed in only one of the two populations.

Eventually, the peripheral cells follow the central cells into the invaginating furrow (Figure 2d). Up to this point, the cells have not moved relative to each other, but maintained their positions within the epithelium. Thus, the first phase of mesoderm invagination is characterized by the folding of an epithelium without any interference by cell migration, division or growth. During the second phase, the epithelial tube of invaginated mesoderm (Figure 2e) disperses into individual cells (Figure 2f), which now divide and migrate outwards along the ectoderm to form the mesodermal cell layer. This process does not depend on cell division, since it also occurs in mutants in which these cells fail to divide (Leptin and Grunewald 1990).

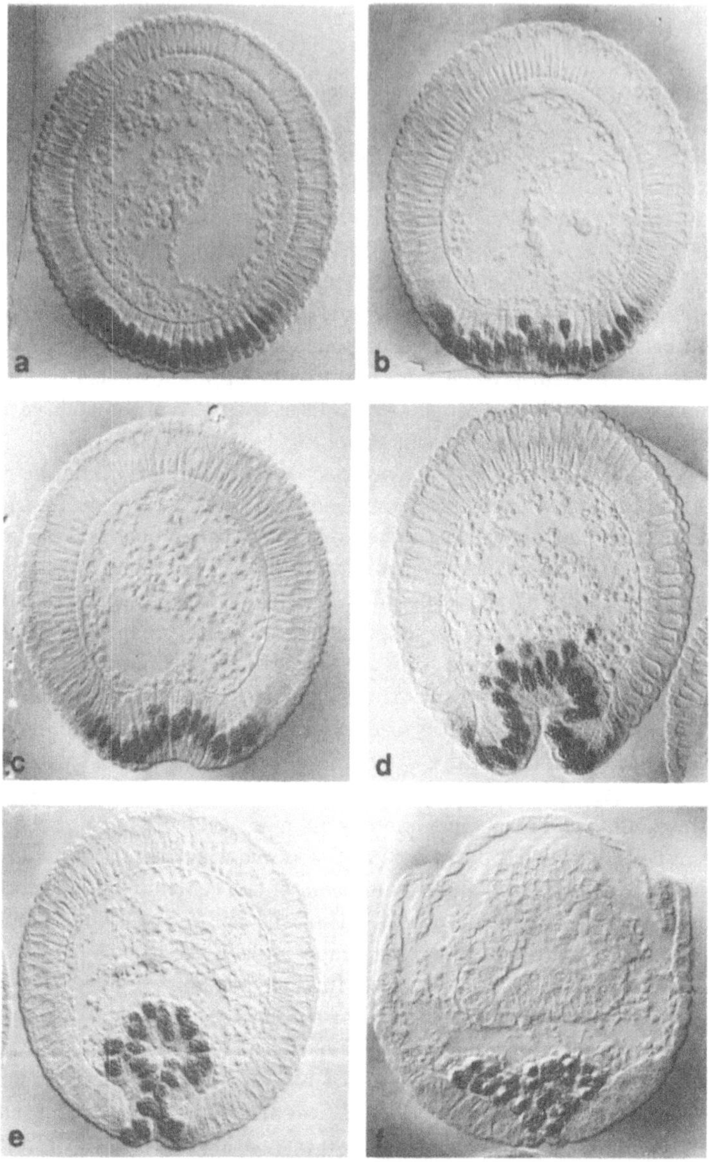

Figure 2. Transverse sections through embryos (stained with anti-*twist* antibodies) depicting six stages of mesoderm invagination and internalization (dorsal is up). **a.** Embryo at the completion of cellularization. **b.** Early stage of ventral furrow formation. The apical (outer) surfaces of the *twist*-expressing cells have become smooth, and the nuclei of the central 8-10 cells have begun to migrate away from the apical cell surface. **c.** The apical sides of the central cells have constricted, and the surface of the ventral epithelium is beginning to indent. **d.** The central cells have invaginated completely, and the peripheral cells are following them into the furrow. **e.** The lips of the ventral furrow have joined but the peripheral cells have not completed their invagination. **f.** The invaginated tube of mesoderm has dispersed into single cells, which are in the process of dividing. The dorsal and lateral epithelia have flattened; the structure in the middle of the embryo is a section through part of the invaginating posterior midgut.

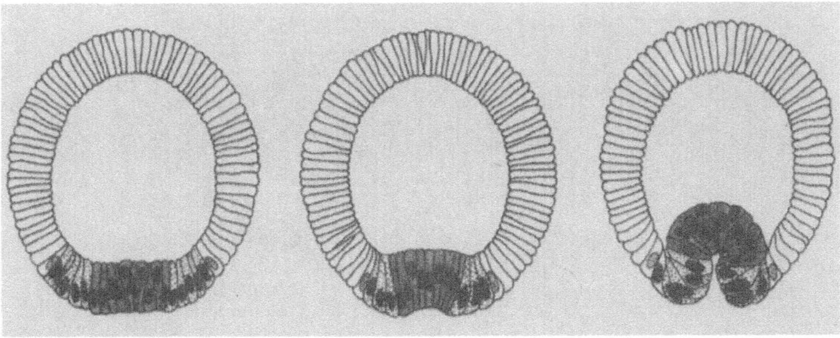

Figure 3. Tracings of three of the sections in Figure 2 (2b,c,d) to show the boundaries of the central and peripheral population in the prospective mesoderm. The populations are at present only defined by their shape and the positions of their nuclei, and at the boundary between central and peripheral population it is not always clear in any particular section whether a cell should be classified as central or peripheral. Cells considered to be within the central population are shaded grey, peripheral cells are stippled. The most lateral *twist*-expressing cell on each side (lightly shaded nucleus in this figure, faint *twist* staining in Figure 2) does not invaginate to become mesoderm, but is a mesectoderm cell, which will lose its *twist* staining completely and become part of the nervous system.

Regional Autonomy of Ventral Furrow Formation

For the rest of this article I will concentrate only on the events during the first phase of mesoderm formation, the actual epithelial invagination. I believe that the forces that drive this process come from within the individual prospective mesoderm cells, with little or no cell-cell communication or contribution from dorsal or lateral cell populations, and I will present the evidence that supports this view.

To study the roles of different cell populations in the formation of the ventral furrow, we genetically changed cell fates in the embryo. The functions of some of the genes determining cell fates along the dorso-ventral axis of the embryo are diagrammed in Figure 4. This figure summarizes the work from a number of labs (Doyle *et al.* 1986; Rushlow *et al.* 1987; Thisse *et al.* 1988; Roth *et al.* 1989; Rushlow *et al.* 1989; Steward 1989), as well as unpublished observations (David Stein, personal communication), and some aspects of it are speculative.

The fates of cells along the dorso-ventral axis are determined by a set of genes transcribed while the egg develops in the ovaries of the mother (maternal effect genes). The product of one of these, the *Toll* gene, is a transmembrane receptor (Hashimoto *et al.* 1988) and seems to be activated in a graded way with its highest activity on the ventral side of the embryo. The activated *Toll* protein is thought to regulate the uptake of another maternal gene product, the *dorsal* protein, into nuclei (Roth *et al.* 1989; Rushlow *et al.* 1989; Steward 1989). The *dorsal* protein (probably a transcription factor; Steward 1987) is initially distributed homogeneously in the egg cytoplasm, but is later taken up mainly by ventral nuclei. Lateral nuclei also take up some *dorsal* protein, but less than ventral ones, and the protein is retained entirely in the cytoplasm on the dorsal side of the embryo. Thus, a gradient of nuclear *dorsal* protein concentration along the dorso-ventral axis is formed. The asymmetric distribution of *dorsal* protein in the blastoderm nuclei results in the asymmetric transcription of zygotically active genes in the blastoderm. Genes that determine dorsal fates are only active in dorsal

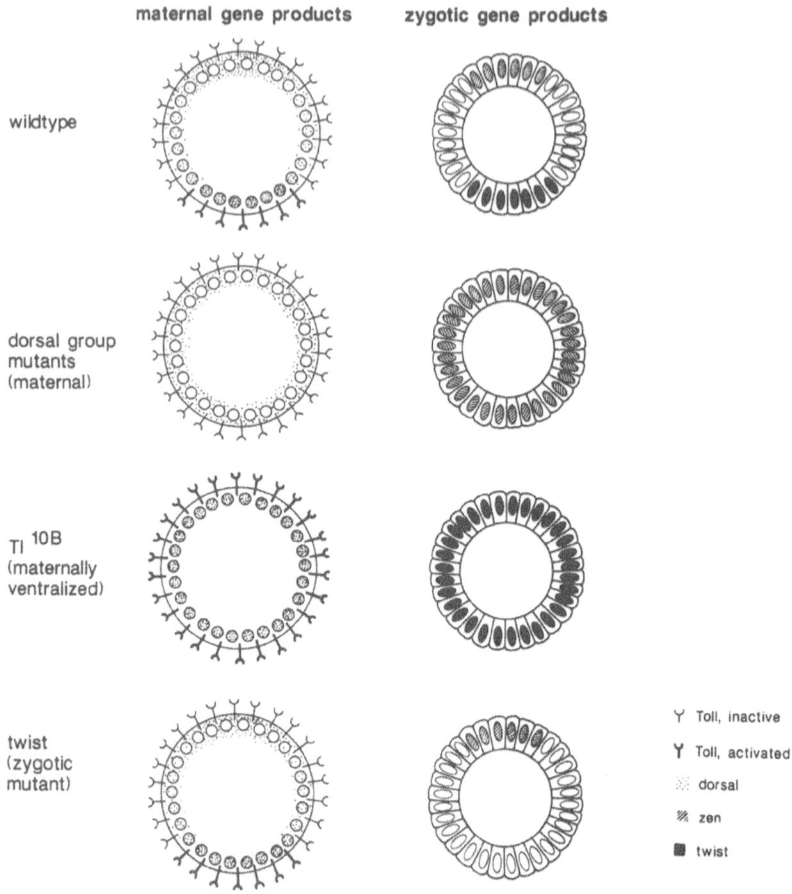

Figure 4. Diagram of the distribution of maternal and zygotic gene products in wildtype embryos and in maternal and zygotic mutant embryos. The left column represents 2.5-3 hr old embryos, before cellularization. The activity of the maternal gene products have already given cells along the dorso-ventral axis different values. The right column shows older embryos (cellular blastoderm) in which zygotic genes have been activated. The distribution of the *dorsal* protein (stippling in cytoplasm or nuclei), and of the *zen* (shaded nuclei) and *twist* (black nuclei) proteins are drawn after Roth *et al.* 1989. (See Figure 5a for photographs of the distribution of these proteins in mutants). The state of activity of the *Toll* transmembrane receptor (inactive: thin lines; active: fat lines) has not been proven directly. It is inferred from experiments by D. Stein (D. Stein, S. Roth, L. Vogelsang, and C. Nüsslein-Volhard, personal communication).

nuclei, which contain little or no *dorsal* gene product (for example, the *zen* gene; Doyle *et al.* 1986). In contrast, the two known genes required for the formation of the mesoderm, *twist* and *snail* (Nüsslein-Volhard *et al.* 1984; Simpson 1983), are transcribed only in nuclei with high levels of *dorsal* protein, *i.e.* in ventral nuclei (Thisse *et al.* 1988; Leptin and Grunewald 1990).

All of these genes can be inactivated by mutation, resulting in cell fate changes along the dorsoventral axis. In the case of maternally active genes, the distribution of all cell fates along the dorso-ventral axis is affected (reviewed in Anderson 1987). For example, in lack-of-function mutants of the *dorsal* group genes, all cells take on dorsal fates (dorsalized embryos) and only those zygotic genes normally expressed in dorsal cells are transcribed, while ventral genes are inactive. Some dominant mutations of the maternal genes (for example $Toll^{10B}$) lead to the opposite phenotype, a ventralization of the embryo, where all cells express mesoderm specific genes. Both types of embryos are defective in many aspects of the gastrulation movements described above, and neither completely dorsalized nor completely ventralized embryos form a ventral furrow.

When the zygotically active genes are mutated, only some cell populations along the dorso-ventral axis (usually those which normally express these genes) are changed in their developmental program. Embryos mutant in dorsally expressed genes form a normal ventral furrow, but have defects in the folding of the dorsal epithelium during germ band extension. Embryos mutant for the ventral genes *twist* and *snail* are normal in all gastrulation movements except for the formation of the ventral furrow (Simpson 1983).

Maternally dorsalized and ventralized mutant embryos can be used to analyze the earliest cellular behavior of dorsal and ventral cells in the absence of neighboring populations with different cellular behaviors. For example, the basal migration of nuclei in the central cells may be an autonomous activity within these cells, or it may be a passive response to pressure from lateral or dorsal cells. The phenotypes of dorsalized and ventralized embryos make the latter interpretation seem unlikely (Leptin and Grunewald 1990). In ventralized ($Toll^{10B}$) embryos, all cells flatten at their apical surfaces, and nuclei at many positions around the circumference move basally, while all cells in dorsalized embryos retain their rounded apices, and the nuclei stay at the apical surface (Figure 5a). This shows that pushing by lateral cells cannot be the only cause of nuclear migration in ventral cells of the wildtype embryo. It is of course possible that all nuclei are initially attached to structures near the apical end of the cell, and are then released only in the ventralmost cells. In this case, only the ventralmost nuclei would be able to respond to a general pressure around the periphery of the egg by moving inwards. But even in this case, the *release* of nuclei would be a process specific to ventral cells, occurring autonomously, independent of neighboring populations. Therefore, apical flattening and release of nuclei are cellular events in ventral cells that occur independent of cells with other fates. The later cellular activities along the dorsoventral axis cannot be analyzed in completely dorsalized or ventralized embryos, because other morphogenetic movements (like germ band extension) make reliable interpretations of events impossible.

The contributions of lateral and dorsal cells to furrow formation were analyzed in zygotic mutants and in embryos from incompletely ventralizing maternal mutants. In embryos mutant for the zygotic genes *twist* or *snail*, or in embryos mutant for both genes, dorsal and lateral gastrulation movements proceed normally, however the ventral epithelium does not fold in the same way as in the wildtype embryo (Leptin and Grunewald 1990). *Twist* mutants form a small and transient furrow; in *snail* mutants, the ventral epithelium flattens and folds slightly, in an irregular pattern, but does not invaginate. Double mutant embryos usually form neither a furrow nor irregular folds. Therefore the activities of dorsal and lateral epithelia are not sufficient to create a ventral invagination, even when the ventral cell sheet has a propensity to fold, as in

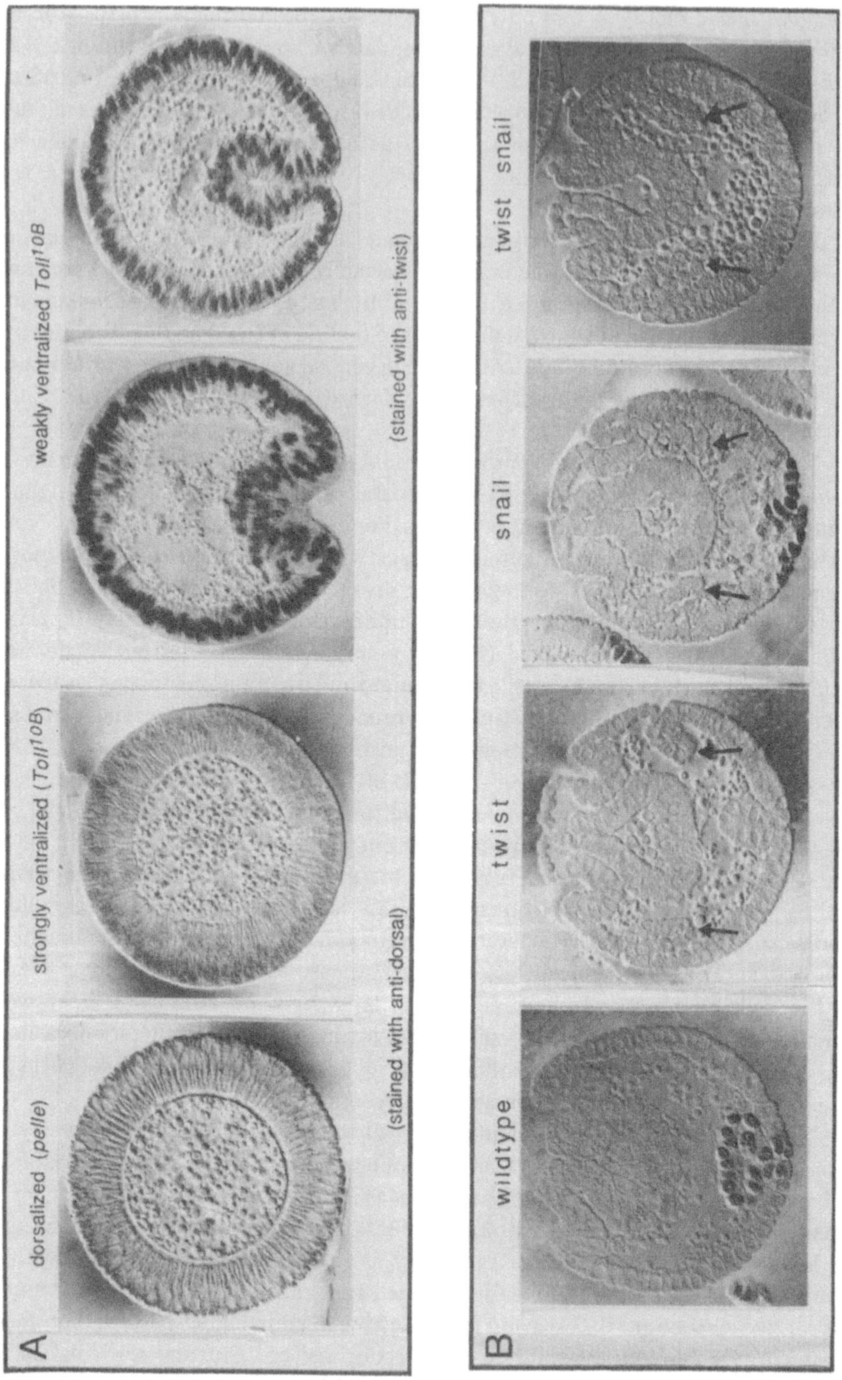
A
dorsalized (pelle)
strongly ventralized (Toll10B)
weakly ventralized Toll10B
(stained with anti-dorsal)
(stained with anti-twist)
B
wildtype
twist
snail
twist snail

snail mutants. Instead, the normal flattening and expansion of the dorsal and lateral epithelia seem to result in extra folds at a more dorsal position (Figure 5b).

The dorsal and lateral epithelia are not only insufficient to create a ventral furrow, but their normal behavior is not even required for furrow formation, as partially ventralized embryos show. Some $Toll^{10B}$ embryos still have some dorsoventral polarity, although they express the *twist* protein in all cells (possibly, the postulated ventral gene that distinguishes the central and peripheral cell populations is not expressed homogeneously in these cases). Thus, these embryos lack cells with lateral or dorsal fates. Nevertheless, they form a ventral furrow (Figure 5; Leptin and Grunewald 1990), showing that cell populations with lateral or dorsal fates are not required for the ventral furrow to invaginate. In summary, cells outside the ventral furrow are not required for the initial cellular events like apical flattening and nuclear migration, and their activities are neither sufficient nor required for the invagination of the furrow as a whole. The ventral genes, expressed in the prospective mesoderm, *are* required for the normal ventral activities that lead to mesoderm invagination. This means that the furrow is formed mainly (and possibly only) by autonomous activities of the invaginating prospective mesoderm cells.

These results do not exclude the possibility that the ventrally expressed genes have only a permissive role, and some other feature of the egg–for example its shape and structure–determines whether and how a furrow is formed (as in amphibians, where the physical properties of the egg partly determine the shape of the invagination at the dorsal lip; Gerhart and Keller 1986; Hardin and Keller 1988). However, this does not seem to be the case, since furrows can be induced at any site in the embryo. In $Toll^-$ maternally dorsalized embryos, in which all cells have dorsal fates, ventral fates can be induced by the injection of wildtype cytoplasm. The ventral cells will always develop only at the site of injection (Anderson *et al.* 1985; Siegfried Roth, unpublished observations). The shape of the region with ventral cells is determined by the way the injected cytoplasm is deposited. Such patches of ventral cells form a furrow, no matter where in the embryo the 'ventral' cells lie. The furrow always follows the longest axis of the patch, and when the patch is round, a pit is formed. This shows that any field of cells expressing ventral genes can organize itself into a furrow, and that there are no epigenetic factors determining this morphogenetic process.

Figure 5. Gastrulation phenotypes in mutant embryos. **A.** Maternal effect genes: embryos from mutant mothers. *pelle* is a recessive dorsalizing mutation. In these embryos, the *dorsal* protein remains in the cytoplasm in all cells. $Toll^{10B}$ is a dominant ventralizing mutation and all nuclei take up the *dorsal* protein. All cells flatten at their apical surfaces and nuclei at many positions around the periphery of the egg move basally in ventralized, but not in dorsalized mutants. This is also true in the weakly ventralized mutants shown on the right. These embryos express the *twist* protein in all cells (that is, all cells have ventral quality), but a residual asymmetry leads to the formation of a ventral furrow (which is larger than in wildtype embryos). **B.** Zygotic mutants. All embryos were stained with anti-*twist* antibodies. The *twist* and double mutant embryos do not express the protein because no protein is produced from the mutated gene. The age of these embryos is between that of the ones shown in Figure 1e and 1f. On the dorsal side of each, the posterior midgut invagination (containing the pole cells) is visible, showing that the invagination of the endoderm occurs normally in these mutants. The arrows point to invaginations of lateral epithelium not seen in wildtype embryos.

Possible Cellular Autonomy in Ventral Cells

Observations of the earliest phase of ventral furrow formation in living embryos suggest that not only the whole region, but also individual cells of the central population undergo their developmental program independent of their neighbors. Computer models indicate that an epithelium could invaginate as the result of a wave of apical constrictions, where the constriction of one cell expands the surfaces of its neighbors and thereby triggers them to constrict (Odell *et al.* 1981). No such wave moves through the central population of ventral furrow cells (*i.e.* those cells that do constrict). However, these cells do not change shape simultaneously either. After the completion of cellularization, individual cells in apparently random positions within the central population begin to change shape (Kam *et al.* submitted). Figure 6 shows the disappearance of nuclei from below the apical cell surfaces in the ventral epithelium, as an indicator for shape change. There is no obvious order in which the central population cells begin to change shape. For example, neighbors of cells which have constricted early do not seem to constrict sooner than any other cell in the epithelium. This stochastic behavior suggests that each cell in the central population undergoes its shape change independent of its neighbors, as part of its intrinsic developmental program. However, this apparent autonomy will have to be proven by making genetic mosaics. Furthermore, we have only analyzed the early phase of ventral furrow formation, and it is possible that later, during the rapid deepening of the furrow cell interactions begin to play a role.

How does each cell determine when to change its shape? I suggest that the timing depends on the accumulation of zygotic gene products in the central population for example of the *twist* and *snail* proteins, and of the product of the postulated gene specific to the central population. This is based on two observations. First, reduction of the amount of *twist* or *snail* protein by half delays the invagination of the furrow, since in embryos with only one copy of the *twist* or the *snail* gene, the ventral furrow appears later than in wildtype embryos (unpublished observations). Second, the stochastic sequence of cell shape changes at the beginning of furrow formation is often overlayed by a bias for cells in certain regions to begin their changes earlier than

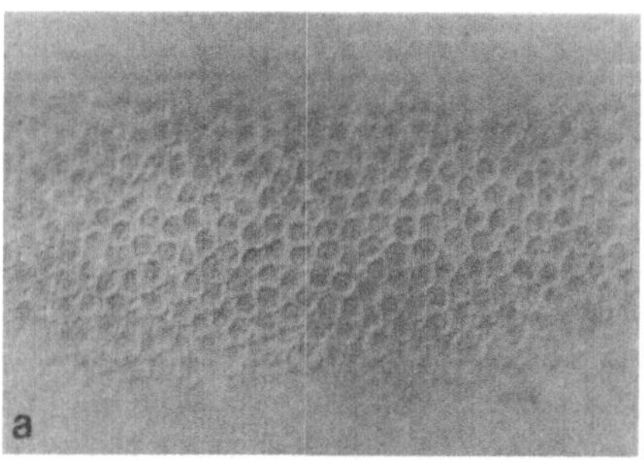

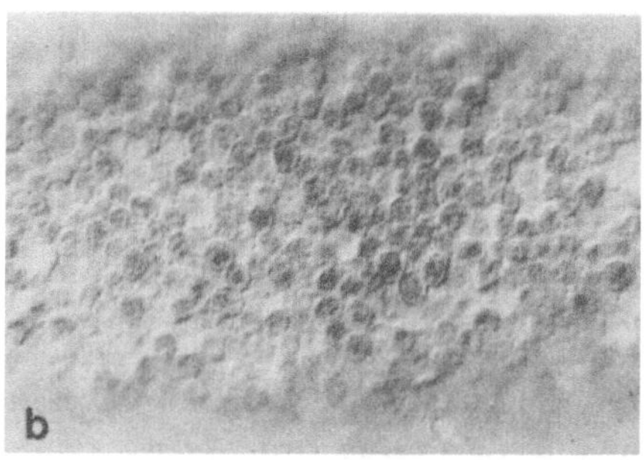

Figure 6. Optical sections just below the ventral surface of embryos at the syncytial blastoderm, cellular blastoderm and beginning ventral furrow formation. The nuclei (dark round structures) are initially all lined up just below the cell membranes on the ventral surface. Ventral furrow formation begins by individual cells of the central population undergoing the changes described in Figure 2, demonstrated here by the disappearance of nuclei from the focal plane below the apical cell surface. The pattern of disappearance does not follow any obvious order, but is apparently stochastic.

others, and this bias mirrors in homogeneities in the early transcription of ventral genes. The ventralmost nuclei often begin transcribing genes slightly earlier than the rest of the prospective mesoderm, and in many stocks the expression of ventral genes along the anterior posterior axis is uneven, with higher expression near the posterior, or in repeating patches about 10-15 cells apart. Similarly, it is not uncommon for the cells closer to the middle of the future furrow to constrict early. This does not mean that in these embryos a wave of constrictions moves from the center outwards, but only that at any given time more cells near the center than the edges of the central population have constricted. Differences in timing along the anterior-posterior axis can also occur, sometimes with patches of early constricting cells in approximately every other segment (see for example Figure 15 in Foe [1989]), and sometimes with a progression of furrow formation from the posterior towards the anterior. In all of these cases however, the final boundaries of the central region are the same (as are the boundaries of ventral gene expression). Thus, the field of cells that constrict is defined independently of the initial pattern of constricting cells. The similarities between patterns of inhomogeneous accumulation of ventral gene transcripts and the timing of the start of ventral furrow formation support the notion that the amount of zygotic gene products in each cell determines when cell shape changes begin. However, more extensive experiments, like inducing overexpression of ventral genes, will be required to prove this.

In summary, the observations described in the last two sections suggest that the ventral furrow is made as the result of each cell following its own developmental program, and that this program is determined by the expression of ventral genes (including *twist* and *snail*, and other, as yet unknown genes). These genes are both required and sufficient to induce furrow formation, since in their absence (in mutants) no normal furrow forms, and when they are ectopically expressed they lead to ectopic invaginations. However, it is important to point out that this does not mean that the products of these genes participate directly in the cellular mechanisms that cause the ventral furrow to invaginate. The *twist* and *snail* proteins are homologous to proteins known to be involved in gene regulation–*twist* has an helix-loop-helix domain (Thisse *et al.* 1988) and *snail* has zinc fingers (Boulay *et al.* 1987). They probably regulate the transcription of other genes whose products mediate the cell shape changes responsible for ventral furrow formation–for example by modifying the state of the cytoskeleton.

Subcellular Events During Ventral Furrow Formation

If the ventral furrow is made by autonomous cell shape changes, then to understand furrow formation, one must understand how individual cells change their shape. Several models for mechanisms by which columnar cells in an epithelium change to wedge or conical shaped cells have been proposed (reviewed in Ettensohn 1985). Some of these mechanisms are unlikely to operate in ventral furrow formation, because predictions of these models do not fit the observations in *Drosophila* ventral furrow cells. For example, 'zippering-up' by increased cell-cell adhesion at the apical end of the cell requires cell-cell interactions and would predict that at least two neighboring cells would undergo simultaneous shape changes, but in the ventral furrow one often sees individual cells change their shape, surrounded by others that do not. Furthermore, in this and other models (cortical tractor; Jacobson *et al.* 1986), the plasma membrane of the apical surface would be pulled into the interface between neighboring cells as these cells change their shapes. This also does not happen in ventral furrow cells, but instead, the excess plasma membrane forms protrusions when the apical side constricts. I therefore favor the idea that cell shape changes in the ventral furrow occur mainly by the rearrangement of components within individual cells, probably mediated by the cytoskeleton, in similar ways as proposed for shape changes in neurulation (Burnside 1973).

In any case, it seems likely that more than one subcellular process directs these changes and that the two microscopically observable components—apical constriction and nuclear migration—are at least partly independent. In *twist* mutant embryos, the ventral cells do not show the strong apical constriction of wildtype cells, but nevertheless the nuclei move away from the apical surface (Figure 7). These embryos do form a ventral furrow but it is less well organized and not as deep as the wildtype furrow. Furthermore, in our video analysis of ventral furrow cells in living wildtype embryos, we occasionally saw nuclei moving before any visible apical constriction of the cell (Kam *et al.* submitted). Thus, apical constriction appears not to be the cause of nuclear migration. On the other hand, the basal migration of nuclei might be a prerequisite for forming the wedge-shaped cells and allowing the complete invagination of the ventral furrow.

It is not yet clear which parts of the cytoskeleton mediate nuclear migration and apical constriction. The actin system probably plays a part in apical constriction since cytoplasmic myosin is redistributed in ventral cells during furrow formation and accumulates at the constricting apical ends (Young *et al.* 1991 and Figure 8). Studies using cytoskeletal drugs, and genetic analysis should allow us to analyze the roles of cytoskeletal components and how they are regulated by the genes that determine ventral fates. We thereby hope to find out how the instructions of the early pattern formation genes are interpreted and translated into the morphogenetic work that shapes the embryo.

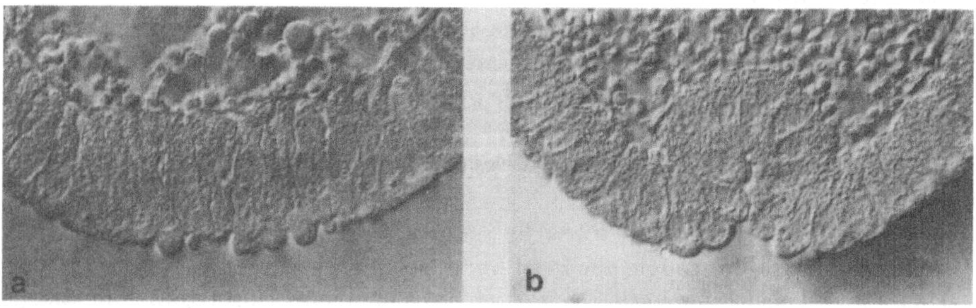

Figure 7. Two stages of furrow formation in *twist* mutant embryos. **a.** Embryo at approximately the same age as the one in Figure 2d. The 10 ventral most cells have undergone shape changes, but differ from wildtype cells. Although their nuclei have migrated away form the apical surfaces and the apical plasma membrane form the same protrusions as in the wildtype, they do not show the same strong apical constrictions. **b.** Slightly older embryo. A furrow has formed, but the cells do not have the typical appearance of wildtype cells. The inner cells are not wedge shaped, and the surfaces of the cells at the lip of the furrow are not stretched toward the center. At later stages, the furrow flattens out and disappears.

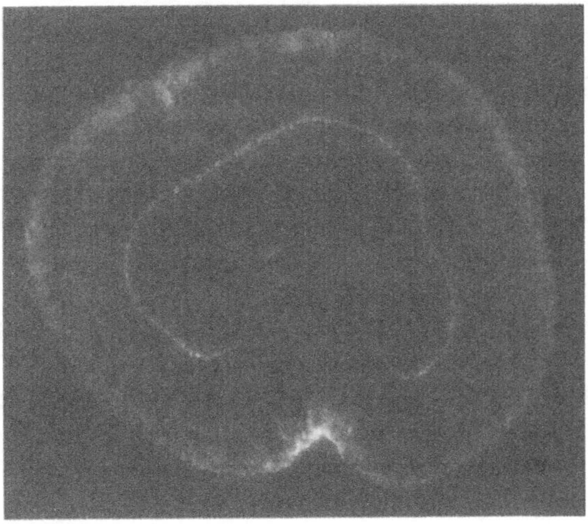

Figure 8. Distribution of cytoplasmic myosin in the ventral furrow (visualized with anti-cytoplasmic myosin antibodies and fluorescent secondary antibodies). At the completion of cellularization, cytoplasmic myosin is located at the bases of the blastoderm cells (Young *et al.* 1991), as is still seen in the dorsal and lateral cells in this section. When ventral furrow formation begins, the myosin disappears from the basal end and accumulates at the apical surfaces of the central cell population.

ACKNOWLEDGEMENTS

I thank Barbara Grunewald for technical assistance and for providing Figure 8, Paul Young and Dan Kiehart for the antibody against cytoplasmic myosin, Rolf Reuter and Mary Mullins for comments on the manuscript, David Stein and Siegfried Roth for allowing me to quote unpublished observations, and Zvi Kam, Jon Minden and Eric Wieschaus for discussions on many of the ideas presented here.

REFERENCES

Anderson, K. 1987. Dorsal-ventral embryonic pattern genes of *Drosophila. Trends Genet.* 3:91-97.

Anderson, K.V., L. Bokla, and C. Nüsslein-Volhard. 1985. Establishment of dorsal-ventral polarity in the *Drosophila* embryo: The induction of polarity by the *Toll* gene product. *Cell* 42:791-798.

Boulay, J.L., C. Dennefeld, and A. Alberga. 1987. The *Drosophila* developmental gene *snail* encodes a protein with nucleic acid binding fingers. *Nature* 330:395-398.

Burnside, B. 1973. Microtubules and microfilaments in amphibian neurulation. *Am. Zool.* 13:989-1006.

Campos-Ortega, J.A. and V. Hartenstein. 1985. *The Embryonic Development of Drosophila melanogaster*. Springer Verlag, Berlin.

Crews, S.T., J.B. Thomas, and C.S. Goodman. 1988. The *Drosophila* single-minded gene encodes a nuclear protein with sequence similarity to the *per* gene product. *Cell* 52:143-151.

Doyle, H.J., R. Harding, T. Hoey, and M. Levine. 1986. Transcripts encoded by a homeo box gene are restricted to dorsal tissues of *Drosophila* embryos. *Nature* 323:76-79.

Ettensohn, C.A. 1985. Mechanisms of epithelial invagination. *Q. Rev. Biol.* 60:289-307.

Foe, V.E. 1989. Mitotic domains reveal early commitment of cells in *Drosophila* embryos. *Development* 107:1-22.

Fullilove, S.L. and A.G. Jacobson. 1978. Embryonic Development: Descriptive. p. 105-227. *In: Genetics and Biology of Drosophila, Vol. 2c.* M. Ashburner and E. Novitski (Eds.). Academic Press, New York.

Gerhart, J. and R. Keller. 1986. Region-specific cell activities in amphibian gastrulation. *Annu. Rev. Cell Biol.* 2:201-229.

Hardin, J. and R. Keller. 1988. The behavior and function of bottle cells during gastrulation of *Xenopus laevis. Development* 103:211-230.

Hashimoto, C., K.L. Hudson, and K.V. Anderson. 1988. The *Toll* gene of *Drosophila,* required for dorsal-ventral embryonic polarity, appears to encode a transmembrane protein. *Cell* 52:269-279.

Jacobsen, A.G., G.F. Oster, G.M. Odell, and L.Y. Cheng. 1986. Neurulation and the cortical tractor model for epithelial folding. *J. Embryol. Exp. Morphol.* 96:19-49.

Kam, Z., J.S. Minden, D.A. Agard, J.W. Sedat, and M. Leptin. 1991. *Drosophila* gastrulation: Analysis of cell behavior in living embryos by three-dimensional fluorescence microscopy. *Development* 112:365-370.

Leptin, M. and B. Grunewald. 1990. Cell shape changes during gastrulation in *Drosophila. Development* 110:73-84.

Nüsslein-Volhard, C., E. Wieschaus, and H. Kluding. 1984. Mutations affecting the pattern of the larval cuticle in *Drosophila melanogaster.* I. Zygotic loci on the second chromosome. *Wilhelm Roux's Arch. Dev. Biol.* 193:267-282.

Odell, G.M., G. Oster, P. Alberch, and B. Burnside. 1981. The mechanical basis of morphogenesis. Epithelial folding and invagination. *Dev. Biol.* 85:446-462.

Rickoll, W.L. 1976. Cytoplasmic continuity between embryonic cells and the primitive yolk sac during early gastrulation in *Drosophila melanogaster. Dev. Biol.* 49:304-310.

Roth, S., D. Stein, and C. Nüsslein-Volhard. 1989. A gradient of nuclear localization of the dorsal protein determines dorsoventral pattern in the *Drosophila* embryo. *Cell* 59:1189-1202.

Rushlow, C., M. Frasch, H. Doyle, and M. Levine. 1987. Maternal regulation of *zerknullt*: A homoeobox gene controlling differentiation of dorsal tissues in *Drosophila. Nature* 330:583-586.

Rushlow, C.A., K. Han, J.L. Manley, and M. Levine. 1989. The graded distribution of the dorsal morphogen is initiated by selective nuclear transport in *Drosophila. Cell* 59:1165-1177.

Simpson, P. 1983. Maternal-zygotic gene interactions during formation of the dorsoventral pattern in *Drosophila* embryos. *Genetics* 105:615-632.

Steward, R. 1987. *Dorsal,* an embryonic polarity gene in *Drosophila,* is homologous to the vertebrate proto-oncogene, c-*rel. Science* 238:692-694.

Steward, R. 1989. Relocalization of the dorsal protein from the cytoplasm to the nucleus correlates with its function. *Cell* 59:1179-1188.

Thisse, B., C. Stoetzel, T.C. Gorostiza, and F. Perrin-Schmitt. 1988. Sequence of the twist gene and nuclear localization of its protein in endomesodermal cells of early *Drosophila* embryos. *EMBO J.* 7:2175-2183.

Turner, F.R. and A.P. Mahowald. 1977. Scanning electron microscopy of *Drosophila* embryogenesis. 2. Gastrulation and segmentation. *Dev. Biol.* 49:403-416.

Young, P.E., T.C. Pesacreta, and D.P. Kiehart. 1991. Dynamic changes in the distribution of cytoplasmic myosin during *Drosophila* embryogenesis. *Development* 111:1-14.

CONVERGENCE AND EXTENSION DURING GERMBAND ELONGATION IN *DROSOPHILA* EMBRYOS

Eric Wieschaus, Dari Sweeton, and Michael Costa

Molecular Biology Department
Princeton University
Princeton, New Jersey 08544

ABSTRACT

After the initial infoldings of gastrulation, the ventral region of the *Drosophila* embryo undergoes a rapid elongation called germband extension. This elongation is produced by intercalation of the more lateral cells as they move toward the ventral midline. In many respects, the process is very similar to the convergent extension which occurs during amphibian gastrulation and to elongation of the archenteron in sea urchins.

Several years ago, Gergen, Coulter and Wieschaus (1986) proposed that the intercalary behavior of cells during germband elongation reflects the adhesive preferences established in individual cells by the anterior-posterior patterning which occurs at the blastoderm stage. This model is formally very similar to the clock model used by French, Bryant, and Bryant (1976) to explain intercalary regeneration in imaginal discs and vertebrate limbs. In that model, positional values within a field are infinitely graded, and cells tolerate only finite differences between themselves and their immediate neighbors. When the discrepancies between adjacent cells are too great, the cells are induced to divide or otherwise fill in the gap. Surgical manipulations and wound healing induce cell proliferation and intercalary regeneration because they juxtapose cells with radically different positional identities.

In our model for germband extension, similarly abrupt juxtapositions of positional values would arise when the graded segmental field is condensed onto the limited number of precursors cells present in each segment at the gastrula stage. In contrast to the clock model, however, inappropriate juxtapositions in the early embryo are not resolved by induced cell proliferation. Instead, cells from the more dorsal regions which by chance have the appropriate intervening positional identities intercalate.

The article below presents a more detailed description of the model for germband extension and describes several tests of the model based on the predicted behavior of cells in embryos with aberrant anterior-posterior and dorsal-ventral patterning.

INTRODUCTION

Germband extension begins in the *Drosophila* gastrula after the mesodermal precursors have moved into the interior of the embryo through the ventral furrow, and the posterior midgut invagination has begun at the posterior pole. At that point the ventral epidermal and mesodermal anlage begin to elongate in the anterior-posterior direction. Within one and a half hours, this "germband" has more than doubled its length and undergone a corresponding reduction in width (Figure 1). The mechanisms underlying the elongation have been investigated by a number of authors (Poulson, 1950; Turner and Mahowald 1977; Rickoll and Counce 1980; Hartenstein and Campos-Ortega 1985). The initial stages of elongation occur at a time when there are no mitoses in the embryo (Foe 1989). Moreover, the elongation is not associated with a corresponding change in the anterior-posterior dimensions of the cells involved (Hartenstein and Campos-Ortega 1985). Instead, the underlying motor for germband elongation appears to be cell movements, specifically intercalation of the more lateral cells as they move toward the ventral midline.

Certain aspects of germband extension in *Drosophila* are very similar to the process of convergent extension which occurs during amphibian gastrulation (Keller *et al.* 1989; Keller and Tibbetts 1989; Wilson *et al.* 1989) and to elongation of the archenteron in sea urchins (Ettensohn 1985; Hardin and Cheng 1986; Hardin 1988). It is possible to explain the behavior of the cells in all these processes by postulating a system of adhesive preferences. In *Drosophila*, for example, cells might exchange neighbors and move in between each other not because they are drawn to the ventral midline, but because by moving in particular directions; they exchange less favorable adhesive contacts for those which are more favorable. The underlying mechanisms which govern those adhesive preference and arrange them in particular patterns in the developing field, however, have proven difficult to investigate.

Several years ago, Gergen, Coulter, and Wieschaus (1986) proposed a model for germband extension in which intercalation was governed by selective adhesion of cells in the early *Drosophila* gastrula. In that model, those adhesive affinities were a direct expression of the anterior-posterior patterning that occurred preceding the blastoderm stage.

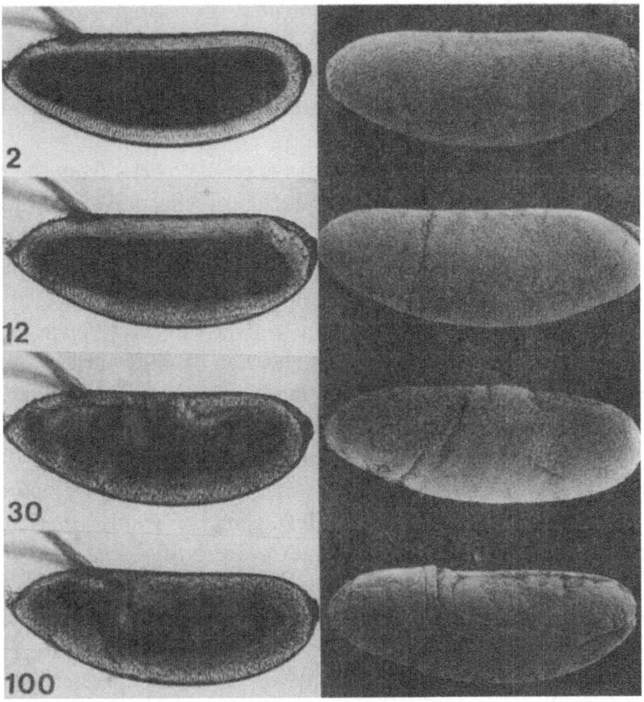

Figure 1. Germband Extension in *Drosophila* Embryos. Living embryos (left) and scanning electronmicrographs (right) of *Drosophila* embryos during the process of germband extension. The time designations in the lower left hand corner of the wild-type embryos indicate the minutes elapsed since the onset of gastrulation.

A General Model for Positional Information

The specific model for germband extension we wish to discuss grew from our studies on patterning along the anterior-posterior axis within each segment. The final pattern of bristles and sense organs in the insect epidermis results from a system of positional information operating in the precursor cells immediately prior to differentiation. Although adjacent cells often form discretely different structures (*i.e.*, cuticle *vs.* bristle *vs.* sensila), surgical manipulation and regeneration studies suggest that the information system which defines these fates is graded. In many models, these differences establish a gradient of positional information which varies along the anterior-posterior axis of the segment (French, Bryant and Bryant 1976). This gradient can be described by assigning numbers (or positional values) to individual cells, such that a given cell is very similar to its neighbors and increasingly different form cells at positions farther and farther away. At differentiation, each embryonic segment is about 12 cells wide. If all the cells are different in their programming, the segment would require a minimum of twelve positional values [If values are graded, then many more values are possible, twelve only indicating the slope of the gradient (*i.e.*, the average difference between any cell and its neighbors)]. In the simplest version of this model, "positional values" might reflect concentrations of a single morphogen. In other versions, positional identities might represent the sum of multiple distributions of different molecules. The essential feature of the model however is that the subtle quantitative differences between adjacent cells are important in determining their final fate.

A second essential feature of the model is derived directly from the clock model of French, Bryant and Bryant (1976) and postulates that cells tolerate only subtle differences between their neighbors (see also Mittenthal, 1981). When the discrepancies between adjacent cells are too great, the cells are induced to divide or otherwise fill in the gap. The final stage of the pattern is stable because the slope of the distribution of positional values is flat enough that no cell is subject to such "uncomfortable" juxtaposition. Such uncomfortable juxtaposition might arise during surgical manipulation or wound healing, and the model was originally proposed to explain intercalary regeneration in imaginal discs, amphibian limbs, and healing in insect epithelia.

One attractive feature of the clock model as it was originally proposed is that it can also be used to explain how in normal development, a small, variable number of adjacent precursor cells can give rise to a much larger final structure of constant size and pattern. It does this by postulating that the same gradient system might operate at earlier stages in development. At these stages however, the reduced number of cells would cause the average difference between adjacent cells to be greater, (assuming the entire range of values 0 to 12 are distributed over the smaller primordia). Cells respond to these differences by growth. As in regeneration following surgical manipulation, this growth smooths out the juxtaposition by supplying cells with intervening positional values. Cell proliferation continues in the normal primordia until each cell is juxtaposed to neighbors who are sufficiently similar to it that the juxtapositions induce no further proliferation. Schematically, the cell proliferation process can be described as flattening an initially steep positional gradient, until a slope is reached that can be tolerated by the individual cells. Because the final distribution is the only stable one (and is fixed by cell tolerance for juxtapositions), precursor cells within a segmental primordium will be driven to the same final pattern from a variety of starting configurations. In particular, the initial number of precursors cells can vary substantially with no effect on the final pattern. Even the precise positional identities assigned to these cells can be varied over a broad range with no effect on the final pattern. It is for this reason that the model allows for a very robust size regulation.

Anterior-posterior Pattern in the Embryonic Segment and its Role in Germband Extension

At the blastoderm stage, segments do not exist as morphological units. The segmental pattern however has already been established by overlapping stripes of segmentation gene expression (for review, see Akam 1987). Predominant among such genes are the seven pair rule genes. Each of the genes in this class which has been analyzed so far is expressed in a distinct pattern of stripes and interstripes. The variable width of the stripes, their graded appearance, and the number of different pair-rule genes means that there are sufficient overlapping molecular patterns within each double segment interval to define each cell with a unique identity.

The phenotypic deletion pattern associated with the different pair-rule genes and their dosage dependence suggests that their products are not only required to establish segments, but that the levels of their products play a direct role in establishing cell identities within the segment (Gergen and Wieschaus 1986). The relationship between these pair-rule values and the final differentiation pattern is complicated by the small size of the blastoderm segmental primordia. At the blastoderm stage, each segmental primordium is only three to four cells wide. Within each primordium, there are not enough cells any given anterior posterior row of cells to accommodate all twelve positional values present at final differentiation. To accommodate the entire field at the blastoderm stage, Gergen and Wieschaus (1985) and Gergen *et al.* (1986) proposed that the same range of values are imposed on the smaller number of cells (Figure 2). In any given anterior-posterior row, the first cell might have any value between 1 and 4, the actual value assumed depending on the levels of different pair-rule genes. The second cell might have values between 3 and 6, the third between 5 and 8, etc. In a slightly more dorsal or ventral row, the first cell would still be assigned a value between 1 and 4, but due to stochastic variability in the levels of the different pair-rule genes, it would not necessarily have the same value assigned the cell in the preceding row. The smaller segment primordium approximates the total range of values present at later stages. Because the same range of positional values is spread over a smaller number of cells, a greater discrepancy exists between each cell and its immediate neighbors.

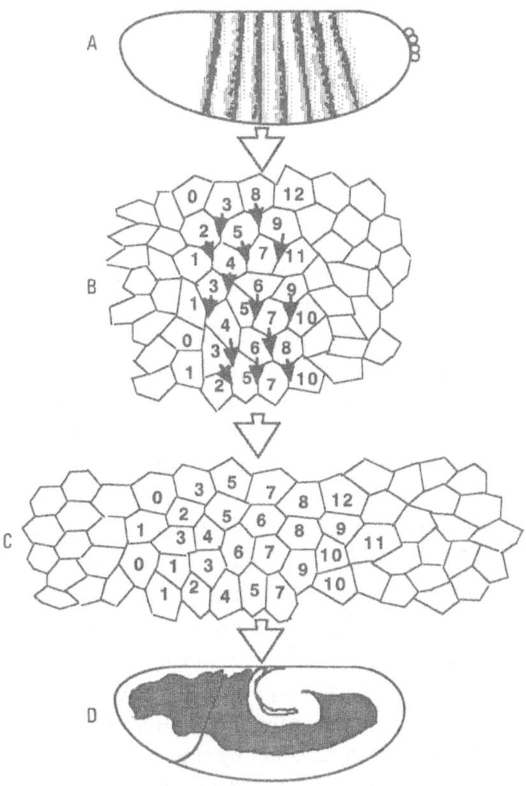

Figure 2. Model for the role of positional identities in controlling cell intercalation during germband extension. At the blastoderm stage, segmental patterning is represented by the overlapping pattern of various pair-rule gene products (**A**). Each segment is about three to four cells wide. When the range of positional identities used to define the pattern in the later segments (see text) is imposed onto the blastoderm segmental primordia (**B**), the reduced number of blastoderm precursors means that many cells are juxtaposed to neighbors very different from themselves. These juxtapositions can be relieved by cell intercalation, since virtually all cells can find more appropriate neighbors by moving ventrally (arrows). This intercalation causes the primordium to elongate in the anterior-posterior direction (**C**), and the germband to extend onto the dorsal side of the embryo (**D**).

In contrast to the random, logarithmic proliferation in imaginal discs, cell division in the *Drosophila* embryo is tightly controlled (Foe 1989). The uncomfortable juxtapositions which arise by superimposing the entire segmental field can therefore not be relieved by intercalary growth or proliferation. Instead we propose that the cells resolve the juxtapositions by slight shifts in their positions. Examination of Figure 2 shows that most cells would find more appropriate adhesive contacts by intercalating between pairs of cells that lie immediately dorsal or ventral to them. This arises because the dorsally and ventrally situated cells have slightly different positional values, due to stochastic error and variability in the assignment of positional values. Such intercalation will only occur in the dorsal-ventral direction, however. In the anterior-posterior direction, the cells are different enough that no intercalation could lead to more comfortable juxtapositions. This arises because the positional values, although variable and subject to error, still reflect the anterior-posterior patterning of the segment primordium.

Tests of the Model: The Effects of Various Patterning Mutations on Germband Elongation

The feature of the model that distinguishes it from most other models for germband extension is that, although the predominant intercalation of cells is in the dorsal-ventral direction, the driving force is anterior-posterior patterning. This means that the model can be tested by observing the effects of mutations that affect anterior-posterior and dorsal-ventral polarity on intercalation and germband extension. One of the advantages of *Drosophila* is that such mutations are available and, in the past ten years, much work has been done on their genetic and molecular nature. This work has shown that by the time the egg is laid, anterior-posterior patterning is controlled by a different set of genes than that controlling dorsal-ventral patterning.

Maternal Effect Mutations Affecting Anterior-posterior Polarity. Mutations in this class cause large deletions in the segmental pattern. Embryos from mutant females have the normal number of blastoderm cells, but lack determinants necessary to specify certain regions or structures. The resultant embryos differentiate a cuticle pattern lacking particular regions. Using molecular markers, this pattern deletion can be traced back to the blastoderm stage, where the epidermal primordium is subdivided into a reduced number of segments of larger size. In the context of the present model, the larger size of the affected segments would demand that the same range of positional values are spread over a larger number of cells (Figure 3). The discrepancy between adjacent cells will be less and the drive for intercalation will be reduced. The germbands of mutant embryos should not elongate to the same extent as in wild-type embryos and, under certain models, the actual rate of elongation should be reduced.

These expectations have been largely born out by an examination of a large number of mutations. Reduced germband lengths are observed in all maternal anterior-posterior mutants examined (Nüsslein-Volhard *et al.* 1987; Schüpbach and Wieschaus 1986). This includes mutations affecting the anterior system, (*bicoid, Bicaudal-D, exuperantia*), the posterior system (*vasa, oscar* and *nanos*) and the terminal system (*torso*). In one extreme case, the triple mutant *bicoid nanos torso-like*, all germband extension

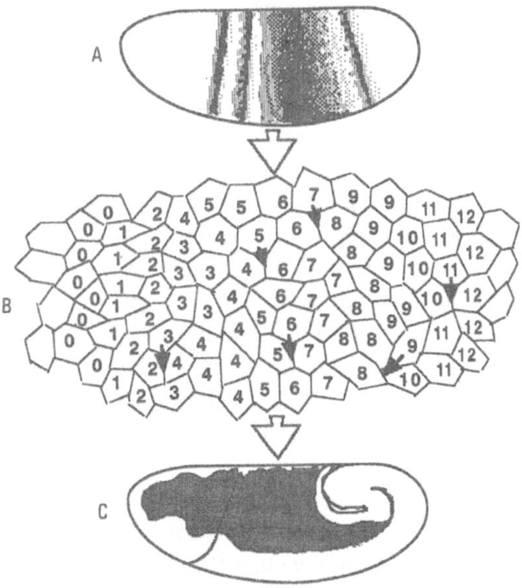

Figure 3. Maternal effect mutations which reduce anterior-posterior pattern show corresponding reductions in germband elongation. Embryos from *vasa* homozygous mothers have a reduced number of abdominal segments. At the blastoderm stage (**A**), cells of the abdominal primordia are allocated to a small number of greatly enlarged segments. If the same field of positional identities is stretched over the enlarged segmental primordium (**B**), the difference between each blastoderm cell and its neighbors is reduced compared to the wild-type blastoderm (see Figure 2). The juxtapositions between cells are less abrupt and fewer cells (arrows) will find more appropriate juxtapositions by ventral intercalation. The reduced intercalation means that the germband does not elongate as much (**C**).

is eliminated. The ventral furrow forms normally along the entire anterior-posterior length, but no extension occurs (unpublished observations). Because none of these mutants have significant effects on dorsal-ventral patterning, their effects on germband extension argue for a major role of anterior-posterior patterning on the process.

Maternal Effect Mutants Affecting Dorsal-ventral Polarity. In the normal embryo, the germband forms on the ventral side of the embryo, and there is an apparent bias in the direction of the intercalation towards the ventral midline. In the model we have proposed, dorsal-ventral polarity is not essential for the elongation process. In its absence, uncomfortable anterior-posterior juxtapositions would still drive cells to intercalate. Without dorsal-ventral polarity, however, the ventral bias would be eliminated and intercalation would occur around the entire circumference of the embryo by cells moving both dorsally and ventrally. In theory, the embryo itself will elongate as intercalation causes the tube to get longer and narrower. In many respects, the process might resemble the elongation of the archenteron during sea urchin development, which also occurs by cell rearrangement (Ettensohn 1985; Hardin and Cheng 1986; Hardin 1988).

The gastrulation observed in dorsalized embryos (Nüsslein-Volhard 1979) is consistent with the above predictions, but it is clearly more complex, probably because elongation is constrained by the surrounding egg shell. The embryos form a normal cellular blastoderm, and at about the time when a normal embryo would begin germband elongation, the cells around the entire circumference are thrown into transverse folds in apparent response to elongation in the anterior posterior direction. As the tube continues to elongate inside the eggshell, it becomes more and more convoluted and folded back on itself. At final differentiation, the embryos consist of a long yolk-filled tube of clearly segmented epidermis. The movements that occur during gastrulation of dosalized embryos can be interpreted as an attempt at intercalation and germband elongation, constrained because the embryo develops in an egg shell and is filled with yolk. On the other hand, fully dorsalized embryos may not provide the best test of the model, since the dorsal epidermis made in such embryos may not normally participate in germband extension. A better test might be provided by recently described mutant combinations which impose lateral cell fates around the entire circumference (Anderson *et al.* 1985).

Zygotic Segmentation Genes. The maternal effect mutations which affect anterior-posterior polarity probably do not control germband elongation directly. The products of these genes are mostly transcription factors (or their regulators) and are thought to control the pattern of zygotic gene expression in the embryo. One obvious set of downstream genes are the zygotically active segmentation genes. It is now known that segmentation is established by a sequence of activities which operate in a hierarchy and establish an increasingly refined segmental pattern (Nüsslein-Volhard and Wieschaus 1980, O'Farrell and Scott 1986; Akam 1987). It is of interest to determine whether and to what extent this set of genes also controls intercalation at gastrulation.

The first genes to operate in this zygotic hierarchy are the "gap" genes which define large contiguous blocks in the segmental pattern. All the gap genes studied so far affect germband elongation, although the effect of single mutations are often subtle (*i.e.*, less than 15% reduction in total length). Embryos which are doubly homozygous for two or more gap mutations show more dramatic effects (Wieschaus unpublished). These gap genes are thought to regulate the expression of pair-rule genes (Carroll and Scott 1986; Frasch and Levine 1987), and mutations in certain pair-rule genes (*e.g.*, even-skipped, Nüsslein-Volhard *et al.* 1986) also appear to reduce germband extension. It is not known, however whether all pair-rule genes behave in this way. The final step in the segmental hierarchy, the segment polarity genes, are activated in late blastoderm and early gastrula embryos (Dinardo and O'Farrell 1987; Ingham *et al.* 1988). Embryos which are mutant for any single one of these loci do not show large scale defects in germ band extension which correspond to their effects on segmentation (Wieschaus *et al.* unpublished). It is possible that subtle defects exist which would only be detected with careful quantitation, or in double or triple mutant combination which might show additivity. Alternatively, segment polarity genes may only be required for the maintenance and elaboration of segmental pattern at later stages, and not determine cellular properties in the early gastrula. This view would place the pair-rule genes at the final patterning step in controlling germband extension.

Cellular Properties Controlling Intercalary Behavior

All the segmentation genes known to affect germband extension encode transcription factors. They presumably affect extension by regulating genes directly responsible for the intercalary behavior. In the model we have proposed, intercalation depends on the adhesive preferences of cells, and thus potentially on the cell surface molecules such as cadherans (Takeichi 1988) and cell adhesion molecules (CAMs) (Edelman 1986) which are known to affect adhesion in other organisms. A relatively small number of different adhesive molecules would be sufficient to account for the behavior described here, particularly if they are distributed in graded, overlapping stripes of expression corresponding to the stripes of pair-rule genes which might control their expression. Although little is known about the molecular basis of cell adhesion in insects, it should be possible to identify the *Drosophila* homologs of vertebrate adhesion protein genes. If the model we have proposed is correct, mutations in some of these genes might reduce the extent of germband elongation. The effect would depend on zygotic activity of these genes in the embryo itself, and the expression should be controlled by segmentation genes in the pair-rule class. Since the adhesive preferences during germband elongation may be only one expression of segmental patterning, the reduced intercalation may not grossly affect the final differentiation of segmental pattern observed in the cuticle. Thus, in contrast to the segmentation genes described above, the genes involved in extension would not necessarily have already been identified based on their segmentation phenotype.

In its simplest version, our model suggests that segmentation genes control the adhesive difference which governs *Drosophila* germband elongation, and that adhesive differences might be present in a segmentally repeated pair rule pattern. A more general version of the model, however, does not require adhesive patterning as precise as that established during segmentation of the *Drosophila* blastoderm. Most of the intercalary behaviors and sorting out will also occur in a non-repeated adhesive gradient, as long as the cells are free to move and exchange neighbors to establish adhesive contact which are more favorable. In fact, in the classical Steinberg differential adhesion model (see Steinberg and Poole 1982), all that would be required for sorting out, is quantitative differences in adhesion. Under certain constraints (*i.e.* if the field is prevented from rounding up into concentric rings, Wieschaus *et al.* in preparation), any field possessing an imprecise adhesive gradient will elongate along the axis of that gradient, as cells that are more adhesive sort out from those which are less adhesive.

ACKNOWLEDGEMENTS

We thank numerous members of the *Drosophila* labs at Princeton for advice, encouragement, discussions, and critical readings of the manuscript. These experiments were supported by National Institutes of Health grant 5RO1HD22780 to E.W.

REFERENCES

Akam, M. 1987. The molecular basis for metameric pattern in the *Drosophila* embryo. *Development* 101:1-22.

Anderson, K.V., G. Jurgens, and C. Nüsslein-Volhard. 1985. Establishment of dorsal-ventral polarity in the *Drosophila* embryo: Genetic studies on the role of the *Toll* gene product. *Cell* 42:779-789.

Carroll, S.B. and M.P. Scott. 1986. Zygotically active genes that affect the spatial expression of the *Fushi-tarazu* segmentation gene during early *Drosophila* embryogenesis. *Cell* 45:113-126.

Dinardo, S. and P.H. O'Farrell. 1987. Establishment and refinement of segmental pattern in the *Drosophila* embryo: spatial control of *engrailed* expression by pair-rule genes. *Genes Dev.* 1:1212-1225.

Edelman, G.M. 1986. Cell adhesion molecules in the regulation of animal form and tissue pattern. *Annu. Rev. Cell Biol.* 2:81-116.

Ettensohn, C.A. 1985. Gastrulation in the sea urchin embryo is accompanied by the rearrangement of invagination epithelial cells. *Dev. Biol.* 112:385-390.

Frasch, M. and M. Levine. 1987. Complementary patterns of *even-skipped* and *fushi tarazu* expression involve their differential regulation by a common set of segmentation genes in *Drosophila*. *Genes Dev.* 1:981-995.

French, V., P.J. Bryant, and S.V. Bryant. 1976. Pattern regeneration in epimorphic fields. *Science* 193:969-981.

Foe, V. 1989. Mitotic domains reveal early commitment of cells in *Drosophila* embryos. *Development* 107:1-22.

Gergen, J.P., D. Coulter, and E. Wieschaus. 1986. Segmental pattern and blastoderm cell identities. *Symp. Soc. Dev. Biol.* 43:195-220.

Gergen, J.P. and E. Wieschaus. 1985. The localized requirements for a gene affecting segmentation in *Drosophila*: Analysis of larvae mosaic for *runt*. *Dev. Biol.* 109:321-335.

Gergen, J.P. and E. Wieschaus. 1986. Dosage requirements for *runt* in the segmentation of *Drosophila* embryos. *Cell* 45:289-299.

Hardin, J. 1988. The role of secondary mesenchyme cells during sea urchin gastrulation studied by laser ablation. *Development* 103:317-324.

Hardin, J. and L.Y. Cheng. 1986. The mechanisms and mechanics of archenteron elongation during sea urchin gastrulation. *Dev. Biol.* 115:490-501.

Hartenstein, V. and J.A. Campos-Ortega. 1985. Fate mapping in wild type *Drosophila melanogaster*. I. The spatio-temporal pattern of embryonic cell divisions. *Wilhelm Roux's Arch Dev. Biol.* 194:181-195.

Ingham, P.W, N.E. Baker, and A. Martinez-Arias. 1988. Regulation of segment polarity genes in the *Drosophila* blastoderm by *fushi-tarazu* and *even-skipped*. *Nature* 331:73-75.

Keller, R., M.S. Cooper., M. Danilchik, P. Tibbetts, and P. Wilson. 1989. Cell intercalation during notochord development in *Xenopus laevis*. *J. Exp. Zool.* 251: 134-154.

Keller, R. and P. Tibbetts. 1989. Mediolateral cell intercalation in the dorsal axial mesoderm of *Xenopus laevis*. *Dev. Biol.* 131:539-549.

Mittenthal, J.E. 1981. The rule of normal neighbors: A hypothesis for morphogenetic pattern regulation. *Dev. Biol.* 88:15-26.

Nüsslein-Volhard, C. 1979. Maternal effect mutations that alter the spatial coordinates of the embryo of *Drosophila melanogaster*. *Symp. Soc. Dev. Biol.* 37:185-211.

Nüsslein-Volhard, C., H.G. Frohnhofer, and R. Lehmann. 1987. Determination of anteroposterior polarity in *Drosophila*. *Science* 238:1675-1681.

Nüsslein-Volhard, C., H. Kluding, and G. Jürgens. 1985. Genes affecting the segmental subdivision of the *Drosophila* embryo. *Cold Spring Harbor Symp. Quant. Biol.* 50:145-154.

Nüsslein-Volhard, C. and E. Wieschaus. 1980. Mutations affecting segment number and polarity in *Drosophila*. *Nature* 287:795-801.

O'Farrell, P. and H.M. Scott. 1986. Spatial programming of gene expression in early *Drosophila* embryogenesis. *Annu. Rev. Cell Biol.* 2:49-80.

Poulson, D.F. 1950. Histogenesis, organogenesis, and differentiation in the embryo of *Drosophila melanogaster* Meigen. p. 168-274. *In: Biology of Drosophila*. M. Demerec (Ed.). John Wiley, New York.

Rickoll, W.L. and S.J. Counce. 1980. Morphogenesis in the embryo of *Drosophila melanogaster*–Germ band extension. *Wilhelm Roux's Arch. Dev. Biol.* 188:163-177.

Schüpbach, G.M. and E. Wieschaus. 1986. Maternal-effect mutations altering the anterior-posterior pattern of the *Drosophila* embryo. *Wilhelm Roux's Arch. Dev. Biol.* 195:302-317.

Steinberg, M.S. and T.J. Poole. 1982. Liquid behavior of embryonic tissues. p. 583-607. *In: Cell Behavior*. R. Bellairs, A. Curtis, and G. Dunn (Eds). Cambridge University Press, Cambridge.

Takeichi, M. 1988. The cadherins: Cell-cell adhesion molecules controlling animal morphogenesis. *Development* 102:639-665.

Turner, F.R. and A.P. Mahowald. 1977. Scanning electron microscopy of *Drosophila melanogaster* embryogenesis II. Gastrulation and segmentation. *Dev. Biol.* 57:403-416.

Wilson, P.A., G. Oster, and R. Keller. 1989. Cell rearrangement and segmentation in *Xenopus*: direct observation of cultured explants. *Development* 105:155-166.

ULTRAVIOLET-SENSITIVE DETERMINANTS OF GASTRULATION AND AXIS DEVELOPMENT IN THE ASCIDIAN EMBRYO

William R. Jeffery

Bodega Marine Laboratory
P.O. Box 247
Bodega Bay, CA 94923* and
Department of Zoology
University of California
Davis, CA 95616

* address for correspondence

ABSTRACT

The axes of the ascidian embryo are determined after fertilization during a series of cytoplasmic movements known as ooplasmic segregation. UV irradiation of the vegetal hemisphere of the fertilized *Styela* egg during ooplasmic segregation prevents gastrulation, axis formation, and brain sensory cell differentiation without affecting subsequent cytoplasmic movements, the early cleavage pattern, or differentiation of muscle, epidermal, and endodermal cells. Proteins synthesized in UV-irradiated (axis-deficient) and normal embryos were compared by 2D-gel electrophoresis and autoradiography. Of 433 polypeptides detected in normal embryos, only about 5% were

Gastrulation, Edited by R. Keller *et al.*
Plenum Press, New York, 1991

missing or decreased in labelling intensity in UV-irradiated embryos. The most prominent of these is a 30,000 molecular weight (pI 6.0) polypeptide (p30). Several lines of evidence suggest that p30 may play a role in embryonic axis formation. First, p30 labelling peaks during gastrulation, when the embryonic axis is being established. Second, p30 labelling and axis formation are abolished by the same threshold UV dose. Third, the UV sensitivity period for abolishing p30 labelling and axis formation are both restricted to ooplasmic segregation. Because the chromophore(s) involved in axis determination exhibits absorption characteristics similar to nucleic acids, rather than proteins, and p30 is not detectibly synthesized until after the conclusion of the UV sensitivity period, p30 cannot be the UV sensitive target localized in the vegetal pole region. *In vitro* translation of egg RNA and subsequent analysis of protein products showed that p30 is encoded by a maternal mRNA. To determine whether UV irradiation inactivates p30 mRNA, RNA from UV-irradiated and normal eggs was translated in a cell-free system and the labelled proteins were compared by gel electrophoresis. The results showed that translation of p30 mRNA, as well as a number of other mRNAs, was abolished in UV-irradiated embryos. These results show that gastrulation and axis formation are controlled by UV-sensitive components transiently localized in the vegetal hemisphere during ooplasmic segregation and that p30 mRNA is a UV target. UV irradiation of the vegetal pole may affect development by inactivating localized mRNAs encoding proteins that function in gastrulation and axis formation.

INTRODUCTION

Embryonic development involves two distinct processes: organization of the embryonic body plan (or axis) and differentiation of specific cell types with respect to the coordinates established during axial development. In most organisms, these processes are sequential. Axis determination begins either during oogenesis, after fertilization, or during cleavage (see Anderson 1987; Gerhart and Keller 1986; van den Biggelaar and Guerrier 1983 for examples), and the body plan is subsequently established by coordinated cell movements during gastrulation. Later, cell fates are specified in their proper spatial positions. In some organisms, the embryonic axis and cell fates are determined simultaneously at early stages of development. Axis and cell fate determination occur within the period encompassing fertilization and the early cleavages in ascidians (Bates and Jeffery 1987b), which may be a consequence of the rapid and determinative pattern of development exhibited by these organisms.

Ascidians are chordates, yet larval development shows distinct invertebrate features: small eggs, determinate cleavage, invariable cell lineages, and low numbers of different cell and tissue types. The most obvious chordate feature of ascidians is the organization of the tadpole larva, which exhibits a notochord and dorsal nervous system (see Katz 1983 for review). The ascidian tadpole consists of a head, which contains a brain, endoderm, and mesenchyme cells, and a tail, which contains the central notochord surrounded by a dorsal spinal cord, lateral rows of muscle cells, and a ventral endodermal strand. The head and tail tissues are covered by an epidermal layer, which secretes the larval test. Swimming movements, mediated by contraction of the muscular tail, lead to dispersal of the tadpole, which eventually selects an attachment site and undergoes metamorphosis. During metamorphosis, the tail is resorbed and undifferentiated cells in the head are converted into adult tissues.

The ascidian egg is composed of at least three different cytoplasmic regions which are segregated into specific cell lineages during cleavage. In *Styela* eggs, these regions contain pigment granules and can be distinguished by color (Conklin 1905b). The colored cytoplasmic regions are distributed in a distinct spatial pattern before fertilization. The egg cortex contains a yellow cytoplasmic region known as the myoplasm, which is segregated to the larval tail muscle cells during embryogenesis. The animal hemisphere contains a clear cytoplasmic region called the ectoplasm, which is derived from the released germinal-vesicle plasm during oocyte maturation. The remainder of the egg is composed of a yolk-filled cytoplasmic region called the endoplasm. Fertilization triggers ooplasmic segregation, an extensive rearrangement of these egg cytoplasmic regions. Because ooplasmic segregation in ascidian eggs has been reviewed recently (Jeffery and Bates 1989), further description will focus on the myoplasm.

The myoplasm shows two phases of translocation after fertilization. During the first phase of ooplasmic segregation (OS 1), yellow cytoplasm moves along the egg periphery into the vegetal hemisphere, where it forms a transient cap near the vegetal pole. OS 1 is mediated by contraction of an actin filament network lying immediately under the egg plasma membrane (Sawada and Osanai 1981; Jeffery and Meier 1983, 1984). During the second phase of ooplasmic segregation (OS 2), myoplasm moves from the vegetal pole region to the sub-equatorial zone of the uncleaved zygote where it spreads out into a yellow crescent (Conklin 1905b). OS 2 appears to be mediated by the formation of a sperm aster in the vegetal hemisphere (Sardet *et al.* 1989; Sawada and Schatten 1988, 1989). The first cleavage furrow bisects the yellow crescent, distributing myoplasm to each blastomere. During the second and third cleavages, however, myoplasm is distributed to only two blastomeres, and finally into the primary muscle cell lineage: 4 cells of the 16-cell embryo, 6 cells of the 32-cell embryo, 8 cells of the 64-cell embryo, and so forth (Nishida 1987).

Bilateral cleavage is equal and regular during the first three divisions; but subsequently becomes unequal and irregular in the animal and vegetal hemisphere, resulting in the formation of four large endodermal blastomeres at the vegetal pole. Gastrulation begins with a flattening of the vegetal pole region followed by elongation and invagination of these endodermal cells (Conklin 1905b; Satoh 1978). The remaining endodermal cells and presumptive notochord, mesenchyme and tail muscle cells are invaginated next; and ectoderm cells eventually spread from the animal hemisphere to cover the surface of the gastrula. The changes in cell shape and movements associated with gastrulation have not been analyzed with modern techniques; however, they are sensitive to cytochalasin B and, therefore, appear to be dependent on the function of microfilaments (Bates, personal communication).

While invagination and epiboly are occurring on the surface, the internalized vegetal cells continue to migrate within the interior of the embryo. Endoderm cells move anteriorly into the future head region, whereas presumptive notochord and muscle cells move posteriorly to form the tail rudiment. Tail formation is driven by rearrangement of the presumptive notochord cells (Conklin 1905b; Cloney 1964; Miyamoto and Crowther 1985), which interdigitate and extend into a rod (Nishida and Satoh 1985). As the embryo elongates anteroposteriorly, the neural plate, which has developed in the vegetal hemisphere, invaginates to form the neural tube. Subsequently, the embryonic axis continues to elongate during the tailbud stage and the lateral bands of tail muscle cells begin to differentiate. Although the embryonic axis is established during and after gastrulation, the future axial coordinates are first apparent during ooplasmic segregation. The vegetal cap of myoplasm becomes the initial site of invagination during gastrulation and the dorsal pole, the animal pole region becomes the ventral pole, the yellow crescent of myoplasm marks the posterior pole, and the region 180° opposite the yellow crescent becomes the anterior pole of the embryo (Conklin 1905b; Bates and Jeffery 1987a; Speksnijder *et al* 1990).

Ooplasmic determinants segregated to various blastomeres during cleavage are responsible for specification of cell fate in ascidian embryos (see Venuti and Jeffery 1989 for review). Although the identity and mode of action of these substances is unknown, evidence for their existence is derived from blastomere deletion (Conklin 1905a) or isolation (Whittaker *et al.* 1977; Deno *et al.* 1985), cleavage-arrest (Whittaker 1973), and cytoplasmic redistribution (Whittaker 1982) experiments. The results of these experiments indicate that morphological and biochemical markers for differentiated epidermal, endodermal, and muscle cells develop autonomously; and in the case of muscle cell markers can be expressed in extra cells containing cytoplasm introduced from the muscle cell lineage. Neural tissues, which are specified by inductive interactions (Rose 1939; Reverberi *et al.* 1960; Nishida and Satoh 1989), are an important exception to this generalization.

In contrast to cell fate specification, little is known about the role of ooplasmic determinants in embryonic axis development. One reason for this situation is that experiments used to establish the importance of determinants in specifying cell fate are not suitable for investigating the problem of axis determination. Isolated blastomeres, cleavage arrested embryos, or embryos manipulated to redistribute cytoplasm do not develop into larvae with a distinguishable body plan. Furthermore, grafting experiments, such as those used to demonstrate the presence of the organizer (Spemann and Mangold 1924) or to identify blastomeres responsible for inducing the embryonic axis (Gimlich and Gerhart 1984) in amphibian embryos, are not easily applied to ascidian embryos because of their small size and fragility. Bates and Jeffery (1987a) tested for localization of axis-forming potential by microsurgical removal of cytoplasmic regions of unfertilized and fertilized *Styela* eggs. Although no deleterious effects were observed in unfertilized eggs (after subsequent fertilization and development), removal of myoplasm from the vegetal pole region of fertilized eggs at the yellow cap stage resulted in specific inhibition of gastrulation and subsequent axis formation. After the completion of OS 2, however, deletion of the vegetal or posterior pole region did not affect gastrulation and axis formation. These results suggest that "axis determining potential" is transiently localized in the vegetal pole region during OS 1.

Further understanding of axis determination in ascidian embryos requires identification and characterization of the molecules responsible for the "axis determining potential." However, microsurgical deletion experiments are unsuitable for obtaining large quantities of axis-deficient embryos required for biochemical studies. This paper describes experiments in which ultraviolet (UV) irradiation is used to inhibit axis development in large numbers of *Styela* embryos (Jeffery 1990a,b). Using this approach, we have discovered a UV-sensitive maternal mRNA encoding a cytoskeletal protein that may play a role in embryonic axis formation.

EXPERIMENTAL

Effect of UV Irradiation on Embryonic Axis Formation

Experiments with different UV doses were conducted to determine the effect of UV light on embryonic axis formation. UV irradiation was carried out using a short-wave Mineralight Lamp (Model UVG-11; UVP, Inc., San Gabriel, CA.), which emits maximally at 254 nm. Unfertilized eggs or zygotes were irradiated at a concentration of about 1000 eggs/ml in a Syracuse dish positioned 5 cm below the UV source. The time of irradiation varied from 15 sec to 3 min. The UV dose was determined with a Blak-Ray Short Wave UV Meter (UVP, Inc.). To expose the entire cell surface to UV light (global irradiation), eggs or zygotes were irradiated while being agitated with a small magnetic stirring bar. Axis development was evaluated by microscopic examination of the extent of tail formation in living or fixed embryos. A tail formation index was developed empirically from these observations. Normal development was scored as Grade 3, embryos that developed with an incompletely-extended tail were scored as Grade 2, embryos that formed no tail but exhibited an elongated anteroposterior axis were scored as Grade 1, and embryos that remained spherical and showed no indication of tail extension were scored as Grade 0. The UV dose-response curve obtained for a clutch of zygotes is shown in Figure 1. At doses of 0.8×10^{-3} J mm^{-2} and below, most zygotes developed as Grade 3 embryos (Figure 2A). These embryos, which exhibited a fully-extended tail containing a single row of notochord cells, hatched into swimming larvae that underwent metamorphosis. Between about 0.8 and 2.3×10^{-3} J mm^{-2}, the proportion of normal embryos gradually decreased. The abnormal embryos were a mixture of Grade 2, 1, and 0 embryos (Figure 2B-D), which lacked a fully-extended tail and notochord, and did not hatch or undergo metamorphosis. Finally, at 2.3×10^{-3} J mm^{-2} or greater, most zygotes developed as Grade 0 embryos (Figure 2D), which lacked any visible indication of an embryonic axis. To determine the UV-sensitive period for axis development, zygotes were irradiated (3.0×10^{-3} J mm^{-2}) at various times between insemination and fourth cleavage. As shown in Figure 3, UV irradiation was effective in inhibiting axis formation only up to the completion of OS 2. Most zygotes irradiated with the same UV dose during the early cleavages developed normally. These results show that UV irradiation during ooplasmic segregation prevents subsequent axis development.

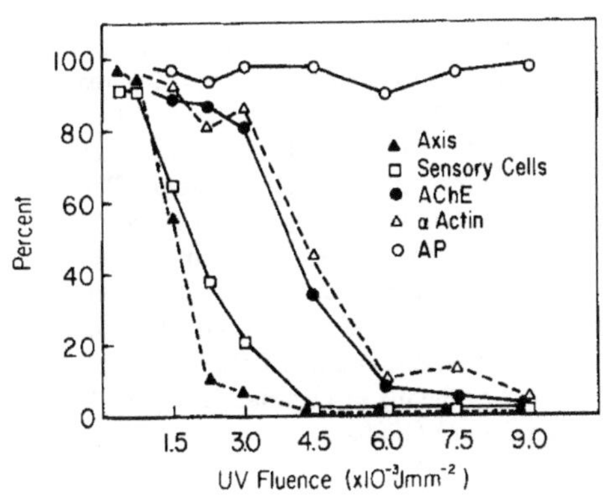

Figure 1. The effect of UV irradiation on embryonic axis formation and cell differentiation. A clutch of zygotes at the yellow cap stage was irradiated globally with different UV doses, cultured until controls reached the late tailbud stage, and assayed for axis development (see Figure 2; filled triangles), melanin pigment granules (open squares), muscle actin mRNA accumulation (open triangles), AChE activity (filled circles), and AP activity (open circles). Percent is the proportion of irradiated zygotes that formed an embryonic axis (Grade 3, 2 or 1 embryos), developed melanized brain sensory cells, or expressed muscle actin mRNA, AChE, or AP, relative to unirradiated controls. From Jeffery (1990a).

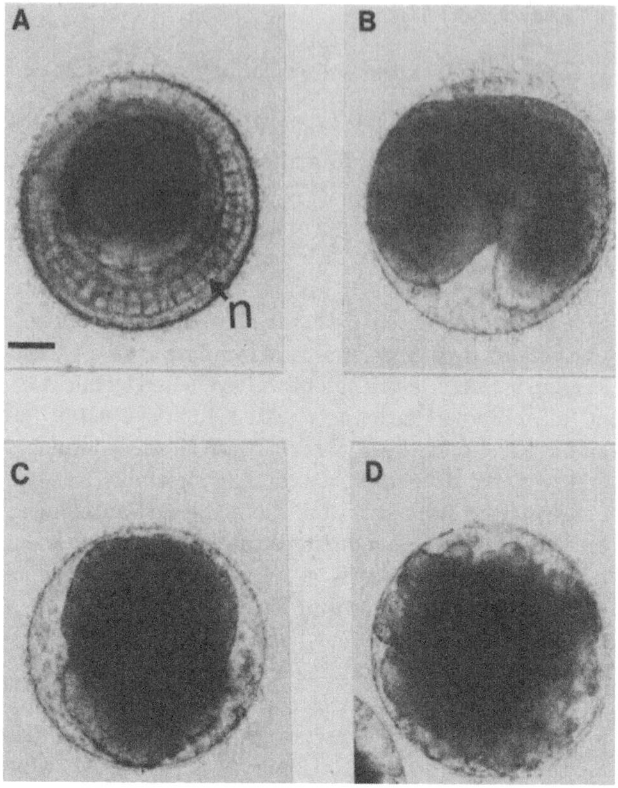

Figure 2. Embryonic axis development in UV-irradiated embryos. **A.** A normal (Grade 3) embryo that developed from a zygote irradiated with 0.8 × 10^{-3} J mm^{-2}. This frame shows a late tailbud embryo with a completed tail curving around the head of the tadpole larva. The single row of tail notochord cells (N) can be distinguished. **B.** A Grade 2 embryo that developed from a zygote irradiated with 2.3 × 10^{-3} J mm^{-2} showing an incompletely extended tail (T) lacking a differentiated notochord. The anterior region of the embryo faces the left of the frame. **C.** A Grade 1 embryo that developed from a zygote irradiated with 2.3 × 10^{-3} J mm^{-2} showing a slight anteroposterior elongation of the body axis, but no tail formation. The anterior region of the embryo faces the top of the frame. **D.** A Grade 0 embryo that developed from a zygote irradiated with 3.0 × 10^{-3} J mm^{-2}. This embryo is spherical and lacks a detectible embryonic axis. The embryos shown in **B-D** were photographed at the same time in development as the embryo shown in A. Scale bar: 20 μm; magnification is the same in each frame. From Jeffery (1990a).

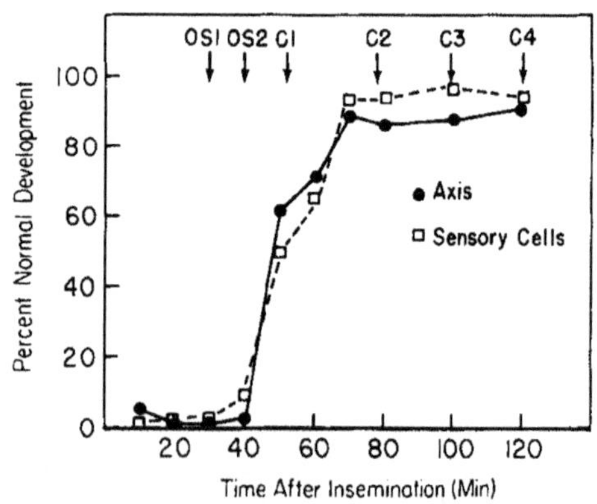

Figure 3. The UV sensitivity periods for axis and sensory cell development. Zygotes were irradiated globally with 3.0 × 10^{-3} J mm^{-2} at various times between insemination and fourth cleavage. The timing of various developmental stages is indicated by the arrows. OS 1: completion of the first phase of ooplasmic segregation. OS 2: completion of the second phase of ooplasmic segregation. C1-C4: cleavages 1-4. Other details are similar to Figure 1. From Jeffery (1990a).

Effect of UV Irradiation on Ooplasmic Segregation, Cleavage, and Gastrulation

Further observations were made to determine whether axis deficient embryos are defective in ooplasmic segregation, cleavage, or gastrulation. Embryos were irradiated with a threshold UV dose of 3.0×10^{-3} J mm^{-2}, which abolished axis formation in more than 95% of the embryos (see Figure 1). To examine the effect of UV light on ooplasmic segregation, unfertilized eggs or zygotes at the yellow cap stage were irradiated and the distribution of myoplasm was followed in sections stained with Milligan's trichrome (data not shown), which specifically stains this region of eggs and embryos (Jeffery 1989). UV-irradiated eggs could not be fertilized, but artificial activation and ooplasmic segregation were induced by ionophore A23187 treatment (Jeffery 1982). Irradiated eggs formed a yellow cap of myoplasm after activation, indicating that OS 1 is not affected by UV light. Likewise, zygotes irradiated at the yellow cap stage formed a yellow crescent, indicating a normal OS 2. The pattern of cleavage and gastrulation in normal and UV-irradiated zygotes was examined by light microscopy. As shown in Figure 4A-E, Grade 0 embryos cleaved normally; a typical bilateral cleavage pattern was evident in vegetal cells at the 16-cell stage (Figure 4D), and small B6.5 blastomeres formed in the posterior region of 32-cell embryos (Figure 4E). Cell counts indicated that Grade 0 larvae contained about 500 cells. Since there are about 2500 cells in most ascidian larvae (Berrill 1935), cell division must be terminated early in Grade 0 embryos. Cleavage also continued in embryos irradiated with higher UV doses (3.0-15.0×10^{-3} J mm^{-2}). Gastrulation, which is normally initiated after the 64-cell stage by extension and invagination of endodermal cells surrounding the vegetal pole (Conklin 1905b; Satoh 1978; Bates and Jeffery 1987b), is shown in Figure 4G-H. In UV-irradiated embryos, the vegetal pole blastomeres did not invaginate (Figure 4F), although there was a slight flattening of the vegetal hemisphere and ectoderm cells underwent epipoly. These results show that UV irradiation during ooplasmic segregation inhibits invagination of vegetal pole cells during gastrulation without affecting ooplasmic segregation or the early cleavage pattern.

Effect of UV Irradiation on Cell Differentiation

After gastrulation, ascidian embryos begin to express specific markers in differentiating larval cells. These markers include melanin pigment granules in brain sensory cells (Dilly 1962; 1964), an epidermal antigen in ectoderm cells (Nishikata *et al.* 1987a), muscle actin (Tomlinson *et al.* 1987b), myosin heavy chain (Nishikata *et al.* 1987b), and acetylcholinesterase (AChE) (Whittaker 1973) in muscle cells, and alkaline phosphatase (AP) in endoderm cells (Whittaker 1977). These markers were used in subsequent experiments to determine the effect of UV light on cell differentiation.

The effect of UV light on sensory cell differentiation was determined by counting the number of irradiated embryos that developed melanin pigment granules. The formation of melanin pigment granules exhibited a UV dose-response curve (Figure 1) and sensitivity period (Figure 3) similar to axis development, suggesting that these events are coordinated during embryogenesis. The effect of UV light on muscle cell development was determined by monitoring muscle actin, myosin heavy chain, and AChE expression. Muscle actin mRNA expression was determined by *in situ*

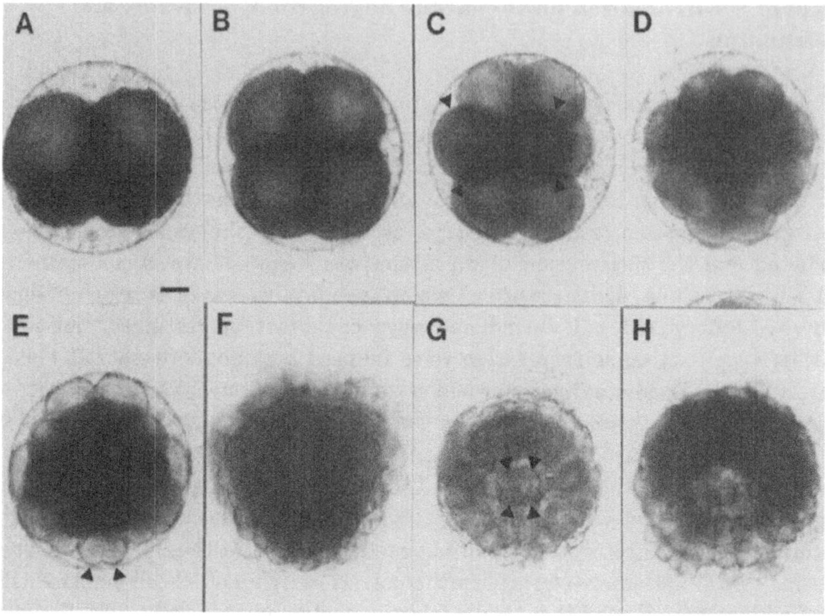

Figure 4. The effect of UV light on cleavage and gastrulation. **A-E.** Cleavage in UV irradiated zygotes. Zygotes were irradiated at the yellow cap stage with a UV dose of 3.0×10^{-3} J mm^{-2}. **A.** A 2-cell embryo viewed from the posterior side. **B.** A 4-cell embryo viewed from the animal pole. **C.** An 8-cell embryo viewed from the vegetal-posterior side showing the opaque B4.1 blastomeres (arrowheads). **D.** A 16-cell embryo viewed from the vegetal pole showing the unequal size and diagonal cleavage planes of the vegetal blastomeres. **E.** A 32-cell embryo viewed from the vegetal pole showing the two small B6.5 blastomeres (arrowheads). **F-H.** UV irradiated and normal embryos viewed from the vegetal pole during the time of gastrulation in normal embryos. **F.** An 64-128 cell embryo that developed from a UV irradiated zygote showing no invagination of vegetal cells. **G.** A normal embryo at the 64-cell stage showing the bases of the invaginating endodermal blastomeres (arrowheads). **H.** A normal embryo at about the 128 cell-stage showing further invagination of the vegetal pole cells. Scale bar: 20 μm; magnification. From Jeffery (1990a).

hybridization of sectioned embryos with a *Styela* muscle actin cDNA probe (Tomlinson *et al.* 1987b) as described previously (Jeffery 1989). Muscle actin mRNA accumulated in distinct cells in embryos irradiated with 3.0×10^{-3} J mm^{-2}, but was absent in embryos irradiated with higher UV doses (Figure 5). The positive cells were identified as muscle cells by staining alternate serial sections with Milligan's trichrome (Jeffery 1989). Most Grade 0 embryos irradiated with 3.0×10^{-3} J mm^{-2}, but not above 6.0×10^{-3} J mm^{-2}, contained cells that expressed myosin heavy chain (Figure 6), as determined by fluorescence microscopy of whole mount embryos stained with the monoclonal antibody Mu-2, which recognizes myosin heavy chain in ascidian embryos (Nishikata *et al.* 1987b). AChE activity was determined by histochemical methods (Karnovsky and Roots 1964). The results showed that axis deficient embryos expressed AChE similarly to normal embryos (Figure 7A-B). Cleavage-arrest experiments conclusively identified the AChE-positive cells of UV-irradiated embryos as part of the muscle lineage. In these experiments, UV-irradiated zygotes were treated with 2 μg/ml cytochalasin B at the 4- or 8-cell stage and processed for AChE histochemistry when untreated controls reached the late tailbud stage. As shown in Figure 7C-D and Table 1, a maximum of 2 cells developed AChE in both UV-irradiated and control embryos arrested at the 4- or 8-cell stage, indicating that a normal muscle lineage forms in axis-deficient embryos (Whittaker 1973). The UV dose-response curves for muscle actin and AChE development are shown in Figure 1. The effect of UV irradiation on epidermal cell differentiation was examined by staining whole mounts with the monoclonal antibody Epi-1, which recognizes an epidermal antigen (Nishikata *et al.* 1987b). As shown in Table 2, the epidermal antigen was produced in most zygotes irradiated with 3.0×10^{-3} J mm^{-2}, but not those irradiated with 6.0×10^{-3} J mm^{-2}. The UV dose-response curve for endodermal AP development, assayed as described by Bates and Jeffery (1987b), is shown in Figure 1. The results show that UV doses 3 times higher than the threshold required to inhibit gastrulation, sensory cell differentiation, or axis formation fail to affect AP activity.

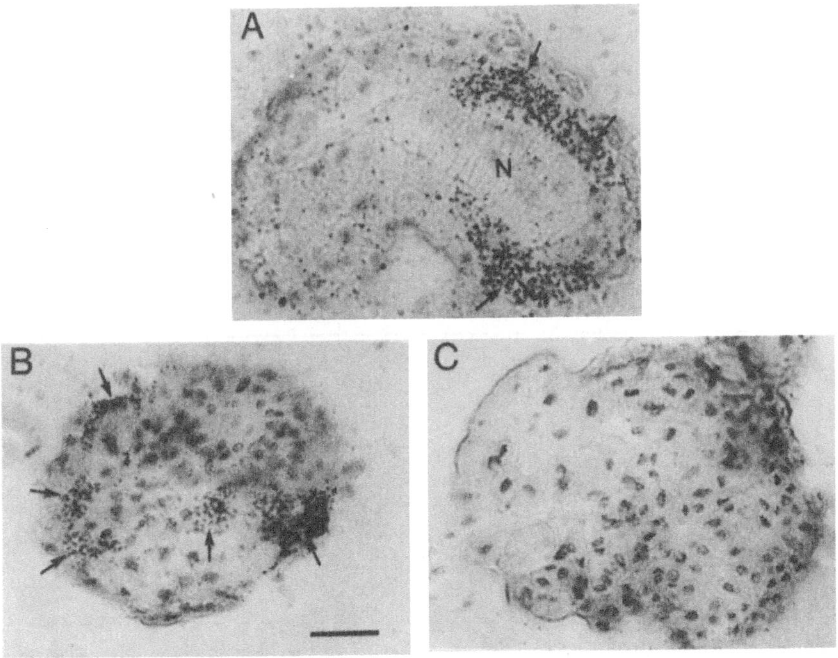

Figure 5. *In situ* hybridization showing the distribution of muscle actin mRNA in normal and UV irradiated embryos. **A.** A normal embryo at the mid-tailbud stage showing muscle actin mRNA in two rows of tail muscle cells (arrows). The row of notochord cells (N) can also be distinguished in this section. **B.** An axis-deficient embryo that developed from a zygote irradiated with 3.0×10^{-3} J mm^{-2} showing muscle actin mRNA in groups of muscle cells (arrows) scattered throughout the embryo. **C.** An axis-deficient embryo that developed from a zygote irradiated with 9.0×10^{-3} J mm^{-2} that has not accumulated detectible amounts of muscle actin mRNA. Sections of UV-irradiated embryos were hybridized with a muscle actin probe (Tomlinson *et al.* 1987b). Scale bar: 20 μm; magnification is the same in each frame. From Jeffery (1990a).

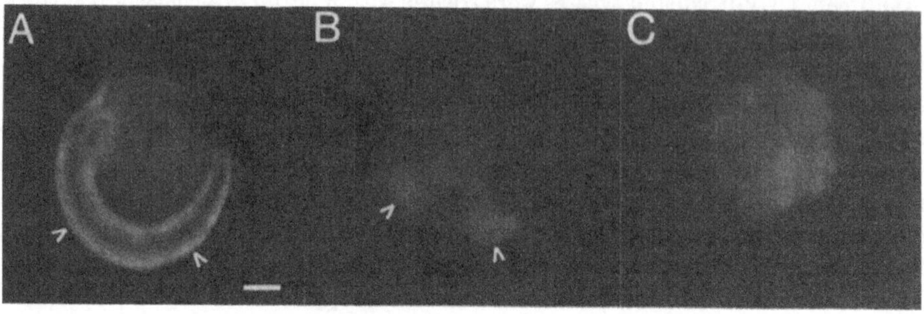

Figure 6. Immunofluorescence micrographs showing the accumulation of myosin heavy chain in normal and UV-irradiated embryos stained with the monoclonal antibody Mu-2. **A.** A normal embryo at the mid-tailbud stage showing fluorescence in the tail muscle cells (arrowheads). **B.** An axis deficient embryo that developed from a zygote irradiated with 3.0×10^{-3} J mm^{-2} showing fluorescence in some embryonic cells (arrowheads). **C.** An axis deficient embryo that developed from a zygote irradiated with 9.0×10^{-3} J mm^{-2} showing background fluorescence. Scale bar: 20 μm; magnification is the same in each frame. From Jeffery (1990a).

Table 1. The effect of UV light on AChE expression in cleavage-arrested embryos developing from zygotes irradiated with 3.0×10^{-3} J mm^{-2} at the yellow cap stage

Stage	Number of AChE positive cells	UV-irradiated[a]	Controls
4-cell	0	59	18
	1	19	43
	2	11	20
	>2	0	0
8-cell	0	16	31
	1	56	29
	2	46	16
	>2	0	0

a. Zygotes at the yellow cap stage were UV irradiated with 3.0×10^{-3} J mm^{-2}.

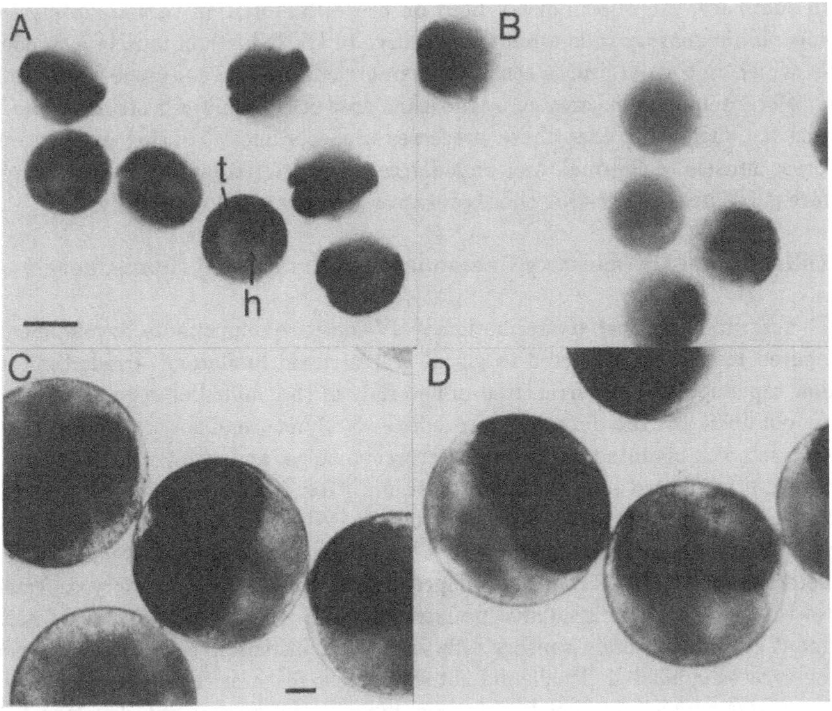

Figure 7. AChE expression in UV-irradiated embryos and their counterparts that were cleavage-arrested by treatment with 2 μg/ml cytochalasin B. **A.** Normal embryos at the mid-tailbud stage showing the tail (t) curling around the head (h). AChE activity (darkly-stained areas) is seen in the tail muscle cells. **B.** Axis-deficient embryos that developed from zygotes irradiated with 3.0 $\times$ 10^{-3} J mm^{-2}. These embryos contain cells with AChE activity (darkly-stained areas). The unstained specimens represent unfertilized eggs that served as a negative control. Scale bar in A: 100 μm; magnification is the same in A-B. **C-D.** Embryos that developed from normal zygotes (**C**) and zygotes irradiated with 3.0 $\times$ 10^{-3} J mm^{-2} (**D**) that were arrested in cleavage at the 4-cell stage. These embryos show AChE activity in 2 cells (darkly-stained cells). Scale bar in C: 20 μm; magnification is the same in C-D.

Table 2. The effect of UV light on the accumulation of an epidermis-specific antigen as determined by staining with the monoclonal antibody Epi-1

Irradiation	Number	Percent stained
None	34	94
3.0×10^{-3} J mm^{-2}	34	61
6.0×10^{-3} J mm^{-2}	30	0

In summary, the effects of UV light on cell differentiation vary according to cell lineage: brain sensory cells are more sensitive to UV light than muscle or epidermal cells, which in turn are more sensitive than endodermal cells. Furthermore, sensory cell differentiation is inactivated by the same dose of UV light as gastrulation and axis formation, suggesting that these processes are coordinated during embryogenesis, whereas muscle, epidermal, and endodermal cell differentiation are inactivated by different UV does, suggesting these processes are controlled differently.

Localization of UV-sensitive Components in the Vegetal Hemisphere

To determine the spatial distribution of UV-sensitive components, development was compared in zygotes subjected to global or restricted (unilateral) irradiation at the yellow cap stage. Zygotes irradiated unilaterally in the animal or vegetal hemisphere were obtained by the following procedure. A Syracuse dish containing the egg suspension was mounted on a stereomicroscope stage, and zygotes at the yellow cap stage were irradiated in a stationary position. After irradiation, the UV source was removed and replaced with a stereomicroscope. Whereas zygotes irradiated globally developed mostly as Grade 0 embryos, a significant proportion of zygotes irradiated unilaterally (3.0×10^{-3} J mm^{-2}) developed into Grade 1, 2, or 3 embryos (Table 3). Likewise, some zygotes irradiated unilaterally (6.0×10^{-3} J mm^{-2}) developed melanin pigment granules in brain sensory cells and AChE in muscle cells; however, most of those zygotes irradiated globally did not show these features. These results show that UV-sensitive components involved in axis, muscle, and sensory cell development are localized in the uncleaved zygote.

The region in which the UV-sensitive components are localized was determined by the following procedure. After unilateral irradiation, zygotes were separated manually according to whether the yellow cap of myoplasm was oriented toward (vegetal hemisphere irradiation) or away (animal hemisphere irradiation) from the UV source. Most zygotes irradiated (3.0×10^{-3} J mm$^{-2)}$ in the animal hemisphere developed normally, or as Grade 1 or 2 embryos; whereas zygotes irradiated in the vegetal hemisphere developed into Grade 0 embryos (Table 3). Similar results were obtained for AChE development. Some of the zygotes irradiated (9.0×10^{-3} J mm^{-2}) in the animal hemisphere later developed AChE activity, but most zygotes irradiated in the vegetal hemisphere lacked AChE (Table 3). Although most zygotes irradiated (3.0×10^{-3} J mm^{-2}) in the animal hemisphere developed melanin pigment in brain sensory cells, only 10% did so when zygotes were irradiated in the vegetal hemisphere (Table 3). Thus, despite their origin from animal hemisphere cells (Nishida 1987), UV-sensitive components required for brain sensory cell differentiation are localized in the vegetal hemisphere. The results show that UV-sensitive components involved in axis formation and sensory and muscle cell development are localized in the vegetal hemisphere during ooplasmic segregation.

Absorption of UV-sensitive Components

The absorption characteristics of the unknown chromophore(s) involved in axis determination were characterized by comparing survival curves generated by UV irradiating standard chromophores with and without a cut-off filter (Epstein and Schiff 1960). UV irradiation with or without the cut-off filter (Stumpf and Shugar 1962) was conducted while zygotes were irradiated globally as described above. The time of irradiation varied from 15 sec to 30 min. The cut-off filter was a #4015 microscope cover glass (Albert A. Henning Co., New York, N.Y.), which transmitted UV light above about 270 nm, as determined by spectrophotometry (Figure 8A). The standard chromophores were DNase I, which absorbs maximally at 280 nm (Setlow 1960), and bacteriophage T2, which exhibits an action spectrum peaking at 260 nm (Winkler *et al.* 1962). DNase I activity was determined by the method of Kunitz (1950). Bacteriophage T2 infectivity was measured using a standard plaque assay using *Escherichia coli* strain B_{S-1} as the dark-reactivation insensitive host (Ellison *et al.* 1960). The survival curves for DNase I, bacteriophage T2, and embryonic axis formation are shown in Figure 8B-D. The dose ratio for axis inactivation (10.7) was similar to that obtained for bacteriophage T2 inactivation (9.5), but distinct from that obtained for DNase I (2.1), suggesting that the UV-sensitive components have absorption characteristics similar to nucleic acids.

Protein Synthesis in Axis-deficient Embryos

The pattern of [^{35}S]-methionine incorporation was compared in UV-irradiated embryos and controls by gel electrophoresis and autoradiography. Zygotes at the yellow cap stage were irradiated with 3×10^{-3} J mm^{-2}. After irradiation, 50 μC/ml [^{35}S]-methionine (1290 C/mmole; Amersham Radiochemicals, Arlington Heights, IL.) was added, and labelling was continued until controls reached the mid-tailbud stage. Proteins were extracted from labelled embryos, separated by two-dimensional gel electrophoresis, and autoradiographed as described by Tomlinson *et al.* (1987a). Most of the 433 polypeptides detected in the gels were not affected by UV irradiation (Figure 9). For example, the three major actin isoforms (Tomlinson *et al.* 1987a) were labelled to the same extent in UV-irradiated embryos and controls. However, about 5% of the

Table 3. The effect of global and unilateral UV irradiation on embryonic axis, muscle, and brain sensory cell development

Irradiation	Axis[a]				AChE[b]		Brain pigment[a]	
	3	2	1	0	Number	%+	Number	%+
Global	0	0	7	42	44	7	58	12
Unilateral	16	19	10	22	57	75	61	53
Animal hemisphere	18	8	10	8	54	61	33	63
Vegetal hemisphere	0	6	2	26	76	17	39	10

a. Irradiation with a UV dose of 3.0×10^{-3} J mm^{-2}.
b. Irradiation with a UV dose of 6.0×10^{-3} J mm^{-2}.

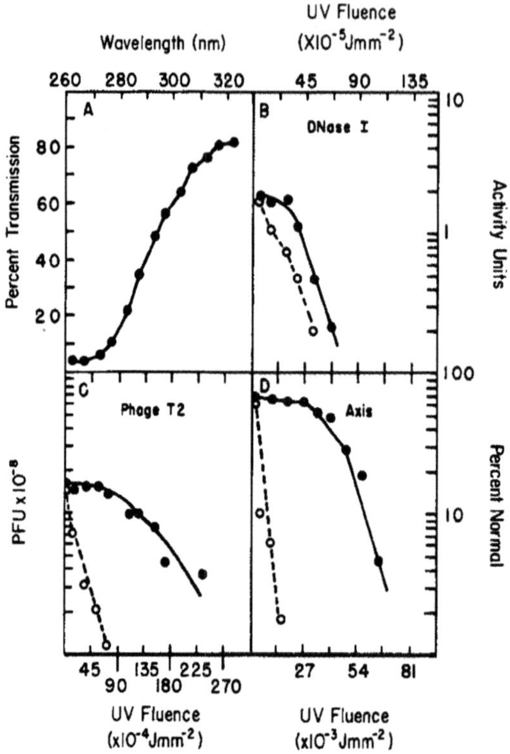

Figure 8. Absorption characteristics of UV-sensitive components. **A.** Percent UV transmission of the cut-off filter. **B-D.** Survival curves for DNase I activity (**B**), bacteriophage T2 (**C**), and the *Styela* embryonic axis (**D**) after irradiation with increasing UV doses in the presence (closed circles) or absence (open circles) of a cut-off filter. PFU: plaque forming units.

detectible polypeptides decreased in labelling or disappeared after UV-irradiation (Figure 9A). Although the effect of UV irradiation on protein synthesis varied between different experiments, the labelling of a 30,000 molecular weight (pI 6.0) polypeptide (p30) was consistently abolished in 15 separate experiments. This polypeptide, which sometimes resolved into two distinct isoforms (see Figure 11), was one of the most prominent proteins labelled between the 2-cell and tailbud stage in normal embryos. Subsequent experiments were designed to determine whether p30 could be identified by mass in eggs, embryos, and adults. Figure 10 shows silver-stained gels of eggs and neurulae. The position of p30 in these gels was established by comparing its electrophoretic mobility to those of the actin isoforms. Although a spot corresponding to p30 was detected in silver-stained gels containing proteins from neurulae (Figure 10B), a similar spot could not be identified in silver-stained gels containing proteins from unfertilized eggs (Figure 10A), indicating that p30 is synthesized only in the embryo. The results suggest that UV irradiation during ooplasmic segregation affects subsequent synthesis of a relatively small number of polypeptides, including p30.

Relationship of p30 Synthesis to Embryonic Axis Formation

The lack of p30 synthesis in UV-irradiated embryos suggests that this protein may be involved in axis formation. If p30 is involved in axis formation, it is expected to be

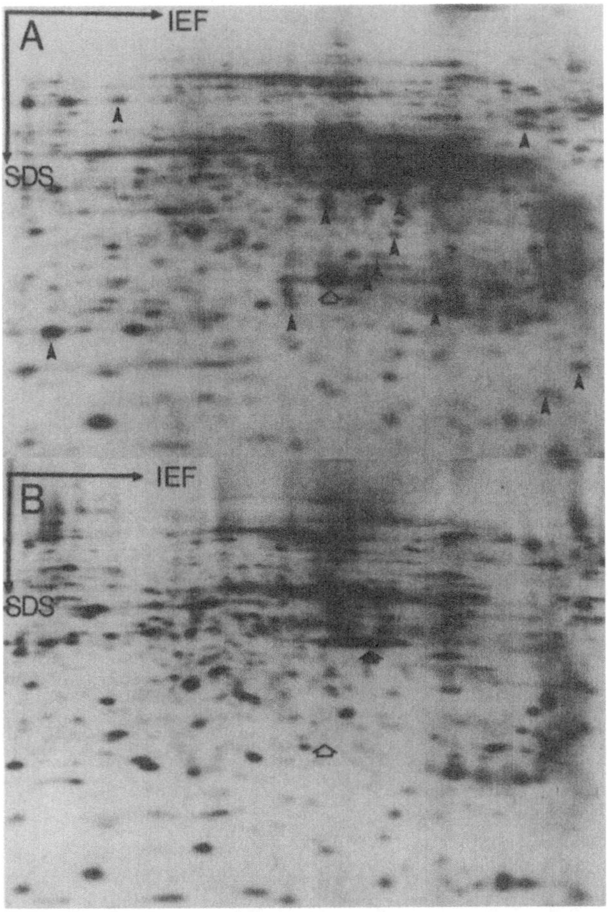

Figure 9. Autoradiographs of gels containing radioactive proteins from normal and UV-irradiated embryos. **A.** Proteins extracted from normal embryos labelled with [^{35}S]-methionine between the 2-cell and mid-tailbud stages. **B.** Proteins extracted from embryos that developed from zygotes irradiated with 3×10^{-3} J mm^{-2} at the yellow cap stage and labelled with [^{35}S]-methionine from the 2-cell stage until controls reached the mid-tailbud stage. Equal counts were applied to the gels. IEF: direction of isoelectric focussing. SDS: direction of electrophoresis through SDS gel. Solid large arrowhead: position of muscle actin. Open large arrowhead: position of p30. Solid small arrowheads: other polypeptides that disappear or decrease in UV-irradiated embryos. From Jeffery (1990b).

synthesized during gastrulation, when the axis is established, and to show the same UV sensitivity as axis formation. To determine the timing of p30 synthesis, normal embryos were exposed to [^{35}S]-methionine for restricted intervals during development, and p30 labelling was assessed by gel electrophoresis and autoradiography. The results are summarized in Table 4. Incorporation of [^{35}S]-methionine into p30 was undetectable in early cleaving zygotes. Synthesis of p30 was first detected between the 32- and 64-cell

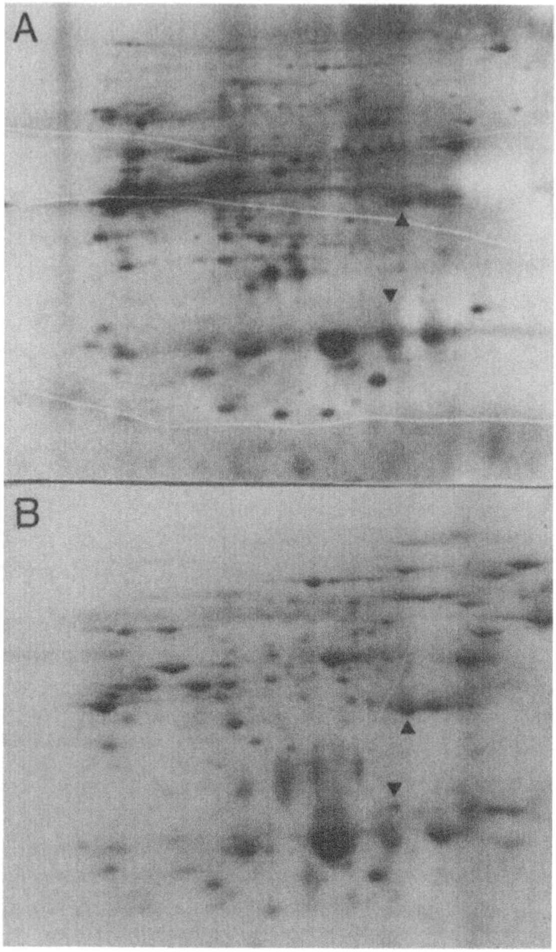

Figure 10. Silver-stained gels containing proteins extracted from eggs (**A**) and neurulae (**B**). Equal quantities of protein were applied to each gel. Upward-facing arrowheads: position of β-actin. Downward-facing arrowheads: position of p30. Electrophoretic conditions were similar to Figure 9. From Jeffery (1990b).

stages; synthesis peaked in gastrulae, was reduced in neurulae, and could not be detected during later embryogenesis. To determine whether p30 synthesis is affected by the same UV dose as axis formation, zygotes at the yellow cap stage were irradiated with increasing UV doses, and p30 synthesis was monitored as described above. The results are shown in Figure 11. Irradiation with a UV dose (1.5×10^{-3} J mm^{-2}) that reduced the number of normal embryos to about 50% (see Figure 1) resulted in a concomitant decrease in p30 labelling (Figure 11A,B). The same UV dose (3×10^{-3} J mm-$^{-2}$) that prevented axis formation in greater than 95% of the embryos virtually abolished p30 labelling (Figure 11C). To determine the UV sensitivity period for p30 synthesis, incorporation of [^{35}S]-methionone was investigated in embryos irradiated at different times during development. Labelling of p30 was suppressed in embryos irradiated at the yellow cap or crescent stages, but not markedly affected when UV irradiation was carried out at the 2- or 8-cell stages (Figure 12). In summary, the timing of synthesis during embryogenesis, the UV dose-response curve, and the UV sensitivity period for p30 are correlated with similar parameters for axis formation.

A UV-sensitive Maternal mRNA Encodes for p30

Although axis formation is correlated with p30 synthesis, this protein cannot be a UV sensitive component in uncleaved zygotes because it is absent in the egg and does not show significant labelling until after the 32-cell stage. Therefore, it is possible that a maternal mRNA, which directs p30 synthesis, is the UV sensitive target. To determine whether p30 is encoded by a maternal mRNA, RNA extracted from fertilized eggs by the method of March *et al.* (1985) was translated in a reticulocyte lysate containing [^{35}S]-methionine (made and used for *in vitro* protein synthesis as described by Pelham and Jackson 1976), and the labeled proteins were examined by gel electrophoresis and autoradiography. The results showed that RNA extracted from fertilized eggs directs the synthesis of a polypeptide of the same molecular weight and isoelectric point as p30 (see Figure 13A), indicating that p30 is encoded by a maternal

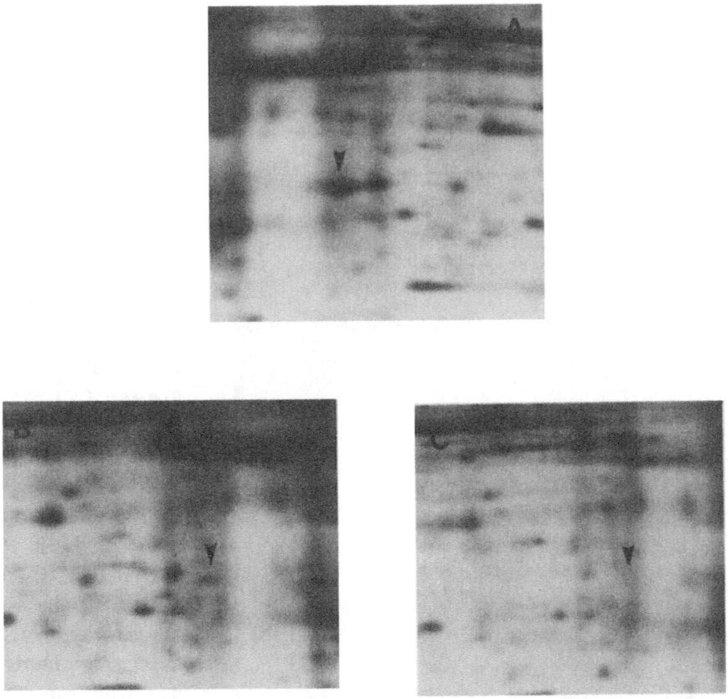

Figure 11. The effect of UV dose on p30 synthesis. **A-C.** Autoradiographs of gels showing incorporation of [^{35}S]-methionine into p30 in **(A)** control embryos and embryos irradiated with **(B)** 1.5×10^{-3} J mm^{-2} or **(C)** 3.0×10^{-3} J mm^{-2} at the yellow cap stage. Downward facing arrowheads: position of p30. Only the p30 region of the gels are shown. Electrophoretic conditions in were similar to Figure 9. From Jeffery (1990b).

Table 4. The timing of [^{35}S]-methionine incorporation into p30 during *Styela* embryogenesis.

Labelling period (hrs after fertilization)	p30 labelling intensity
0-2	–
2-3	+
3-4	++
4-5	+++
5-6	++
6-14	–

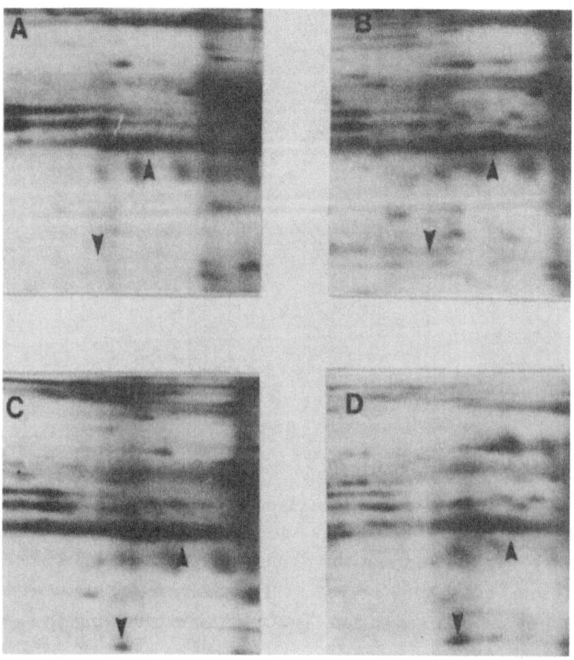

Figure 12. Autoradiographs of gels containing proteins extracted from zygotes irradiated at (**A**) the yellow cap stage, (**B**) the yellow crecent stage, (**C**) the 2-cell stage (60 min after insemination), and (**D**) the 8-cell stage (90 min after insemination). Incorporation of [^{35}S]-methionine into p30 was undetectable in (**A-B**), but observed in (**C-D**). Upward-facing arrow-heads: position of muscle actin. Downward-facing arrow-heads: position of p30. Only the p30 region of each gel is shown. Embryos were irradiated with 3×10^{-3} J mm^{-2}. Electrophoretic conditions were similar to Figure 9. From Jeffery (1990b).

mRNA. If p30 mRNA is a UV sensitive component, its translation should be inactivated when zygotes are irradiated with the same UV dose that prevents axis formation. Zygotes at the yellow cap stage were irradiated with 3×10^{-3} J mm^{-2}, their RNA was extracted and translated *in vitro*, and the translation products were identified as described above. Figure 13 shows the translation products of RNA extracted from control (Figure 13A) and UV-irradiated (Figure 13B) zygotes. RNA from UV-irradiated zygotes did not direct the translation of p30 and a number of other polypeptides detected in the gel (Figure 14B). In contrast, the translation of most proteins, including the actins (Figure 14B), was not affected by UV irradiation. The results suggest that UV irradiation affects p30 synthesis by inactivating p30 mRNA translation.

Association of p30 with the Cytoskeleton

Since invagination during gastrulation does not occur in UV-irradiated embryos and is dependent on microfilaments that initiate shape changes in vegetal pole blastomeres (Bates, personal communication), it was reasoned that p30 might be associated with the cytoskeleton. To test this possibility, [^{35}S] methionine-labelled gastrulae were extracted with the non-ionic detergent Triton X-100 as described by Jeffery and Meier (1983). After detergent extraction, proteins were prepared, and the distribution of p30 labelling in the soluble and insoluble fractions was determined by gel electrophoresis and autoradiography. As shown in Figure 14A, labelled p30 was recovered in the detergent-insoluble residue, suggesting that this protein is associated with the cytoskeleton. Further experiments were conducted to determine whether there is an interaction between p30 and microfilaments. Labelled gastrulae were extracted with Triton X-100 in the presence of DNase I, which depolymerizes F-actin in *Styela* embryos (Jeffery and Meier 1983), or bovine serum albumen (BSA) as a control, and the

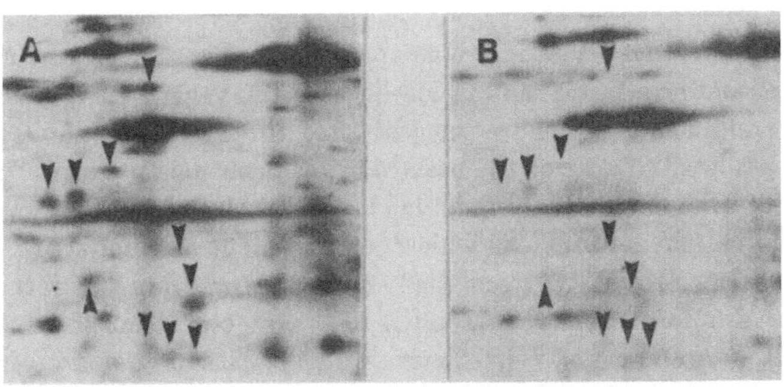

Figure 13. The effect of UV irradiation on the ability of RNA extracted from zygotes to direct protein synthesis in a reticulocyte lysate. **A.** Protein synthesis directed by RNA from normal zygotes. **B.** Protein synthesis directed by RNA from UV-irradiated zygotes. p30 and a number of other polypeptides are missing or decreased in B. Upward-facing arrowheads: position of p30. Downward-facing arrowheads: position of other polypeptides whose translation was affected by UV irradiation. Only the p30 region of the gels is shown. Zygotes that were UV-irradiated (3×10^{-3} J mm^{-2}) at the yellow cap stage. Equivalent radioactivity was applied to each gel. Electrophoretic conditions were similar to Figure 9. From Jeffery (1990b).

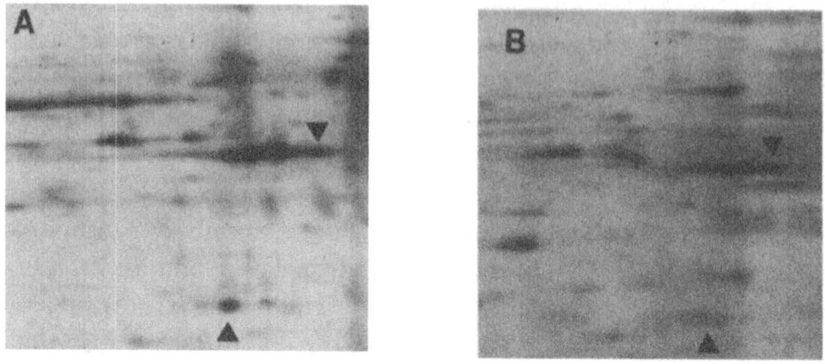

Figure 14. Association of p30 with the Triton X-100-insoluble residue of gastrulae and its release after DNase I treatment. **A-B.** Autoradiographs of gels showing the distribution of labelled p30 in the detergent-insoluble residues of gastrulae extracted with Triton X-100 in the presence of 10 μg/ml BSA (**A**) or 10 μg/ml DNase I (**B**). Equivalent volumes of detergent-insoluble residue were applied to each gel. Downward-facing arrowheads: muscle actin. Upward-facing arrowheads: position of p30. Only the p30 region of the gels are shown. Other details are similar to Figure 9.

distribution of p30 was analyzed by gel electrophoresis and autoradiography. As shown in Figure 14B, p30 was released from the detergent-insoluble residue of embryos extracted in the presence of DNase I, but not BSA (Figure 14A), suggesting that p30 association with the cytoskeleton is dependent on the integrity of microfilaments.

DISCUSSION

UV Irradiation Prevents Gastrulation, Sensory Cell Induction, and Embryonic Axis Formation

The results show that UV-sensitive components that specify gastrulation, sensory cell induction, and axis formation are localized in the vegetal hemisphere of *Styela* zygotes. Several lines of evidence suggest that these components are the same substances postulated to exist on the basis of cytoplasmic deletion experiments (Bates and Jeffery 1987b). First, UV irradiation and cytoplasmic deletion prevent axis formation when applied during the yellow cap stage, but are ineffective after yellow crescent formation. Second, UV irradiation and cytoplasmic deletion at the yellow cap stage prevent axis formation when applied to the vegetal hemisphere, but not the animal hemisphere. Third, UV irradiation and cytoplasmic deletion suppress the invagination of vegetal pole cells during initiation of gastrulation and morphogenetic movements in notochord cells during tail morphogenesis, but do not have significant effects on ooplasmic segregation, cleavage, or the development of muscle, epidermis, or endoderm.

The results define a critical period coincident with ooplasmic segregation in which the components involved in axis determination are sensitive to UV light. Previous studies showed that deletion of the vegetal pole region at the yellow cap stage inhibited gastrulation and axis development, but there was no effect when the same experiment was done at the yellow crescent stage (Bates and Jeffery 1987b). Therefore, it was concluded that either the factors responsible for these events are no longer required after the yellow cap stage or that their localization is changed so that they are unable to be removed by cytoplasmic deletion. The present results show that UV irradiation can inactivate gastrulation and axis formation when administered as late as the yellow crescent stage, supporting the possibility that cytoplasmic deletion between the yellow cap and crescent stage was unsuccessful in preventing gastrulation and axis development because the factors mediating these events spread throughout the vegetal surface of the zygote during OS 2. It is unlikely the UV-sensitive components ingress to a more internal position in the zygote because there is no concomitant increase in the effective dose required to inhibit gastrulation and axis development between the two phases of ooplasmic segregation. The UV-sensitive components presumably spread out in the vegetal hemisphere, as the myoplasm shifts posteriorly to form the yellow crescent during OS 2.

The results suggest that gastrulation, sensory cell induction, and axis formation may be controlled by a common UV-sensitive component localized in the vegetal hemisphere. Axis formation in ascidians is defined as the process in which presumptive notochord cells intercalate at the midline of the embryo and extend posteriorly to form the larval tail (Conklin 1905b; Cloney 1964; Miyamoto and Crowther 1985; Nishida and Satoh 1985). Because of the absence of biochemical markers, notochord differentiation could not be studied quantitatively, but no extended notochord was recognized morphologically in axis deficient embryos. Although UV doses that prevent axis formation had little or no effect on the development of specific markers in the muscle, epidermal and endoderm cells, and melanin pigment deposition was inhibited in sensory cells. Also, the UV-sensitive periods for axis and sensory cell development are the same. In contrast to muscle cells, the brain sensory cells appear to be specified by an inductive process involving interactions between presumptive endoderm and notochord blastomeres in the vegetal hemisphere and sensory cell precursors in the animal hemisphere (Rose 1939; Reverberi and Minganti 1946; 1947; Reverberi *et al.* 1960; Ortolani 1961; Nishida and Satoh 1989). The simplest explanation for these results is that the UV-sensitive components are necessary to initiate cell movements leading to invagination during gastrulation and notochord extension during tail morphogenesis. The lack of cell movements during gastrulation may prevent contact between the vegetal and animal hemisphere cells that is required for sensory cell induction.

The yellow pigmentation of *Styela* eggs has served as an excellent marker for an egg cytoplasmic region destined to enter presumptive muscle cells during development, but the purpose of this pigmentation has been a mystery since its discovery near the turn of the century (Conklin 1905b). Cytological studies have shown that all ascidian eggs contain myoplasm, but it is not pigmented in eggs of most species (Berg and Humphrey 1960). The sensitivity of axis determination to UV irradiation suggests that the myoplasmic pigment granules of *Styela* eggs may be a natural shield for potentially harmful effects of sunlight sustained during development in the intertidal zone. Thus, *Styela* eggs may have evolved a pigmented myoplasm because development is especially sensitive to natural UV radiation. Consistent with this possibility, is the fact that unlike other ascidians, whose eggs are shed at dawn and develop in daylight (Whittingham 1967), *Styela* eggs are shed at dusk and develop during the evening (West and Lambert 1975). This relationship suggests that colored cytoplasmic regions in other invertebrate eggs may be indicative of underlying UV-sensitive substances that are critical for embryonic development.

Role of UV-sensitive Maternal mRNA in Axis Determination

An important advantage of the UV irradiation approach is that it can be used to inactivate axis determination simultaneously in large numbers of zygotes. Using this approach, candidates for some of the UV-sensitive components involved in axis determination have been identified. The UV absorption characteristics described here suggest that the critical components may be nucleic acids. These nucleic axis, presumably mRNAs, probably function indirectly by directing the synthesis of proteins involved in axis formation. One such protein may be p30. Evidence supporting the possibility that p30 functions in axis formation stems from three different observations. First, p30 is most intensely labelled during gastrulation, when the embryonic axis is beginning to be established. Labelling during gastrulation suggests that p30 is likely to be involved in an early event of axis formation, such as invagination of vegetal blastomeres, rather than a late event, such as notochord extension. Second, p30 labelling and axis formation are both sensitive to UV irradiation during ooplasmic segregation, but not during cleavage. Third, p30 labelling is curtailed by the same dose of UV light that inhibits gastrulation and axis formation without affecting yellow crescent formation, cleavage, or cell differentiation. In principle, any of the other proteins affected by UV irradiation could also be involved in axis formation.

p30 itself cannot be a UV-sensitive target for axis formation because its synthesis is not appreciable at least until the 32-cell stage, about 2 hrs after the conclusion of the UV-sensitive period, and the absorption characteristics of the relevant chromophore(s) suggests that the target is a nucleic acid. Since p30 is directed by a maternal mRNA, the latter is a likely UV-sensitive target. This was confirmed by experiments in which p30 mRNA translation was inactivated by UV irradiation of zygotes. The reason that p30 mRNA and other messages whose translation was abolished are particularly sensitive to UV irradiation may be explained by inability of UV light to penetrate very far into the interior of eggs (Youn and Malacinski 1980). It follows that the set of mRNAs shown to be UV-sensitive must be localized near the egg surface. Localized maternal mRNAs have been previously described in the myoplasm of *Styela* eggs (Jeffery *et al.* 1983), the vegetal hemisphere of *Xenopus* eggs (Melton 1987), and the anterior pole region of *Drosophila* eggs (Berleth *et al.* 1988). The localization of p30 mRNA near the cell surface is consistent with the possibility that it may be an axial determinant since these determinants are thought to be localized in the cortex during ooplasmic segregation (Bates and Jeffery 1987b). Recently, it has been proposed that RNA localized in the peripheral region of the vegetal hemisphere of the *Xenopus* oocyte may also be an axial determinant (Elinson and Pasceri 1989).

The axis of the ascidian embryo is initiated after invagination of vegetal pole blastomeres during gastrulation. The descendants of these and adjacent cells migrate through the embryo and eventually reach a position where they induce animal hemisphere blastomeres to elaborate neural structures (Nishida and Satoh 1989) or are extended in an anteroposterior direction to form the notochord (Conklin 1905b). The results of the present investigation suggest that p30 is associated with the cytoskeleton, and that this association is dependent on the integrity of microfilaments. Therefore, p30 may be an actin-binding protein that organizes microfilaments involved in changing the shape and/or motility of the invaginating vegetal pole cells.

The results support the hypothesis that axis determination is mediated by the translation of a maternal mRNA encoding p30, a cytoskeletal protein. During OS 1, p30 mRNA may segregate with the myoplasm to the vegetal pole region of the uncleaved zygote. During OS 2, p30 mRNA probably remains in the vegetal pole region as the myoplasm shifts to form the yellow crescent. If so, the behavior of this mRNA during ooplasmic segregation may be similar to that of cortical components known to translocate with the myoplasm during OS 1, but not OS 2 (see Jeffery and Bates 1989 for review). The localization of p30 mRNA near the vegetal pole would place it in the proper spatial position to be inherited and translated in endodermal cells that are first to invaginate and initiate axis formation during gastrulation. Future studies with specific probes for p30 mRNA and protein will be required to determine the spatial distribution of these molecules during development.

ACKNOWLEDGEMENTS

This research was supported by grants from the NSF (DCB-9196041) and NIH (HD-13970). I thank P. Kemp and M. Badgett for technical assistance, Dr. B.J. Swalla for conducting the immunofluorescence studies, Drs. T. Nishikata and N. Satoh for providing monoclonal antibodies, and Dr. M.M. Winkler for providing reticulocyte lysate.

REFERENCES

Anderson, K.V. 1987. Dorsal-ventral pattern genes of *Drosophila*. *Trends Genet.* 3:91-97.

Bates, W.R. and W.R. Jeffery. 1987a. Alkaline phosphatase expression in ascidian egg fragments and andromerogons. *Dev. Biol.* 119:382-389.

Bates, W.R. and W.R. Jeffery. 1987b. Localization of axial determinants in the vegetal pole region of ascidian eggs. *Dev. Biol.* 33:197-212.

Berg, W.E. and W.J. Humphrey. 1960. Electron microscopy of four-cell stages of the ascidians *Ciona* and *Styela*. *Dev. Biol.* 2:42-60.

Berleth, T, M. Burri, G. Thoma, D. Bopp, S. Richstein, G. Frigerio, M. Noll, and C. Nüsslein-Volhard. 1988. The role of localization of *bicoid* mRNA in organizing the anterior pattern of the *Drosophila* embryo. *EMBO J.* 7:1749-1756.

Berrill, N.J. 1935. Studies in tunicate development. III. Differential retardation and acceleration. *Phil. Trans. Roy. Soc. Lond. Ser. B* 225:266-326.

Cloney, R.A. 1964. Development of the ascidian notochord. *Acta Embryol. Morphol. Exp.* 7:111-130.

Conklin, E.G. 1905a. Mosaic development in ascidian eggs. *J. Exp. Zool.* 2:145-223.

Conklin, E.G. 1905b. The organization and cell lineage of the ascidian egg. *J. Acad. Nat. Sci. Phila.* 13:1-119.

Deno, T., H. Nishida, and N. Satoh. 1985. Histospecific acetylcholinesterase development in quarter ascidian embryos derived from each blastomere pair of the 8-cell stage. *Biol. Bull.* 168:239-248.

Dilly, P.N. 1962. Studies on the receptors in the cerebral vesicle of the ascidian tadpole. 1. The otolith. *Q. J. Microsc. Sci.* 103:393-398.

Dilly, P.N. 1964. Studies on the receptors in the cerebral vesicle of the ascidian tadpole. 2. The ocellus. *Q. J. Microsc. Sci.* 105:13-20.

Elinson, R.P. and P. Pasceri 1989. Two UV-sensitive targets in dorsoanterior specification of frog embryos. *Development* 106:511-518.

Ellison, S.A., R.R. Feiner, and R.F. Hill. 1960. A host effect on bacteriophage survival after ultraviolet irradiation. *Virology* 11:294-296.

Epstein, H.T. and J. Schiff. 1960. A simple and rapid method for estimating an action spectrum where nucleic acids (or, in principle, other materials) are the chromophores involved. *Exp. Cell Res.* 23:623-624.

Gerhart, J.C. and R.E. Keller. 1986. Region-specific activities in amphibian gastrulation. *Mod. Cell Biol.* 2:483-507.

Gimlich, R.L. and J.C. Gerhart. 1984. Early cellular interactions promote embryonic axis formation in *Xenopus laevis*. *Dev. Biol.* 104:117-130.

Jeffery, W.R. 1982. Calcium ionophore polarizes ooplasmic segregation in ascidian eggs. *Science* 216:545-547.

Jeffery, W.R. 1989. Requirement of cell division for muscle actin expression in the primary muscle cell lineage of ascidian embryos. *Development* 105:75-84.

Jeffery, W.R. 1990a. Ultraviolet irradiation prevents gastrulation, sensory cell induction, and axis formation in the ascidian embryo. *Dev. Biol.* 140:388-400.

Jeffery, W.R. 1990b. An ultraviolet-sensitive maternal mRNA encoding a cytoskeletal protein may be involved in axis formation in the ascidian embryo. *Dev. Biol.* 141:141-148.

Jeffery, W.R. and W.R. Bates. 1989. Ooplasmic segregation in the ascidian *Styela*. p. 341-367. *In: The Molecular Biology of Fertilization*. H. Schatten and G. Schatten (Eds.). Academic Press, New York.

Jeffery, W.R. and S. Meier. 1983. A yellow crescent cytoskeletal domain in ascidian eggs and its role in early development. *Dev. Biol.* 96:125-143.

Jeffery, W.R. and S. Meier. 1984. Ooplasmic segregation of the myoplasmic actin network in stratified ascidian eggs. *Wilhelm Roux's Arch. Dev. Biol.* 193:257-262.

Jeffery, W.R., C.R. Tomlinson, and R.D. Brodeur. 1983. Localization of actin messenger RNA during early ascidian development. *Dev. Biol.* 99:408-417.

Karnovsky, M.J. and L. Roots. 1964. A "direct-coloring" thiocholine method for cholinesterase. *J. Histochem. Cytochem.* 12:219-221.

Katz, M.J. 1983. Comparative anatomy of the tunicate tadpole, *Ciona intestinalis*. *Biol. Bull.* 164:1-27.

Kunitz, M. 1950. Crystalline desoxyribonuclease. I. Isolation and general properties. Spectrophometric method for the measurement of desoxyribonuclease activity. *J. Gen. Physiol.* 33:349-362.

March, C.J., B. Mosley, A. Larsen, D.P. Cerretti, G. Braedt, V. Price, S. Gillis, C.S. Henney, S.R. Kronheim, K. Grabstein, P.J. Conlon, T.P. Hopp, and D. Cosman. 1985. Cloning, sequence, and expression of two distinct human interleukin-1 complementary DNAs. *Nature* 315:641-647.

Melton, D.A. 1987. Translocation of a localized maternal mRNA to the vegetal pole of *Xenopus* oocytes. *Nature* 328:80-82

Miyamoto, D.M. and R.J. Crowther. 1985. Formation of the notochord in living ascidian embryos. *J. Embryol. Exp. Morphol.* 86:1-17.

Nishida, H. 1987. Cell lineage analysis in ascidian embryos by intracellular injection of a tracer enzyme. III. Up to the tissue restricted stage. *Dev. Biol.* 121:526-541.

Nishida, H. and N. Satoh. 1985. Cell lineage analysis in ascidian embryos by intracellular injection of a tracer enzyme. II. The 16- and 32-cell stages. *Dev. Biol.* 110:440-454.

Nishida, H. and N. Satoh. 1989. Determination and regulation in the pigment cell lineage of the ascidian embryo. *Dev. Biol.* 132:355-367.

Nishikata, T., I. Mita-Miyazawa, T. Deno, K. Takamura, and N. Satoh. 1987a. Expression of epidermal-specific antigens during embryogenesis in the ascidian, *Halocynthia roretzi*. *Dev. Biol.* 121:408-416.

Nishikata, T., I. Mita-Miyazawa, T. Deno, K. Takamura, and N. Satoh. 1987b. Muscle cell differentiation in ascidian embryos analyzed with a tissue-specific monoclonal antibody. *Development* 99:163-171.

Ortolani, G. 1961. L'evocazione del cervello nelle larve di Ascidie. *Ric. Sci.* 31:157-162.

Pelham, H.R.B. and R.J. Jackson. 1976. An efficient mRNA-dependent translation system for reticulocyte lysates. *Eur. J. Biochem.* 67:247-256.

Reverberi, G. and A. Minganti. 1946. Fenomeni di evoczione nello sviluppo dell'uovo di Ascidie. Risultati dell'indagine sperimentale sull'uvovo di *Ascidiella aspersa* e di *Ascidia malaca* allo stadio di 8 blastomeri. *Pubbl. Stne. Zool. Napoli I Mar. Ecol.* 20:199-152.

Reverberi, G. and A. Minganti. 1947. La distribuzione delle potenze nel germ di Ascidie sllo stadio di otto blastomeri, analizzata mediante le combinazioni e i trapianti di blastomeri. *Pubbl. Stne. Zool. Napoli I Mar. Ecol.* 21:1-35.

Reverberi, G., G. Ortolani, and N. Farinella-Ferruza. 1960. The causal formation of the brain in the ascidian larva. *Acta Embryol. Morphol. Exp.* 3:296-336.

Rose, S.M. 1939. Embryonic induction in the Ascidia. *Biol. Bull.* 77:216-232.

Sardet, C., J.E. Speksnijder, S. Inoué, and J.F. Jaffe. 1989. Fertilization and ooplasmic movements in the ascidian egg. *Development* 105:237-249.

Satoh, N. 1978. Cellular morphology and architecture during early morphogenesis of the ascidian egg: An SEM study. *Biol. Bull.* 155:608-614.

Sawada, T. and K. Osanai. 1981. The cortical contraction related to ooplasmic segregation in *Ciona intestinalis* egg. *Wilhelm Roux's Arch. Dev. Biol.* 190:208-214.

Sawada, T. and G. Schatten. 1988. Microtubules in ascidian eggs during meiosis, fertilization, and mitosis. *Cell Motil. Cytoskeleton* 9:219-230.

Sawada, T. and G. Schatten. 1989. Effects of cytoskeletal inhibitors on ooplasmic segregation and microtubule organization during fertilization and early development in the ascidian *Molgula occidentalis*. *Dev. Biol.* 132:331-342.

Speksnijder, J.E., C. Sardet, and L.F. Jaffe. 1990. The activation wave of calcium in the ascidian egg and its role in ooplasmic segregation. *J. Cell Biol.* 110:1589-1598.

Spemann, H. and H. Mangold. 1924. Uber Induktion von Embryonanlagen durch Implantation artfremder Organisatoren. *Wilhelm Roux's Arch. Dev. Biol.* 100:599-638.

Setlow, R.B. 1960. Ultraviolet wave-length-dependent effects on proteins and nucleic acids. *Radiat. Res. Suppl.* 2:276-289.

Stumpf, E. and D. Shugar. 1962. Microscope cover glasses as cut-off filters in the medium ultraviolet. *Photochem. Photobiol.* 1:337-338.

Tomlinson, C.R., W.R. Bates, and W.R. Jeffery. 1987a. Development of a muscle actin specified by maternal and zygotic mRNA in ascidian embryos. *Dev. Biol.* 123:470-482.

Tomlinson, C.R., R.L. Beach, and W.R. Jeffery. 1987b. Differential expression of a muscle actin gene in muscle cell lineages of ascidian embryos. *Development.* 101:751-765.

van den Biggelaar, J.A.M. and P. Guerrier. 1983. Origin of spatial organization. p. 179-213. *In: The Mollusca, Vol. 3, Development.* N.H. Verdonk, J.A.M. van den Biggelaar, and A.S. Tomba (Eds.). Academic Press, New York.

Venuti, J.M. and W.R. Jeffery. 1989. Cell lineage and determination of cell fate in ascidian embryos. *Int. J. Dev. Biol.* 33:197-212.

West, A.B. and C.C. Lambert. 1975. Control of spawning in the tunicate *Styela plicata* by variations in a natural light regime. *J. Exp. Zool.* 195:263-270.

Whittingham, D.G. 1967. Light-induction of shedding of gametes in *Ciona intestinalis* and *Molgula manhattensis. Biol. Bull.* 132:292-298.

Whittaker, J.R. 1973. Segregation during ascidian embryogenesis of egg cytoplasmic information for tissue specific enzyme development. *Proc. Natl. Acad. Sci. USA* 70:2096-2100.

Whittaker, J.R. 1977. Segregation during cleavage of a factor determining endodermal alkaline phosphatase development in ascidian embryos. *J. Exp. Zool.* 20:139-154.

Whittaker, J.R. 1982. Muscle lineage cytoplasm can change the developmental expression in epidermal lineage cells of ascidian embryos. *Dev. Biol.* 93:463-470.

Winkler, U., H.E. Johns, and E. Kellenberger. 1962. Comparative study of some properties of bacteriophage T4D irradiated with monochromatic ultraviolet light. *Virology* 18:343-358.

Youn, B.W. and G.M. Malacinski. 1980. Action spectrum for ultraviolet irradiation of a cytoplasmic component(s) required for neural induction in the amphibian egg. *J. Exp. Zool.* 211:369-377.

RAPID EVOLUTION OF EARLY DEVELOPMENT: REORGANIZATION OF EARLY MORPHOGENETIC PROCESSES IN A DIRECT-DEVELOPING SEA URCHIN

Rudolf A. Raff, Jonathan J. Henry⁺, and Gregory A. Wray*

Institute for Molecular and Cellular Biology, and
Department of Biology, Indiana University
Bloomington, IN 47405, U.S.A.

⁺Present address: Department of Cell and Structural Biology
University of Illinois
Urbana, IL, 61801, U.S.A.
*Present address: Friday Harbor Laboratories
University of Washington, Friday Harbor, WA 68250, U.S.A.

EVOLUTIONARY DEVELOPMENTAL BIOLOGY

A century ago Wilhelm Roux proposed that the underlying processes of development could be revealed by experimental interference with specific developmental events (Roux 1895). Most of our current understanding of developmental processes has been achieved by experimental studies along the lines propounded by Roux. Modern practice uses experimental manipulations at cellular, molecular, and genetic levels, with primary emphasis on gaining an understanding of mechanisms of development. Evolutionary questions are seldom asked, and have had little influence on the mainstream of modern developmental biology. Nevertheless, some 20th Century

Gastrulation, Edited by R. Keller *et al.*
Plenum Press, New York, 1991

investigators have maintained an interest in the study of the role of developmental processes in evolution (see Bonner 1982; Garstang 1922; deBeer 1958; Gould 1977; Raff and Kaufman 1983). An appreciation of the role of evolution in developmental biology has begun to take hold during the past decade with a growing interest in the attempt to fuse these two very dissimilar disciplines in a workable manner.

The current recognition of the potential importance of evolutionary aspects of developmental biology is analogous to the end of the long separation between development and genetics (Raff and Kaufman 1983). Students immersed in the exciting findings of developmental genetics are probably little aware that developmental genetics has only flowered during the past 15 years. The synthesis of developmental genetics required a shift in focus from transmission genetics to gene expression, and it required an appreciation on the part of developmental biologists that development is governed by gene action. The establishment of an evolutionary developmental biology requires an analogous fusion of two fields that have distinct sets of problems, concepts, terminologies, and approaches.

A set of hypotheses has been generated about the role of development in evolution, and feasible experimental approaches to test these hypotheses have begun to appear (see Nitecki 1990; Raff 1990; Stern 1990). Experimental studies will provide answers to questions about the role of development in evolution, but there are other, equally significant reasons for such studies. Hypotheses and data on evolution of development have the potential to provide important information about developmental processes *per se*. Comparisons of evolutionary differences can reveal successful alternate developmental mechanisms; this is different from the general case in developmental genetics where mutations are exploited to disrupt the process being studied. Lastly, evolutionary comparisons make it possible to identify developmental processes or structures which can change readily *versus* those which are more resistant to change.

Defining developmental stages highly resistant to evolutionary modification provides a way to identify functionally important processes. The limited changes that occur in conservative features may define the critical boundaries for function. Exploitation of the same evolutionary logic has been used profitably in molecular biology. A number of upstream sites that control gene transcription (for instance the TATA box) have been revealed by DNA sequence comparisons to be present in many different genes. In other cases, highly specific sites for binding transcriptional regulatory proteins have been found in homologous genes in several organisms, or in genes expressed in a common tissue (Darnell *et al.* 1990). Comparisons of sequences of RNAs such as the ribosomal RNAs (Woese *et al.* 1983) or RNase P RNA (Pace *et al.* 1989; Waugh *et al.* 1989) have been vital in revealing major secondary structure elements and domains necessary for function.

Some of the major current questions about development and evolution deal with basic mechanistic issues. One important question is whether developmental processes constrain evolution. This is a major aspect of the old question in evolutionary biology, raised by biologists critical of the sufficiency of natural selection, of whether internal factors produced by organismal organization affect or determine the course of evolution. The concept of developmental constraint predicts that the course of evolution is limited or channeled by the genetic constitution or epigenetic interactions that govern developmental processes. If that is true, evolution cannot be a result of selection operating on a completely flexible responding system (Alberch 1982; Maynard Smith *et al.* 1985). If developmental mechanisms rule out some directions of change and favor others, then the study of development becomes crucial to understanding evolution. Developmental constraints appear probable, but at present examples of suggested constraints come mainly from our reading of evolutionary histories and from examples of conserved developmental features. It will be vital to investigate the existence of developmental constraints experimentally.

A second major issue arises from the question of whether there are preferred mechanisms for evolutionary changes in development. The current favorite, heterochrony (evolutionary changes in relative timing between developmental events) has been widely held to be the primary means by which evolution occurs (*e.g.*: Gould 1977; Buss 1987). Little is presently known about the mechanisms that underly heterochronies, but it is clear that they are complex in cause, and that heterochromic results often stem from underlying mechanisms that are not related to control of developmental timing (Raff and Wray 1989). Other mechanistic changes in development, such as spatial alterations (heterotopy), are less well understood.

A third issue also has a long history. Phylogenetic histories indicate that some features of development are retained over immensely long periods of geologic time. These features are most dramatic in earlier stages of development, and in some cases must have been retained from the times of origins of the phyla in which they occur (over 500 million years). Observations of this kind have been interpreted to mean that early development cannot accept evolutionary changes because all later events of development depend in a hierarchical way upon those that have occurred earlier (Arthur 1988). But, an alternate view can also be suggested. That is that developmental processes are quite distinct during different stages of development. Thus, maternally-stored information can play a dominant role in early developmental decisions, but processes such as cell-cell interactions may do so later. Thus, we should expect different developmental stages to respond very differently to selective pressures in evolution (Raff *et al.* 1991). If this proves to be so, evolution will have provided a major clue to the overall organization of ontogeny.

Finally, in more general terms, one can ask questions about mechanisms without posing specific theoretical issues to be tested. For example, how many processes are affected by evolutionary changes in development, and how do they change? We have addressed all of these questions in our studies of the evolution of early development in sea urchins. We focus here on changes in morphogenetic processes and their consequences.

EVOLUTIONARY CONSERVATION OF MORPHOGENESIS IN SEA URCHINS

Most of the roughly 1000 living species of sea urchins spawn large numbers of relatively small eggs (*ca.* 100 μm in diameter), and develop via a feeding pluteus larva. The echinus rudiment that will give rise to the juvenile at metamorphosis grows within the pluteus over a period of a few weeks of feeding. The basic attributes of morphogenesis of these typical indirect developing sea urchin embryos are well known (Okazaki 1975). Cleavage is radial and equal for the first two cell cycles. The third cleavage is equatorial, producing two tiers of equal sized cells. At the fourth cleavage, the animal tier of four cells again cleaves equally (Figure 1). The vegetal tier cleaves unequally to produce four large macromeres and four small micromeres. The micromeres will divide unequally again at the fifth cleavage. The machinery that sets

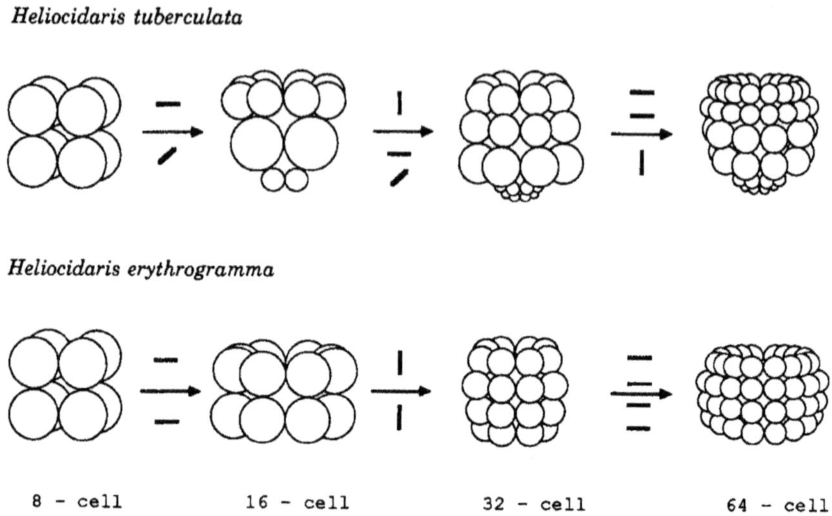

Figure 1. Comparisons of cleavage patterns of indirect developing sea urchins with that exhibited by *H. erythrogramma*. The animal pole is oriented upwards. In indirect developing species, such as *H. tuberculata*, an unequal division of vegetal blastomeres at the fourth cleavage produces a three-tiered, 16-cell embryo. In subsequent divisions, cells of the three tiers divide in different orientations. The heavy bars indicate the axes of the mitotic spindles in the indicated cell tiers at each cleavage. In *H. erythrogramma* there is no unequal cleavage. From Wray and Raff 1989. Used with permission of Academic Press.

up the unequal cleavage is apparently maternal in origin (Jenkinson 1911; Maruyama *et al.* 1985). Unequal cleavage results from the migration of the nucleus to a site at the vegetal pole in each vegetal cell shortly before the fourth cleavage (Dan 1979). The asters of the mitotic apparatus that forms in the vegetal cells are located asymmetrically, with one pole of the mitotic apparatus free in the cell, and the other anchored to the cortex of the vegetal end of the cell (Schroeder 1980). The fates of the micromeres are set at the time of their appearance by as yet undefined maternal factors (Harkey 1983; Maruyama *et al.* 1985). The fates of the other blastomeres are not rigidly specified by maternal factors, but require inductive interactions (Hörstadius 1973; Henry *et al.* 1989; Davidson 1989).

At about the 256-cell stage, the cells become arrayed around the blastocoelar space to form a hollow ball of ciliated cells, the typical coeloblastula. Gastrulation is initiated when the daughters of the micromeres situated at the vegetal pole enter the blastocoel and give rise to primary mesenchyme cells. This event is soon followed by the invagination of the archenteron and the ingression of secondary mesenchyme, both derived from the macromeres (Hörstadius 1973; Cameron *et al.* 1987). The cellular mechanisms which underly gastrulation in indirect-developing sea urchins are relatively well defined (Ettensohn 1984, 1985; Hardin and Cheng 1986; Hardin 1989). Invagination of the archenteron begins as the vegetal plate buckles inward to produce a short, wide archenteron. Further extension is produced by cell rearrangements and cell shape changes that transform a short wide tube into a long narrow tube *via* isovolumetric extension. During the final phase of gastrulation, filopodia guide the tip of the archenteron to the roof of the blastocoel, which forms the prospective oral face of the pluteus (Hardin and McClay 1990).

The basic cellular mechanisms of gastrulation are similar among the typical indirect-developing sea urchins of the two extant subclasses, the Cidaroidea and the Euechinoidea (Ettensohn 1985; Hardin and Cheng 1986; Hardin 1989). These two subclasses separated about 250 million years ago (Kier 1984). Since indirect development is the ancestral mode within echinoderms, including sea urchins (Strathmann 1978; Raff 1987) the sequence of events just described was likely shared by the common ancestor of cidaroids and euechinoids. There is also evidence that isovolumetric cell rearrangements drive gastrulation in starfish (Ettensohn 1984). Since starfish and sea urchins diverged 490-530 million years ago (Smith 1988a), this mechanism is clearly very ancient.

The last phase of embryonic development in typical sea urchins is the product of a set of complex morphogenetic events by which the gastrula is transformed into the pluteus. The various cell lineages of the embryo derive from the three distinct and characteristic blastomere types of the 16-cell embryo. Mesomeres, the most animal pole cells, yield ectodermal cells and the larval nervous system. The most vegetal pole cells, the micromeres yield the skeleton-secreting primary mesenchyme cells as well as some coelomic precursors. Macromeres, the fourth cleavage sister cells to the micromeres yield the endodermal cells of the gut, secondary mesenchyme cells, including pigment cells, some coelom precursors and some ectoderm. The characteristic shape of the pluteus is largely the result of the secretion and extension of the larval skeleton to produce the initial four arms of the pluteus. Eventually eight arms are generated. The complex ciliated band important in feeding is formed along the arm surfaces and a larval nervous system appears near the base of the arms. The dorsal (aboral) ectoderm cells become squamous epithelium and cover the greatly expanded dorsal surface. The archenteron tip makes contact with the oral ectoderm, resulting in the opening of the mouth and the archenteron differentiates into a functioning digestive system. The first steps in adult differentiation also occur during the early development of the pluteus, with the budding of coelomic pouches from the walls of the gut. In combination with the overlying ectoderm on the left side of the pluteus these pouches will later give rise to the echinus rudiment (Okazaki 1975).

THE UTILITY OF DIRECT DEVELOPMENT

To investigate evolutionary changes in early development, a suitable experimental system is necessary. Ideally, organisms used for such studies should differ in substantive ways in their early development from their closest relatives. They should be as closely related as possible to the species with which they will be compared, and the genetic distance should be known. Finally, the phylogenetic history of the experimental species and its relatives is required so that the direction of evolutionary change can be inferred. These conditions are met by some species of sea urchins, which develop from the egg to juvenile adult without producing a pluteus larva. Because such species have (to a greater or lesser degree) an abbreviated early ontogeny, they are said to exhibit direct development.

The species discussed in this chapter are the two Australian species *Heliocidaris tuberculata* and *H. erythrogramma*. *Heliocidaris* species are camarodont euechinoids, sufficiently closely related to *Strongylocentrotus purpuratus* to allow close developmental and molecular comparisons to this well studied species (Smith 1988b).

Both *Heliocidaris* species are common subtidal animals on the New South Wales coast in the vicinity of Sydney. They are similar in morphology and habits, with *H. tuberculata* being larger and having a more robust test and spines. The reproductive behavior of the two species is strikingly different. *H. tuberculata* produces fertile gametes over the entire year, whereas *H. erythrogramma* is fertile only during the summer months of December through February (Laegdsgaard 1989). There are also profound differences in gametes. The eggs of *H. tuberculata* are 90 μm in diameter, whereas those of *H. erythrogramma* are 430 μm in diameter. These larger eggs, although well supplied with protein, lack typical yolk proteins (Scott *et al.* 1990). The eggs are extremely rich in oil and float.

Development of the two *Heliocidaris* species is also strikingly different (Raff 1987). *H. tuberculata* develops indirectly and forms a feeding pluteus. *H. erythrogramma* produces no pluteus, but develops directly from egg to juvenile adult without feeding (Figure 2) (Williams and Anderson 1975; Raff 1987). These two species provide an ideal basis for study of the evolution of early development for several reasons. First, the two species are sufficiently closely related that differences in gene expression and cellular behavior in development should be related to developmental mode *per se*. Second, we can recognize the direction of evolutionary change because we know that the indirect development of *H. tuberculata* represents the ancestral mode. Third, it is possible to demonstrate homologies of cell types in the two species by molecular probes, cell position, cellular organization, and cell fate. Finally, because *Heliocidaris* is closely related to *Strongylocentrotus*, the great amount of molecular and developmental data collected for *S. purpuratus* can be readily utilized as a basis for comparisons.

Briefly, *H. erythrogramma* differs from indirect developers in that it undergoes a modified cleavage in which no micromeres are formed. It produces an unusual wrinkled blastula and generates a very large number of primary mesenchyme cells, rather than the small number produced in indirect development. Gastrulation is also modified to produce a short archenteron and a highly accelerated initiation of coelom formation. No larval skeleton is produced, and the larva is elongated along the animal-vegetal axis. The larva is brightly pigmented and possesses a prominent ciliated band. Adult structures develop rapidly and metamorphosis takes place about four days after fertilization. Superficially, it appears that *H. erythrogramma* development is the product of a simple set of heterochronies in which some features of the pluteus have been deleted, and adult development has been accelerated (Raff 1987). Deeper examination shows that this is an oversimplified picture. In fact, the entire program of early development has been remodeled, and the *H. erythrogramma* larva is the product of a novel developmental program.

Although this chapter will be concerned with morphogenesis in *H. erythrogramma*, it is worth mentioning that direct development is a common developmental strategy among sea urchins and other echinoderms (as well as among other groups of animals better known for having feeding larvae–for example, frogs). Direct development has arisen independently in at least six orders of sea urchins, and an estimated 20% of sea urchin species are direct developers. Although only indirect developers occur in North America and Europe, direct developers are very prevalent in some faunas, such as those of Australia and Antarctica (Emlet *et al.* 1987; Raff 1987). The vast amount of data collected on indirect development over the past century provides an enormous database for comparisons of data collected from direct developers, and aids greatly in the design

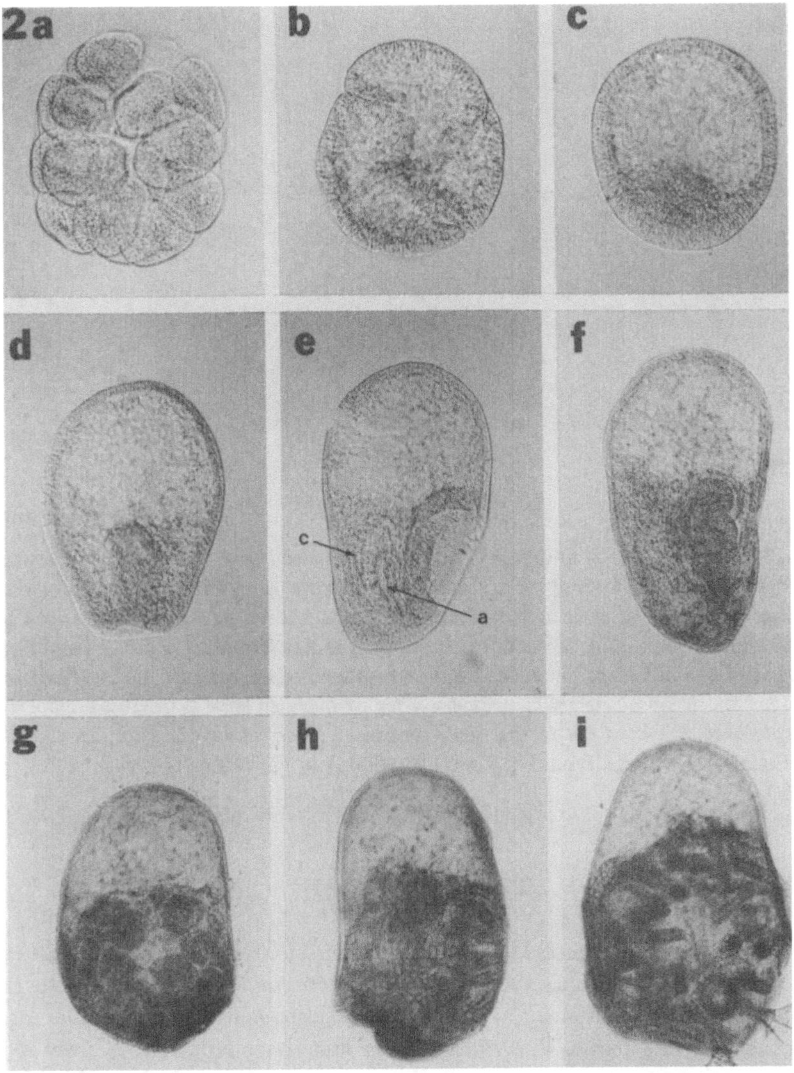

Figure 2. Development of *H. erythrogramma*. Embryos were cleared with xylene. **(a)** Sixteen-cell stage. **(b)** 8-hour wrinkled blastula. **(c)** 12-hour mesenchyme blastula. **(d)** 19-hour gastrula. **(e)** 28.5-hour early larva with vestibule (on right). **(f)** 37-hour larva with tube feet forming in vestibule. **(g)** 58-hour embryo viewed from vestibule. **(h)** 70-hour larva with heavily calcified echinus rudiment filling most of larva. **(i)** 81-hour larva shortly prior to hatching, with tube feet and spines extending from vestibule. All embryos are oriented with the animal pole upward. c = coelom; a = archenteron. Scale bar = 120 μm. From Bisgrove and Raff (1989) with permission of *Dev. Growth & Differ.*

of studies with direct developers. Direct developing sea urchin larvae are quite diverse in form (Figure 3), a product of their independent evolutionary origins. Thus, the larvae of closely related species can be strikingly different, reflecting the multiple possible evolutionary solutions to the achievement of direct development.

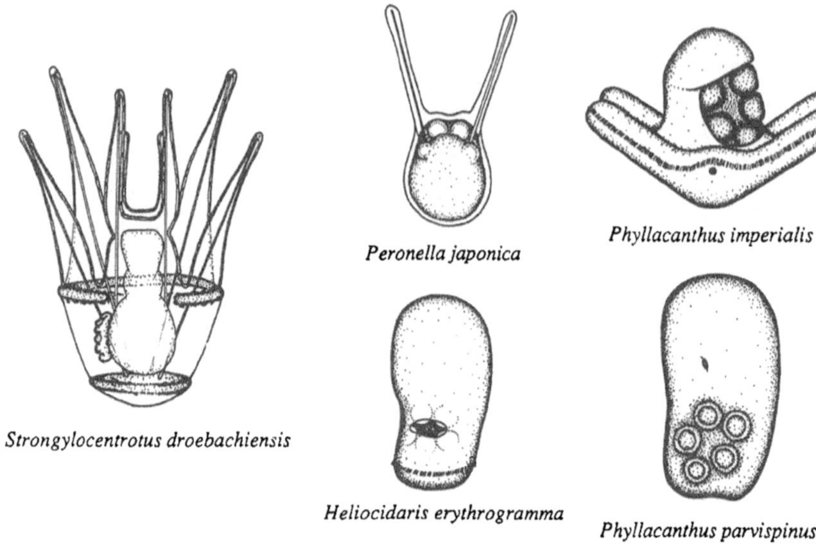

Figure 3. Diversity of sea urchin larvae. The pluteus of *Strongylocentrotus droebachiensis* represents the ancestral feeding larval form. All direct-developing forms evolved from ancestors developing by means of a pluteus. Independently evolved direct developers show various degrees of modification of the pluteus larval form. Some, such as *Phyllacanthus imperialis* and *Peronella japonica* produce non-feeding pluteus-like larvae. In other direct developers, such as *Phyllacanthus parvispinus* and *Heliocidaris erythrogramma*, a highly modified larva is produced. Note that the two *Phyllacanthus* species are in the same genus but possesses two distinct direct-developing larvae. Scale bar = 200 μm. From Wray and Raff (1991a), by permission of *Trends in Ecology and Evolution*.

A RAPID RATE OF EVOLUTIONARY CHANGE

The fossil record suggests that the genus *Heliocidaris* diverged about 30 million years ago from related genera such as *Strongylocentrotus* (Phillip 1965; Smith 1988b). Because indirect development *via* the pluteus represents the primitive mode of development in this genus, *H. erythrogramma* must have acquired its novel mode of development during or subsequent to its divergence from *H. tuberculata*. We have used two approaches to dating this divergence: total single copy nuclear DNA divergence (Smith *et al.* 1990) and divergence of mitochondrial DNA (McMillan *et al.* 1991). Both methods agree in giving inferred dates of separation of the two species from a common ancestor of about 10 million years.

Unfortunately, such an estimate of divergence time of the two lineages does not give the time at which direct development evolved in the *H. erythrogramma* lineage. Differences in the mitochondrial DNAs of East and West coast populations of *H. erythrogramma* provide some further narrowing of this window. The adult morphology of the two populations are distinct from each other as well as from south coast populations (McMillan *et al.* 1991). This genetic differentiation is probably a consequence of the short developmental time and resulting low dispersal capacity. However, the widely separated populations develop identically (Raff and Wray, unpublished observations). Mitochondrial DNA differences are consistent with a separation of about 2 million years for east and west coast populations (McMillan *et al.* 1991). Direct development thus arose before genetic isolation. This narrows the time window for evolution of direct development to 8 million years. The critical changes were probably much faster, but direct evidence of timing of these changes is presently not available. Given the long conservation of development *via* a pluteus (over 250 million years), even a rough maximum estimate of 8 million years still emerges as very rapid for the radical and pervasive changes in early development we document below.

CLEAVAGE AND RADICAL REORGANIZATION OF FATE MAP

The first and second cleavages of *H. erythrogramma* are parallel to the animal-vegetal axis, and the third cleavage produces two equal tiers of cells (Figure 1) (Williams and Anderson 1975; Wray and Raff 1989, 1990). The typical echinoid unequal fourth and fifth cleavages, however, fail to occur. Instead these cleavages are equal. Later cleavages are also equal; thus, unequal cleavages are not simply delayed to later division cycles. Since the positions of the mitotic spindles in the vegetal blastomeres of *H. erythrogramma* have a new and precise cellular orientation and location, a different spatial control has replaced that producing the unequal cleavages of indirect development. This change in cleavage pattern sets up a new set of cell precursors out of which the cell types of the embryo will arise. At a minimum, such a change will alter the timing of cell lineage decisions. Changes in cell lineage specification and commitment, however, have occurred in *H. erythrogramma* that go beyond what might be expected from a simple modification in cleavage pattern (Wray and Raff 1989, 1990).

Modified cleavage poses an interesting problem in establishing homologies between cell types of the two *Heliocidaris* species, since the differentiated cell types of each arise from cell precursors with different cell division patterns. To trace cell lineages, we injected fluorescently-labeled dextran into individual *H. erythrogamma* blastomeres and tracked the fates of labeled descendant cells into larval stages (Wray and Raff 1989, 1990). Since cleavages are constant with respect to embryonic axes, and equal and synchronous to the 7th cleavage, the cell cleavage pattern lends itself to ready analysis of cleavage pattern and cell lineage.

The 32-cell fate maps for *H. erythrogramma* and for indirect development are diagrammed in Figure 4 (Wray and Raff 1990). In both modes of development, all of the animal tier blastomeres, along with some of the vegetal blastomeres give rise to ectoderm. There are, however, four major differences in origins of ectodermal cells and structures, which have taken place in the shift to direct development. First, there is a major shift in symmetry such that ectodermal fates extend into the most vegetal blastomeres on the dorsal side. Second, the ectoderm of the *H. erythrogramma* embryo undergoes a dramatic distortion during gastrulation and formation of the vestibule (see below). Much of the ventral blastomere of the two-cell embryo contributes to internal cell fates, with the vestibule arising from the animal descendants of this blastomere. Whereas only part of two left side animal blastomeres in the 32-cell embryo of indirect developers gives rise to vestibule, fully eight blastomeres produce vestibular ectoderm in *H. erythrogramma*. Third, the ciliated band has shifted from the animal to the vegetal hemisphere. And fourth, the larval mouth has been lost.

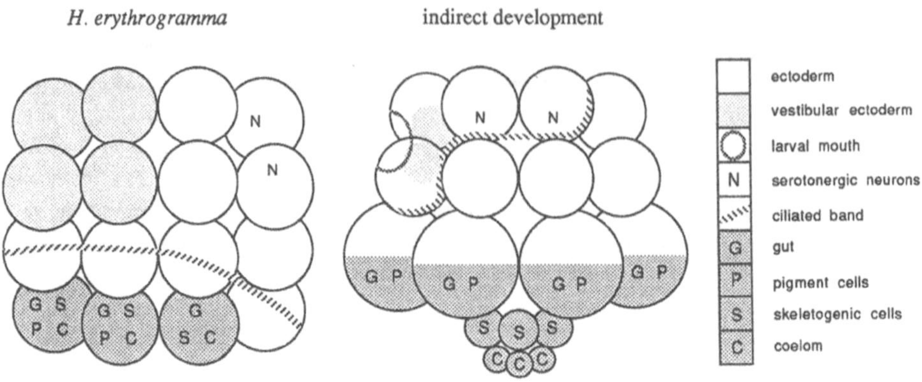

Figure 4. Fate maps of 32-cell indirect developing sea urchin and *H. erythrogramma*. The embryos are viewed from their left sides. Blastomeres giving rise to ectodermal structures are shown in white and light stipple. Blastomeres with internal fates are shown in heavy stipple. The first cleavage in *H. erythrogramma* divides the embryo into dorsal and ventral halves with respect to both the larval and adult axes, which correspond. In the indirect developing embryo, the adult axis is offset 45° from the larval axis. Prospective vestibular ectoderm is symmetrically distributed in *H. erythrogramma*, and arises from four left blastomeres and four right blastomeres. In indirect development the vestibule arises asymmetrically, with only a portion of two left blastomeres giving rise to vestibule. Most vegetal fates are symmetrical in indirect development, but strikingly asymmetric in *H. erythrogramma*. Other differences are also apparent. The full description of the cell lineage of *H. erythrogramma* is given by Wray and Raff (1990). From Wray and Raff (1990) by permission of Academic Press.

Changes in ectodermal cell fates are extensive and underly many of the modifications evident in the fate map (see Figure 4). In indirect development only 16 cells produce ectoderm exclusively, whereas, in *H. erythrogramma*, 26 cells give rise to ectoderm and ectodermal derivatives. Of these, eight cells, constituting one quarter of the embryo, form the vestibule. Thus, ectodermal cells have switched to new fates. A possible adaptive explanation is that in the non-feeding *H. erythrogramma* larva a full sized vestibule must be formed rapidly. The more proximal cause of these switches in cell fate is not yet known, although it is clear that extensive genetic controls govern cell lineage behavior (Sternberg and Horvitz 1984; Ambros 1989). The evolutionary translocation of the ciliated band is also dramatic and poses problems for identifying homologies. In the pluteus the ciliary band defines the margin of the oral face (Cameron *et al.* 1990). The ciliated band of the pluteus is roughly oval in shape when the larval mouth forms, and becomes highly contorted into long narrow loops running along the larval arms surrounding the mouth. The origins of the ciliated band lie in major part in oral and lateral animal pole blastomeres. In contrast, the ciliated band in *H. erythrogramma* circles the blastopore, and arises entirely from vegetal blastomeres. This new location may be related to the mode of swimming of the larva (Wray and Raff 1990).

Equally dramatic changes have evolved in internal cell fates. First, the internal cell types, (gut, pigment cells, skeletogenic cells, and coelomic cells) are all clearly homologous to those of indirect developers such as *S. purpuratus*, yet the changed cleavage pattern of *H. erythrogramma* has resulted in a very different set of cellular precursors for them. These differences are reflected in the timing at which founder cells are established. For example, coelomic precursors are established one cleavage later in *H. erythrogramma* than in indirect development, and skeletogeneic cell precursors at least two cell cleavages later than in indirect development. Second, there is a shift from a radial symmetry in the origins of internal cell types to a pronounced asymmetry (see Figure 4).

There are also profound changes in order of founder cell origins. The sequence in which founder cells for internal larval cell types split from each other is significantly different from the sequences observed in indirect development. Thus, for example, in indirect development the division of each micromere at the fifth cleavage produces a skeletogenic founder cell and a coelomic founder cell. In *H. erythrogramma*, coelomic founder cells arise at the sixth cleavage, but their sister blastomeres are not skeletogenic founder cells. Those founder cells are produced in later divisions from the progeny of sixth cleavage cycle sister cells which contain endoderm, skeletogenic and secondary mesenchyme fates (Wray and Raff 1990). These extensive changes indicate that only a loose determinative relationship exists between cell lineage and cell fate in sea urchins. Thus, the skeletogenic cells of *H. erythrogramma* arise from a different cell lineage than those of indirect development. Nevertheless, they possess the same morphology, carry out the same activities, and express the same skeletogenic cell-specific antigen msp130, which are characteristic of skeletogenic cells in indirect development (Wray and Raff 1990). Other results point to a great potential flexibility for evolutionary modification of cell lineages in sea urchins. For example, embryos of more or less normal morphology containing cell types such as skeletogenic mesenchyme can also be obtained from isolated animal blastomeres of indirect-developing sea urchin embryos (Henry *et al.* 1989; Raff *et al.* 1991).

DETERMINATION OF AXES

Development proceeds as an organized process within a framework of axial systems which are established either maternally or through epigenetic events. Sea urchin embryos have provided experimental embryologists with a model system for studying various aspects of developmental axis formation. Until recently, however, these studies have concentrated on indirect developing species.

One such axis, the animal-vegetal axis, is established maternally (Boveri 1901; Hörstadius 1973; Maruyama *et al.* 1985). The animal pole is defined as the site where the meiotic maturation divisions that produce the polar bodies take place (Lindahl 1932). The vegetal pole, directly opposite the animal pole, marks the site where gastrulation takes place. The early cleavage divisions are oriented in a characteristic fashion relative to the animal-vegetal axis, and subsequent animal (ectodermal) and vegetal (endodermal and mesodermal) developmental potential is organized relative to this axis. Echinoderm eggs are surrounded by a transparent jelly coat which contains a small conical micropile at the animal pole in many species (Boveri 1901; Hörstadius 1973; Schroeder 1980). The animal-vegetal axis is distinctly marked in the eggs of the indirect developing sea urchin *Paracentrotus lividus* by a pigment band which surrounds the egg just below the equator within the vegetal hemisphere.

Jenkinson (1911) demonstrated that the animal-vegetal axis in sea urchins is related to the orientation and attachment of the oocyte within the ovary, and therefore appears to be established during oogenesis. Experimental work has demonstrated that animal-vegetal developmental potential is governed in part by maternal factors which are localized in the unfertilized egg Maruyama *et al.* (1985). In general, the animal-vegetal axis in the eggs and embryos of echinoids appears to be very stable and is not easily altered by various experimental treatments (Morgan and Spooner 1909; Hörstadius 1973; Harvey 1956; Henry *et al.* 1990).

Although Williams and Anderson (1975) reported the presence of an animal pole micropile in the jelly coat surrounding the eggs of *H. erythrogramma*, we have not been able to detect it. However, a small hillock in the plasma membrane is present at the animal pole, which marks the site where polar body formation has taken place in *H. erythrogramma* (Henry *et al.* 1990). Other maternal indices of animal-vegetal polarity are present in *H. erythrogramma* eggs. For instance, lipids are more concentrated in the vegetal hemisphere; as a consequence eggs float with the vegetal pole upwards. In addition, first cleavage in *H. erythrogramma* usually takes place in a unipolar fashion (Wray and Raff 1989), with cleavage furrow formation beginning at the animal pole.

Unlike the animal-vegetal axis, the dorsoventral axis appears to be set up later in development, and is far more labile in indirect developing sea urchins. The studies of Hörstadius and Wolskey (1936) and Kominami (1988) indicated no definite relationship between the first two cleavage planes and the larval dorsoventral axis in the sea urchin *Paracentrotus lividus* and *Hemicentrotus pulcherrimus*. In these cases the dorsoventral axis appears to be set up some time after the four-cell stage. A similar situation appears to occur in the embryos of the starfish, *Asterina pectinifera* (Kominami 1983). Contrasting results were obtained by Cameron *et al.* (1987, 1989) during fate mapping studies of *Strongylocentrotus purpuratus*. These investigators observed a fairly consistent relationship (90% of the cases) between the first two cleavage planes and the larval dorsoventral (oral-aboral) axis.

Regardless of the relationship between the early cleavage planes and the larval dorsoventral axis in these indirect developing species, the results of blastomere isolation experiments indicate that the dorsoventral axis in these indirect developing species is determined epigenetically after the four-cell stage. Blastomeres isolated through the four-cell stage are able to display complete regulation (Hörstadius 1973; Henry and Raff 1990). On the other hand, the dorsoventral axis of the direct developing sea urchin, *H. erythrogramma* is specified prior to first cleavage and is likely to be established maternally. Fate mapping studies in this species indicate that the first cleavage plane always bisects the embryo along the frontal plane (Wray and Raff 1989, 1990). In addition, the two blastomeres isolated after first cleavage display distinct differences in dorsal and ventral developmental potential (Henry and Raff 1990). By experimentally re-orienting the first cleavage spindle, Henry *et al.* (1990) have demonstrated that the specification of the dorsoventral axis in *H. erythrogramma* is not tied to the first cleavage division. Therefore, this axis must be specified prior to first cleavage, most likely in the unfertilized egg.

It has been proposed that the evolution of a more direct mode of development may entail the accelerated establishment of axial systems so that development can proceed more rapidly (Freeman 1982). Although the information discussed above appears to support this hypothesis, other reports suggest that accelerated axial establishment is not pivotal to the evolution of direct development. Gastropods display both indirect and direct forms of development. *Ilyanassa obsoleta*, an indirect developing gastropod mollusc displays a very early determination of the dorsoventral axis (Clement 1952), whereas direct developing species, such as *Lymnaea palustris* (van den Biggelaar and Guerrier 1979; Arnolds, *et al.* 1983; Martindale, *et al.* 1985) display a much later determination of this axis. Not all direct developing echinoids may display an accelerated establishment of the dorsoventral axis. In fact, evidence suggests that this may be the case for the direct developing sea urchin, *Peronella japonica*. Okazaki and Dan (1954) did not report any differences in the developmental potential of blastomeres isolated at either the two- or four-cell stages in this species. Direct development in echinoids is derived from an ancestral condition which involved the formation of a feeding pluteus larva, and it is likely that direct developing molluscs also evolved from forms with feeding larvae (Strathmann 1978). Therefore, it is more likely that a variety of developmental solutions have been utilized in establishing direct development (Wray and Raff 1991a), which has arisen many times within these two groups of organisms (Strathmann 1978).

THE WRINKLED BLASTULA

Early cleavage divisions of echinoderms are generally radial and holoblastic (Hörstadius 1973). As a result of these early cell divisions and the arrangement of the resulting blastomeres, a central fluid-filled cavity develops which becomes the blastocoel (Figure 5a-c). In most species of echinoderms which display indirect development, subsequent cleavage leads to the formation of a rounded smooth-walled epithelium containing a single layer of cells surrounding the blastocoel (Figure 5c). There are many echinoderms, however, which form a wrinkled blastula instead (Inaba 1968; Williams and Anderson 1975; Mladenov 1979; Strathmann 1987; Parks, *et al.* 1989; Henry *et al.* 1991). In these species, pronounced wrinkles or furrows appear in the epithelial wall as the blastula forms (Figure 2b). Wrinkled blastula formation is most common in direct developing species, which develop from large lecithotrophic eggs. However, formation of a wrinkled blastula has been noted in a few indirect developing species which produce small eggs.

The furrows of the wrinkled blastula appear along one, two, or three orthogonal planes. It has been noted that the furrows superficially subdivide the embryos so that they resemble two-cell, four-cell, and eight-cell cleavage stages (Gemmill 1912, 1916; Hörstadius 1939; Kubo 1951; Strathmann 1987; Okazaki and Dan 1954; Williams and Anderson 1975). Previous investigators have suggested that the wrinkles form by active invagination of the blastula epithelium (Inaba 1968; Hayashi 1972). However, we have found in *H. erythgrogramma* that the wrinkles actually begin to form during the early cleavage stages as the blastomeres undergo partial separation along the first, second, and in some cases the third cleavage planes (Henry *et al.* 1991). Cell separation begins during the four-cell stage (Figure 6a). Subsequent cell divisions are oriented such that the blastula epithelium is formed with furrows along these planes of separation (Figure 6b). We have verified that the major wrinkles correspond to these planes of cleavage through injections of fluorescent tracer dye (Henry *et al.* 1991). If one blastomere in a two-, four-, or eight-cell embryo is injected, label is distributed in halves, quarters, or octants, respectively, during wrinkled blastula stages with the boundaries between labeled and unlabeled regions occurring at the wrinkles (Figure 7a-c).

There is another factor which contributes to the formation of the wrinkled blastula. As cell divisions continue to generate smaller cells, there is a tremendous expansion of the surface area of the blastula epithelium. The embryos of indirect-developing sea urchins are generally surrounded by a fertilization envelope in which there is ample room for expansion of the blastocoel and surrounding epithelium (Figure 5). On the other hand, the larger embryos of *H. erythrogramma* are tightly surrounded by a

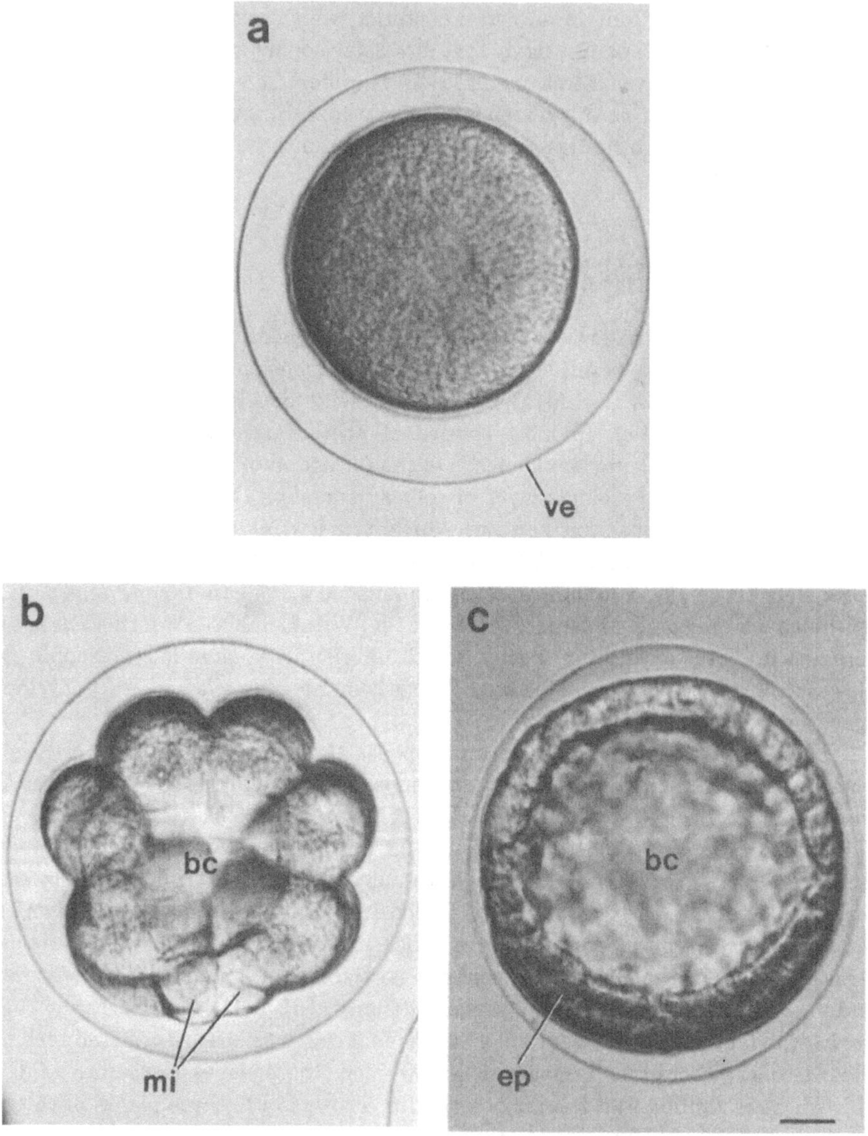

Figure 5. Light micrographs of living *Lytechinus variegatus* embryos. **(a)** Fertilized egg with elevated fertilization envelope. **(b)** 16-cell embryo. Blastocoel formation has clearly begun. **(c)** Blastula stage embryo. A smooth-walled blastula epithelium has formed as the embryo has expanded within the vitelline envelope. ve, vitelline envelope; bc, blastocoel; ep, blastula epithelium; mi, micromeres. Scale bar equals 20 μm.

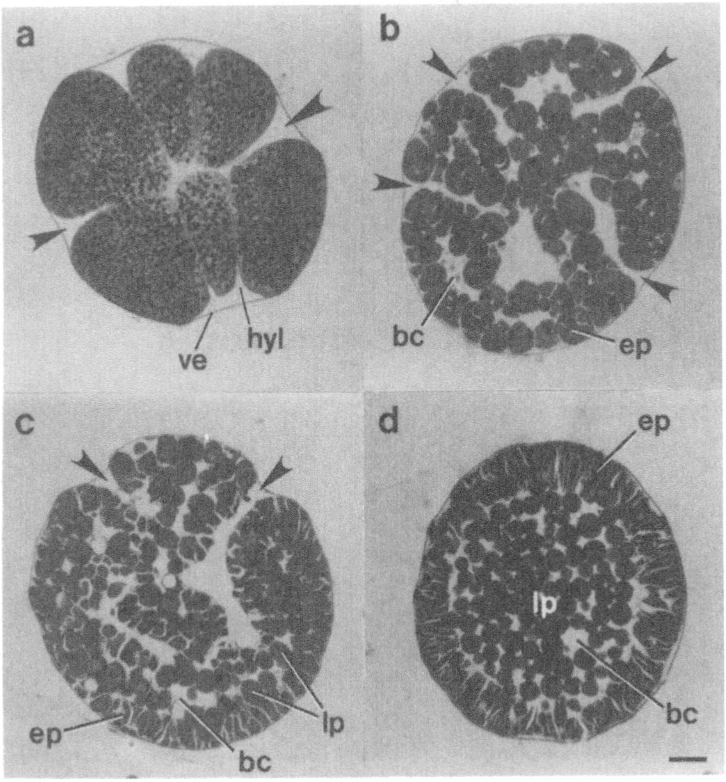

Figure 6. Light micrographs of sections prepared at various stages of wrinkled blastula formation in *H. erytrhogramma*. **(a)** Eight-cell embryo showing partial separation of the blastomeres along the first cleavage plane (arrows). The fertilization envelope surrounds the embryo. **(b)** Newly formed wrinkled blastula stage embryo; arrows indicate sites of wrinkles. **(c)** Later stage of wrinkled blastula formation showing the secretion of lipid droplets into the blastocoel. Arrows indicate wrinkles. **(d)** Smooth-walled blastula stage after wrinkle formation. The embryo has not yet hatched from its fertilization envelope. The cells of the blastula epithelium have become highly elongated and more tightly arrayed. The central blastocoel is now filled with secreted lipid droplets, each bounded by a thin layer of cytoplasm and a plasma membrane. bc, blastocoel; lp, lipid droplets; ve, vitelline envelope; ep, blastula epithelium. Scale bar equals 50 μm.

fertilization envelope (Figure 2a). As the blastula epithelium increases in area, there is no room for expansion of the embryo, therefore the epithelium must assume a highly folded state (Figure 2c, 6b-c). It is therefore not necessary to invoke a mechanism in which the blastula wall undergoes active invagination (Henry *et al.* 1991). Just prior to the time of hatching the wrinkles disappear and the blastula epithelium becomes smooth and rounded (Figure 6d). The wrinkles are lost because the blastula epithelium undergoes a significant increase in thickness, and the cells become more tightly packed (Figure 6d). Another important process takes place during these stages which may contribute to the smoothing of the blastula epithelium. The eggs of *H. erythrogramma* are filled with numerous lipid droplets which serve as nutrient reserves and provide buoyancy to the eggs and embryos. During wrinkled blastula stages, these lipids are

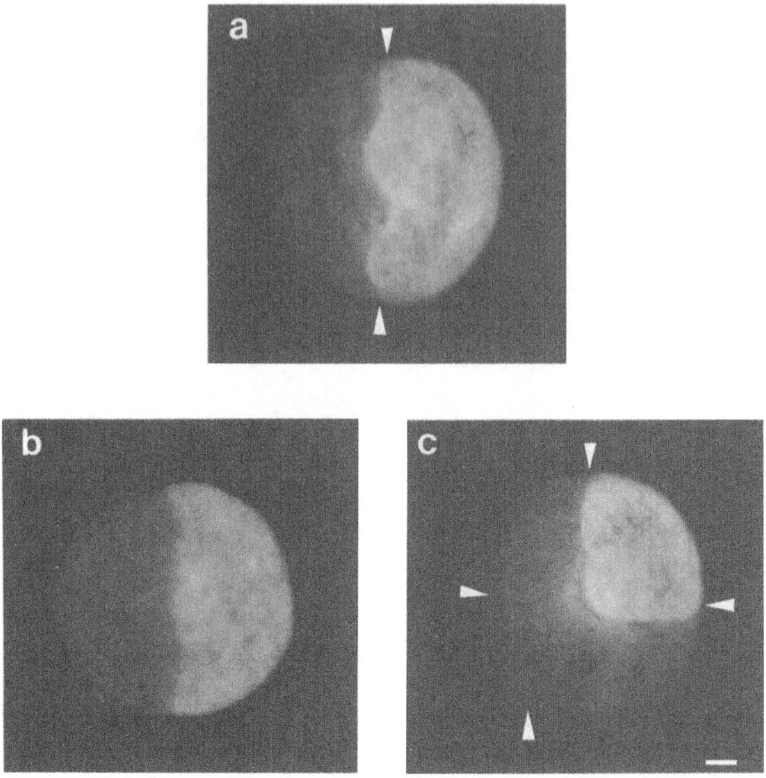

Figure 7. Fluorescence micrographs showing the relationship between the early cleavage furrows and the furrows present in the blastula epithelium during wrinkled blastula stages in *H. erythrogramma*. **(a)** Wrinkled blastula stage embryo resulting from the injection of a single blastomere at the two-cell stage. The boundary between labeled and unlabeled regions (arrow heads) corresponds to one of the furrows. **(b)** An older blastula stage embryo showing the symmetrical distribution of label in one hemisphere after the wrinkles have disappeared. In this case one blastomere was injected at the two cell stage. **(c)** Wrinkled blastula stage embryo resulting from the injection of a single blastomere at the four-cell stage. The labeled region is bounded by the two major furrows which are located at right angles to one another. Scale bar equals 50 μm. From Henry *et al.* (1991), with permission of *Dev. Growth & Differ.*

secreted into the blastoceol (Figures 6c,d) through an apocrine process in which they are extruded in membrane bound sacs (Henry *et al.* 1991). A large amount of cell membrane could be removed from the cells of the blastula epithelium as a result of this process, and lead to a decrease in the surface area of the blastula. It should be mentioned, however, that this process of apocrine secretion does not take place in most species which form a wrinkled blastula, therefore the formation and loss of wrinkles must be a completely independent process from lipid secretion, although they are coincident in *H. erythrogramma*.

There does not appear to be any clear taxonomic trend among species which form wrinkled blastulae (Hayashi 1972; Henry *et al.* 1991). Wrinkled blastula are formed by members of most echinoderm classes including the Echinoidea (sea urchins), Asteroidae (starfish), Holothuroidea (sea cucumbers), and Ophiuroidea (brittle stars). However, wrinkled blastula formation is more commonly seen among direct-developing species with large lecithotrophic eggs. Since direct development has arisen from an ancestral condition in which extended planktotrophic larval development took place (Strathmann 1978; Raff 1987), the widespread occurrence of wrinkled blastulae in direct-developing species indicates that this process has arisen independently on many occasions and must be readily achieved. This would seem to indicate that wrinkled blastula formation probably results form simple physical parameters such as those discussed above. It is important to note, however, that the process of wrinkled blastula formation is not strictly tied to direct development since there are direct developing species which do not form a wrinkled blastula, and wrinkled blastula formation has been noted to occur in a few indirect-developing species (Henry *et al.* 1991).

A NOVEL MODE OF GASTRULATION

Gastrulation in *H. erythrogramma* begins in much the same manner as that in indirect development. The first overt sign of morphogenesis is the ingression of primary mesenchyme cells from the vegetal plate at about 12 hours (Williams and Anderson 1975). Embryos of indirect developers produce only 32 to 64 primary mesenchyme cells, but in *H. erythrogramma* some 2000 primary mesenchyme cells ingress into the blastocoel (Parks *et al.* 1988). This dramatic change in cell number is in keeping with the generally higher number of cells present in the *H. erythrogramma* embryo: the gastrula contains 10,000 to 14,000 cells instead of the approximately 1000 cells in the gastrula of an indirect developer. Three to four additional cell cycles must occur during cleavage to account for this general increase in cell number (Parks *et al.* 1988).

Gastrulation proper begins by 15 hours, as the vegetal plate buckles inwards. This first phase of gastrulation in *H. erythrogramma* produces a short, wide archenteron (Wray and Raff 1991b). Cell marking experiments demonstrate that this first phase of gastrulation is a radially symmetrical process with respect to the animal-vegetal axis (Figure 8a). In both respects, this first phase of gastrulation in *H. erythrogramma* closely resembles that in indirect development (Hörstadius 1973; Ettensohn 1984). As gastrulation proceeds, however, substantive differences become apparent. The animal-vegetal axis moves ventrally during the second phase of gastrulation (Wray and Raff 1991b). Dorsal ectoderm expands to cover the majority of the exterior of the embryo, and the former animal pole is displaced ventrally. In concert with these cell sheet movements, ventral cells move to the interior of the embryo (Figure 8e,f). Differences in external dorsal and ventral clone sizes are apparent by 18 hours and become progressively more pronounced until about 24 hours, when extension of the archenteron is complete.

Fate mapping studies of 36 hour larvae have demonstrated that the dorsal clone in *H. erythrogramma* contributes the majority of larval ectoderm and a minority of internal fates, whereas the ventral clone makes reciprocal contributions (Wray and Raff 1989, 1990). Observations of late gastrulae (24 hours) confirm that cells derived from the ventral side of the embryo extend from the ventral lip of the blastopore, over the tip of the archenteron, and half way down its dorsal side (Figure 8f; Wray and Raff 1991b). Gastrulation in *H. erythrogramma* thus involves a gradual shrinking of the ventral ectoderm, coupled with a progressive increase in internal contributions by ventral cells. During this period, no net increase in dorsal contribution to internal cell volume occurs.

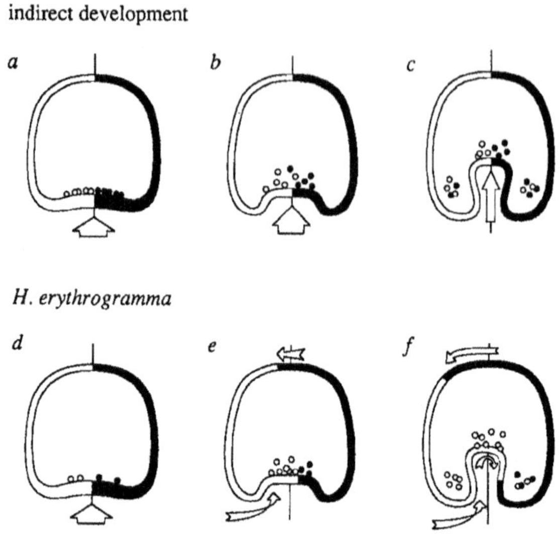

Figure 8. Comparison of gastrulation between indirect-developing sea urchins and *H. erythrogramma*. Diagrammatic saggital sections are oriented with the animal pole up and the ventral side on the left. During indirect development **(a-c)**, invagination begins with a symmetrical buckling of the vegetal plate. Cell rearrangements result in a narrowing and lengthening of the archenteron. Both archenteron and mesenchyme cells are symmetric in origin. In *H. erythrogramma* **(d)**, gastrulation begins with a radially symmetric invagination. As gastrulation proceeds **(e)** ventral cells (the white clone) involute over the tip of the blastopore, producing a ventral cell bias in the archenteron. Primary mesenchyme cells, which continue to ingress during this time also become ventrally biased. During the remainder of gastrulation **(f)**, ventrally biased ingression continues, and the archenteron tip becomes entirely ventral in origin. Secondary mesenchyme cells have an exclusively ventral origin. There is a concomitant stretching of dorsal ectoderm. From Wray and Raff (1991b), Used with permission of *Evolution*.

The most reasonable explanation for these observations is that ventral cells involute over the ventral lip of the blastopore during gastrulation. Because labeling reveals the cell movements to be gradual, this strongly asymmetrical involution probably continues throughout the second phase of gastrulation. It is likely that some cell rearrangements occur during blastopore narrowing and shaping of the archenteron in *H. erythrogramma*. The large cellular volume that involutes during gastrulation would, however, probably swamp out any role that cell rearrangements might play in oriented extension of the archenteron. The second phase of gastrulation in *H. erythrogramma* has thus been modified from the ancestral condition in two important regards: a different type of cellular movement drives archenteron extension, and a strong dorsoventral asymmetry is imposed on cell movements.

The asymmetrical origin of archenteron cells correlates with another unusual feature of development in *H. erythrogramma*. Most of the primary mesenchyme cells and all of the secondary mesenchyme cells in this species are derived from ventral vegetal precursors (Wray and Raff 1989, 1990; Henry and Raff 1990). At the time primary mesenchyme cells are ingressing, the site of ingression is derived in part from dorsal cells, but by the time secondary mesenchyme cells ingress the tip of the archenteron is entirely ventral in origin (Wray and Raff 1991b). Involution during gastrulation may thus participate in generating these cell fate asymmetries by placing presumptive regions for the ingression of mesenchyme cell populations in the right place at the right time.

Because the larva of *H. erythrogramma* does not feed, a functional mouth and gut are not required, and none are produced (Williams and Anderson 1975). The archenteron of *H. erythrogramma* does not completely traverse the blastocoel, but instead halts midway and forms a precocious coelom (Williams and Anderson 1975; Wray and Raff 1991b). Thereafter, the archenteron plays no direct role in premetamorphic development. The blastopore closes by 36 hours, leaving a closed vesicle of internal cells that probably acts as an imaginal source of cells for the functional adult digestive system (Williams and Anderson 1975).

The functional requirements for a generally accelerated pace of development may explain several of the unusual features of gastrulation in *H. erythrogramma*. The coelom, which forms the lining of the adult body cavities, must form very quickly in direct developers, as its derivative, the hydrocoel, is an important component of the adult rudiment (Raff 1987). In *H. erythrogramma* two important modifications ensure that the hydrocoel forms rapidly. First, the coelom arises just a few hours after gastrulation begins (Williams and Anderson 1975), and much earlier than in indirect developers (Okazaki 1975). Second, a much larger proportion of cellular volume is committed to coelomic fates (Wray and Raff 1990). During indirect development, the coelom forms from relatively few cells: the small micromere daughters and an undefined set of veg_2 descendants (Pehrson and Cohen 1986; R.A. Cameron, personal communication). Thereafter, the coelom grows gradually as the pluteus feeds. In *H. erythrogramma* no net growth occurs because the larva does not feed, and thus the coelom forms at full size instead of growing from a rudiment. Perhaps because this additional cellular volume must be packed into the interior, involution rather than cell rearrangements have become an important component of gastrulation. The net result of all these changes is that a large number of cells are made available for development of the coelom and hydrocoel immediately following gastrulation.

ECTODERMAL DISTORTIONS SHAPE THE EMBRYO

The accelerated development of adult structures during gastrulation in *H. erythrogramma* has had a second profound affect on morphogenesis in this embryo. In *H. erythrogramma* the coelom rapidly gives rise to the hydrocoel (Williams and Anderson 1975). The hydrocoel gives rise to the water vascular system of the developing adult. The echinus rudiment of sea urchins arises from the morphogenetic interactions of the hydrocoel with an invagination of the ectoderm called the vestibule (Okazaki 1975). In indirect development an initially small vestibule invaginates after the development of a definitive pluteus, long after gastrulation. In *H. erythrogramma*, however, the vestibule forms at its full size a few hours after the completion of gastrulation (Williams and Anderson 1975). Our cell lineage tracing studies (Wray and Raff 1989, 1990) show that all of the cells that will be included in the vestibule arise from the progeny of the ventral cell of the two-cell embryo (see Figure 4).

As shown in Figure 9, the larvae of both indirect-developing sea urchins and *H. erythrogramma* are shaped by distortions of the ectoderm. The cell movements responsible, however, are quite different. In indirect development, gastrulation is symmetrical. Thus, archenteron formation removes cells from the surface, but does not distort the surface. The striking distortions of the ectoderm that later appear in formation of the pluteus come from the extension of the arms. Extension accompanies elongation of the skeletal rods by calcium deposition by primary mesenchyme cells. The aboral ectoderm stretches to provide most of the surface of the larva. In *H. erythrogramma*, no spicular skeletal rods form. Later in larval development, primary mesenchyme cells secrete adult skeletal elements, but these have little affect on larval surface morphology until metamorphosis (Parks *et al.* 1988). The larva of *H. erythrogramma* retains the bullet-shape of the late gastrula. Nevertheless, drastic remodeling of the surface is revealed by cell-tracing studies. Asymmetric gastrulation initiates a process of ingression of cells derived from the ventral blastomere of the two-cell embryo. This process reduces the amount of the surface covered by ectodermal

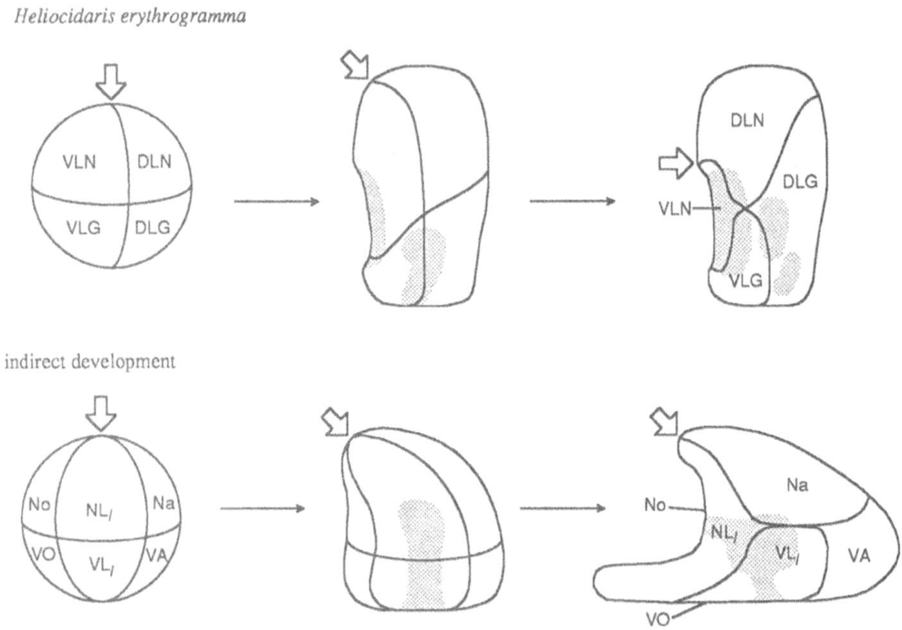

Figure 9. Distortions of clonally-derived ectodermal regions in indirect-developing sea urchin (*S. purpuratus*) and in *H. erythrogramma*. Embryos are shown in three successive stages, blastula, gastrula, and early larva. Views are left lateral, such that the plane of larval bilateral symmetry coincides with the page. Eight-cell clonal boundaries are shown as solid lines. Animal poles are indicated by the arrows. The archenteron, coelom, and vestibule are stippled. Distortions in ectodermal clones are caused by extension of the larval arms in indirect development, and primarily from the invagination of the vestibule in *H. erythrogramma*. Abbreviations designate clone names; those for *H. erythrogramma* are from Wray and Raff (1989, 1990), and those for indirect development are from Cameron *et al.* (1987). From Wray and Raff (1990), used with permission of Academic Press.

cells derived from the ventral blastomere. At the same time, the dorsal ectoderm stretches over the animal pole to cover part of the ventral surface. Ingression of the vestibule continues the process of surface distortion with a massive removal of ectoderm derived from the animal half daughters of the ventral blastomere from the surface. By the time the larva is fully formed, essentially all of these cells are incorporated into the vestibule, and only a fraction of the ventral, vegetal surface is covered with ventral blastomere ectoderm (Wray and Raff 1990). The dorsal blastomere-derived ectoderm has stretched to cover the majority of the larval surface. As in indirect-developing sea urchins, ectodermal distortions in *H. erythrogramma* result from stretching of a coherent ectodermal sheet. The only other surface cells are pigment cells, which as in indirect developing sea urchins arise from secondary mesenchyme and migrate into the ectoderm (Wray and Raff 1989).

CELL-CELL INTERACTIONS

Cell-cell inductive interactions appear to play an important role in directing cell determination and subsequent morphogenesis in *H. erythrogramma*. It is quite clear from cell lineage studies that the vestibule in the intact *H. erythrogramma* embryo is formed entirely by daughter cells of the animal-ventral blastomeres (Wray and Raff 1990). Either this capacity is endogenous to these cells, or the vestibule is induced by underlying vegetally-derived internal cells, such as the coelom. We have tested these alternatives (Henry and Raff 1990). Isolated animal half embryos of *H. erythrogramma* prepared at the eight-cell stage generally form only ectodermal derivatives. Significantly, these half embryos produce well formed invaginations of vestibular ectoderm. Thus, the animal ventral ectodermal cells have an endogenous capacity to form a vestibule.

Nevertheless, the potential for vestibule formation does not appear to be restricted to animal-ventral blastomeres. Isolated vegetal half embryos regulate sufficiently that they differentiate all cell types including a vestibule. The autonomous formation of a vestibule in embryos derived from the animal hemisphere indicates that inductive influences from adjacent endodermal and coelomic tissues are not required for vestibule formation. This result correlates well with the findings of Runnström (1917) and Czihak (1960, 1965). In indirect-developing embryos invagination of a vestibule is not dependent on the presence of an underlying hydrocoel, as would be expected if the vestibule required induction from internal cells with which it later interacts to form the echinus rudiment. However, one can not rule out the possibility that inductive interactions are responsible for the regulation of vestibule formation in other experimental situations. Embryos derived from isolated 8-cell embryo vegetal halves or from isolated dorsal 2-cell embryo blastomeres also produce vestibules despite the fact that these cells do not contribute to vestibular ectoderm in the intact embryo. Thus, we have the possibility that tissues formed in vegetal or dorsal half embryos (such as endodermal, coelomic or certain mesodermal tissues) may produce inductive signals capable of directing the formation of a vestibule in competent ectodermal cells. One the other hand, animal cells which normally contribute to the formation of the vestibule may inhibit other cells from assuming a vestibular ectodermal fate in the intact embryo. In either case, in the absence of the animal-ventral daughter cells, new cell fates would be generated through a novel, synthetic cell lineage. The differentiation of pigmented mesenchyme in half embryos derived form dorsal blastomeres isolated at the two cell stage (Henry and Raff 1990) poses similar questions. During normal development pigmented mesenchyme cells are formed only by progeny of the ventral blastomere (Wray and Raff 1990).

DISCUSSION AND CONCLUSIONS

Developmental mode has major consequences on dispersal, energetics of reproduction, population genetic structure, and speciation and extinction (McMillan *et al.* 1991; Wray and Raff 1991a for fuller discussions and references). The switch from indirect development *via* a feeding larval stage to direct development *via* a non-feeding larva is common. No simple model for such a change explains the complexity of developmental modifications which have occurred in *H. erythrogramma*. An increase in egg size is required for independence from larval feeding, but is not sufficient in itself to change developmental pattern (Wray and Raff 1991a). The evolutionary changes which result in direct development do not simply cause the loss of some larval developmental programs, as would be suggested from superficial appearances. As discussed above, it is clear that the entire program of morphogenetic processes of embryonic and larval development has been remodeled. The changes include an acceleration of axial determination to maternal specification, acceleration of dorsoventral fate specification, modifications in cell cleavage geometry, extensive reorganization of cell lineages and modification of the fate map, formation of a wrinkled blastula and massive extrusion of lipid vesicles into the blastocoel, substitution of a new form of gastrulation, and accelerated adult morphogenesis accompanied by extensive cell movements.

Although development of indirect-developing sea urchins is stereotypic, and cell lineage patterns are invariant in normal development (Davidson 1989), regulation occurs readily in response to perturbations (Hörstadius 1973; Ettensohn and McClay 1988; Davidson 1989; Henry *et al.* 1989; Henry and Raff 1990). Regulation indicates that whatever the functional relationship between cell lineage and cell fate determination in sea urchin embryos, it is labile. In nematode development cell lineage patterns are highly conserved, and evolutionary modifications appear predominantly, but certainly not exclusively, in postembryonic cell lineages (Sternberg and Horovitz 1981, 1982; Sulston *et al.* 1983; Ambros and Fixsen 1987). Other groups with deterministic development also have experienced evolutionary modifications of cell lineage in the evolution of highly modified larval forms. The clam *Unio* provides a classic example (Lillie 1895). The greatly altered larval form of these clams results from changes in cleavage rates and sizes of founder cells. There is no reshuffling of cell lineages: the basic spiralian pattern of cell lineage and fate remains unchanged (Raff and Kaufman 1983).

Deterministic developmental mechanisms are far better understood than regulative mechanisms. Classical embryology demonstrated the existence of maternally-localized, "determinants," which specify developmental fates of particular blastomeres in embryos as diverse as ctenophores, annelids, nemerteans, snails, ascidians, and sea urchins. Recent developmental genetic studies on the specification of anterior-posterior and dorso-ventral axes in *Drosophila* have begun to reveal the molecular details of how such maternal determinant systems operate (Anderson 1989). Regulation, on the other hand, although also widely noted has been more difficult to approach experimentally. Regulation occurs when, despite an experimental or environmental perturbation, an embryo goes on to correct for the perturbation and develop normally. A classic example is the isolation of one of the first two blastomeres of a sea urchin embryo. Despite missing the other half, that would give rise to different part of the intact embryo, the half embryo develops into a whole, albeit smaller embryo. Explanations of regulation have been based on concepts of graded distributions of materials (Hörstadius 1973; Runnström 1975), in which the isolated blastomere contains a sufficient part of the graded pattern of the intact embryo to still form a polarized entity. A model tying regulation to the localized activation of transcription factors has been presented by Davidson (1989).

A second aspect of regulatory phenomena is that induction is required for successful regulation. Outstanding examples demonstrating this point in sea urchins come from the work of Hörstadius (1973). Hörstadius changed the relative cellular compositions of embryos by removing or adding particular blastomeres to complete or partial embryos from which tiers of cells had been removed. These experiments show that micromeres have at least two inductive activities. They induce their neighbors, the macromeres, not to generate sketetogenic mesenchyme, and in cell transplant experiments, they can induce mesomeres to form endoderm. The experiments of Henry *et al.* (1989) on cultured isolated mesomeres indicate inductive interactions among these cells as well. Competence of cells to respond to inductive interactions among sea urchin embryo blastomeres are, as in other systems, time limited (Hörstadius (1973). In some cases inductions occur very early in development (Ettensohn and McClay 1988; Henry *et al.* 1989).

Regulation may be the key to evolutionary flexibility in early development. The ability of indirect-developing sea urchin embryos to regulate extensively despite an invariant cell lineage pattern provides crucial raw material for evolutionary modifications (Wray and Raff 1990). Experiments that demonstrate regulation in indirect-developing embryos show how flexible cell determination is. For instance, sketetogenic mesenchyme cells can arise from three completely different cellular precursors. In normal development, these cells arise from the micromeres, as outlined earlier. In experimental situations in which micromeres have been deleted (Hörstadius 1973; Ettensohn and McClay 1988), macromeres give rise to skeletogenic cells. In the experiments of Ettensohn and McClay (1988) in which primary mesenchyme cells were removed from the blastocoel, skeletogenic cells arose by a process of "fate conversion" on the part of the secondary mesenchyme cells. Conversion was blocked by the readdition of primary mesenchyme cells. It is not clear by what novel lineage route sketetogenic mesenchyme cells are produced when micromeres are deleted prior to ingression. The third route to skeletogenic cells is from cultured isolated pairs of mesomeres (Henry *et al.* 1989). The evolutionary point of such experiments is that they demonstrate that a particular final cell type can be experimentally generated by a variety of cellular precursors and lineage routes. These final cell fates will thus remain accessible to the embryo as evolutionary changes take place that modify cell cleavage and cell lineage behavior.

H. erythrogramma also provides clues to the nature of the changes involved in the evolution of its morphogenetic processes. The extensive changes in early development involve genetically remodeled processes probably. The maternally-specified dorsoventral asymmetry exploits the pre-existing vegetal localization system found in indirect development. The determinants of vegetal fates, however, have been shifted toward the ventral side and relocalized (Henry and Raff 1990). Presumably the machinery of localization is unchanged. Fate map changes result in part from loss of the unequal cleavages of indirect development and their replacement by the spindle controls of equal cleavage. However, more complex changes have also occurred (Wray and Raff 1989, 1990). Some cells have switched fate: the great increase in vestibular precursors in the 32-cell embryo is an obvious instance. At this point we have data that suggest that vestibular fate is autonomous. We do not know how the fate change is achieved, because we do not yet know the basis for this autonomous determination. Changes in cell lineage splitting patterns to produce founder cells also suggest significant but unknown genetic changes. Gastrulation too apparently has evolved a new way of producing an archenteron (Wray and Raff 1991b). Involution of a cell sheet has become a prominent component of gastrulation. The mechanism of change is likely to have required a different set of cellular instructions.

There have also been what might be described as "passive" changes. These are processes that evidently have not been changed directly, but as a consequence of other changes. For example, cleavage divisions continue significantly longer in *H. erythrogramma* than in indirect development. The result is a gastrula with about 10 times the number of cells. The probable cause is the larger cytoplasmic volume of the egg, which assures a larger ratio of cytoplasm to nucleus. Cleavage controls *per se* require no change (Parks *et al.* 1988). Some aspects of wrinkled blastula formation may owe to analogous physical modifications combined with genetic changes in cellular behavior. Finally, in indirect development, the distortions of the ectoderm which produce the arms and stretch the dorsal (aboral) ectoderm are primarily the passive result of the interaction of ectoderm with the mesenchyme cells clustered at the tips of the skeletal spicules (Hörstadius 1973). The result of an evolutionary loss of the larval skeleton should result in an armless ovoid larva lacking extensions and concomitant distortions of its ectoderm. In the embryo of *H. erythrogramma*, the dorsal ectoderm is extensively stretched, but from a quite different cause. The involution processes of the ventral ectoderm produces the archenteron and the vestibule, with concomitant stretching of the dorsal ectoderm to cover most of the surface as the majority of ventral-derived cells involute.

At present, we do not have a very clear idea of the regulatory changes at the molecular level that underly the suite of evolutionary changes in morphogenesis in *H. erythrogramma*. We are approaching this problem by investigating three sets of potential regulatory change. The first is to examine genes that exhibit cell lineage-restricted expression. The sketetogenic mesenchyme-expressed gene *msp130* has undergone a striking heterochromic change in expression in *H. erythrogramma*, although it has retained its cell type-specific expression pattern (Parks *et al.* 1988). We are comparing regulatory domains from the 5' upstream sequence for the *S. purpuratus* gene (Parr *et al.* 1990) to the homologs from the two *Heliocidaris* species. The use of *H. tuberculata* in comparisons against *S. purpuratus* is important because it makes it possible to separate changes related to evolutionary distance from those related to change in developmental mode. The actin gene family offers another valuable gene set because of the spatially-restricted expression patterns of the family members in *S. purpuratus* (Cox *et al.* 1986), and their direct relevance to cellular motility. The actin gene family is similar in the *Heliocidaris* species to *S. purpuratus*, and 5' upstream regulatory regions of the best studied family member are very similar in sequence in all three species, implying that the same trans-acting factors are used to control expression (Hahn and Raff, unpublished).

Regulatory protein-encoding genes offer more direct approaches to evolutionary change in gene regulation. The potential inducer SpEGF2 (Yang *et al.* 1989) in *Heliocidaris* is similar to its homolog in *Strongylocentrotus* (Ferkowicz and Raff, unpublished). Finally, the homeobox genes of *Heliocidaris* are very similar to their homologs in other sea urchins (Popodi, Kissinger and Raff, unpublished). It is of interest to note that the SpEGF2 mRNA has an expression pattern localized similarly to one of the homeobox genes in sea urchins (Yang *et al.* 1989). Homeobox-containing genes offer a direct look at potential evolutionary changes in timing or localization of expression of transcription factors, and a potential correlation with evolutionary changes in other genes expressed under spatial controls. Ultimately, it is the evolution of these genetic controls of cellular behavior that must be explained.

ACKNOWLEDGEMENTS

We thank Donald T. Anderson for his encouragement and hospitality during our work at the University of Sydney. We also thank Maria Byrne, Andy Cameron, and Pia Laegdsgaard for discussions of their unpublished work. This study was supported by NIH Grant HD21337 to R.A.R. and NIH Postdoctoral Fellowship GM12495 to G.A.W.

REFERENCES

Alberch, P. 1982. Developmental constraints in evolutionary processes. p. 313-332. *In: Evolution and Development*. J.T. Bonner (Ed.). Springer-Verlag, Berlin.

Ambros, V. 1989. A hierarchy of regulatory genes controls a larva-to-adult developmental switch in *C. elegans*. *Cell* 57:49-57.

Ambros, V. and W. Fixsen. 1987. Cell lineage variation among nematodes. p. 139-159. *In: Development as an Evolutionary Process*. R.A. Raff and E.C. Raff (Eds.). Alan R. Liss, New York.

Anderson, K.V. 1989. *Drosophila*: the maternal contribution. p. 1-37. *In: Genes and Embryos*. D.M. Glover and B.D. Hames (Eds.). IRL Press, Oxford.

Arnolds, W.J.A., J.A.M. van den Biggelaar, and N.H. Verdonk. 1983. Spatial aspects of cell interactions involved in the determination of dorsoventral polarity in the equally-cleaving gastropods and regulative abilities of their embryos, as studied by micromere deletions in *Lymnaea* and *Patella*. *Wilhelm Roux's Arch. Dev. Biol.* 192:281-295.

Arthur, W. 1988. *A Theory of the Evolution of Development*. John Wiley, Chichester.

Bisgrove, B.W. and R.A. Raff. 1989. Evolutionary conservation of the larval serotonergic nervous system in a direct developing sea urchin. *Dev. Growth & Differ.* 31:363-370.

Bonner, J.T. (Ed.). 1982. *Evolution and Development*. Springer-Verlag, Berlin.

Boveri, T. 1901. Die Polarität von Oocyte, Ei, und Larve des *Strongylocentrotus lividus*. *Zool. Jahrb. Abt. Anat. Ontog. Tiere* 14:630-653.

Buss, L.W. 1987. *The Evolution of Individuality*. Princeton University Press, Princeton.

Cameron, R.A., S.E. Fraser, R.J. Britten, and E.H. Davidson. 1989. The oral-aboral axis of a sea urchin embryo is specified by first cleavage. *Development* 106:641-647.

Cameron, R.A., S.E. Fraser, R.J. Britten, and E.H. Davidson. 1990. Segregation of oral from aboral ectoderm precursors is completed at 5th cleavage in the embryogenesis of *Strongylocentrotus purpuratus*. *Dev. Biol.* 137:77-85.

Cameron, R.A., B.R. Hough-Evans, R.J. Britten, and E.H. Davidson. 1987. Lineage and fate of each blastomere of the sea urchin embryo. *Genes Dev.* 1:75-84.

Clement, A.C. 1952. Experimental studies on germinal localization in Ilyanassa. I. The role of the polar lobe in determination of the cleavage pattern and its influence on later development. *J. Exp. Zool.* 121:563-626.

Cox, K.H., L.M. Angerer, J.J. Lee, E.H. Davidson, and R.C. Angerer. 1986. Cell lineage-specific programs of expression of multiple actin genes during sea urchin embryogenesis. *J. Mol. Biol.* 188:159-172.

Czihak, G. 1960. Untersuchungen uber die Coelomanlagen und die Metamorphose des Pluteus von *Psammechinus miliaris* (Gmelin). *Zool. Jahrb. Abt. Anat. Ontog. Tiere.* 78:235-256.

Czihak, G. 1965. Entwicklungsphysiologische Untersuchungen an Echiniden. Experimentelle Analyse der Coelomentwicklung. *Wilhelm Roux' Arch. Entwicklungsmech. Org.* 155:709-729.

Dan, K. 1979. Studies on unequal cleavage in sea urchins. I. Migration of the nuclei to the vegetal pole. *Dev. Growth & Differ.* 21:527-535.

Darnell, J., H. Lodish, and D. Baltimore. 1990. *Molecular Cell Biology*. 2nd ed. Scientific American Books, New York.

Davidson, E.H. 1989. Lineage-specific gene expression and the regulative capacities of the sea urchin embryo. *Development* 105:421-445.

deBeer, G.R. 1958. *Embryos and Ancestors*. 3rd edition. Clarendon Press, Oxford.

Emlet, R.B., L.R. McEdward, and R.R. Strathmann. 1987. Echinoderm larval ecology viewed from the egg. p. 55-136. *In: Echinoderm Studies, vol. 2.* M. Jangoux and J.M. Lawrence (Eds.). A.A. Balkema, Rotterdam.

Ettensohn, C.A. 1984. Primary invagination of the vegetal plate during sea urchin gastrulation. *Am. Zool.* 24:571-588.

Ettensohn, C.A. 1985. Gastrulation in the sea urchin embryo is accompanied by the rearrangement of invaginating epithelial cells. *Dev. Biol.* 112:383-390.

Ettensohn, C.A. and D.R. McClay. 1988. Cell lineage conversion in the sea urchin embryo. *Dev. Biol.* 125:396-409.

Freeman, G. 1982. What does a comparative study of development tell us about evolution? p. 155-167. *In: Evolution and Development.* J.Y. Bonner (Ed.). Springer-Verlag, Berlin.

Garstang, W. 1922. The theory of recapitulation: A critical restatement of the biogenetic law. *J. Linn. Soc. Lond. Zool.* 35:81-101.

Gemmill, J.F. 1912. The development of the starfish *Solaster endeca* Forbes. *Trans. Zool. Soc. Lond.* 20:1-71.

Gemmill, J.F. 1916. Notes on the development of starfishes *Asterias glacialia* O.F.M.; *Cribrella oculata* (Linck) Forbes; *Solaster endeca* (Retzius) Forbes; *Stichaster roseus* (O.F.M.) *Sars. Proc. Zool. Soc. Lond.* 39:553-565.

Gould, S.J. 1977. *Ontogeny and Phylogeny*. Harvard University Press, Cambridge.

Hardin, J. 1989. Local shifts in position and polarized motility drive cell rearrangements during sea urchin gastrulation. *Dev. Biol.* 136:430-445.

Hardin, J. and L.Y. Cheng. 1986. The mechanisms and mechanics of archenteron elongation during sea urchin gastrulation. *Dev. Biol.* 115:490-501.

Hardin, J. and D.R. McClay. 1990. Target recognition by the archenteron during sea urchin gastrulation. *Dev. Biol.* 142:86-102.

Harkey, M.A. 1983. Determination and differentiation of micromeres in the sea urchin embryo. p. 131-155. *In: Time, Space, and Pattern in Embryonic Development.* W.R. Jeffery and R.A. Raff. (Eds.). Alan R. Liss, New York.

Harvey, E.B. 1956. *The American Arbacia and Other Sea Urchins.* Princeton University Press, Princeton.

Hayashi, R. 1972. On the relations between the breeding habits and larval forms in asteroids, with remarks on the wrinkled blastula. *Proc. Jpn. Soc. Syst. Zool.* 8:42-48.

Henry, J.J., S. Amemiya, G.A. Wray, and R.A. Raff. 1989. Early inductive interactions are involved in restricting cell fates of mesomeres in sea urchin embryos. *Dev. Biol.* 136:140-153.

Henry, J.J. and R.A. Raff. 1990. Evolutionary change in the process of dorsoventral axis determination in the direct developing sea urchin, *Heliocidaris erythrogramma. Dev. Biol.* 141:55-69.

Henry, J.J., G.A. Wray, and R.A. Raff. 1990. The dorsoventral axis is specified prior to first cleavage in the direct developing sea urchin, *Heliocidaris erythrogramma. Development* 110:875-884.

Henry, J.J., G.A. Wray, and R.A. Raff. 1991. Mechanism of an alternate type of echinoderm blastula formation: The wrinkled blastula of the sea urchin *Heliocidaris erythrogramma. Dev. Growth & Differ.* 33:317-328.

Hörstadius, S. 1939. Über die Entwicklung von *Astropecten aranciacus L. Pubbl. Stne. Zool. Napoli.* 17:221-312.

Hörstadius, S. 1973. *Experimental Embryology of Echinoderms.* Clarendon Press, Oxford.

Hörstadius, S. and A. Wolskey. 1936. Studien über die Determination der Bilateralsymmetrie des jungen SeeigelKeimes. *Wilhelm Roux's Arch. Dev. Biol.* 135:69-113.

Inaba, D. 1968. Holothuria. p. 316-329. *In: Invertebrate Embryology.* M. Kúme and K. Dan (Eds.). Nolit, Belgrade.

Jenkinson, J.W. 1911. On the origin of the polar and bilateral structure of the egg of the sea urchin. *Wilhelm Roux's Arch. Dev. Biol.* 32:699-716.

Kier, P.M. 1984. Echinoids from the Triassic (St. Cassian) of Italy, their lantern supports, and revised phylogeny of Triassic echinoids. *Smithson. Contrib. Paleobiol.* 56:1-41.

Kominami, T. 1983. Establishment of embryonic axes in larvae of the starfish, *Asterina pectinifera. J. Embryol. Exp. Morphol.* 75:87-100.

Kominami, T. 1988. Determination of dorsoventral axis in early embryos of the sea urchin *Hemicentrotus pulcherrimus. Dev. Biol.* 127:187-196.

Kubo, K. 1951. Some observations on the development of the sea-star *Leptasterias ochotensis similspinis* (Clark). *J. Fac. Sci. Hokkaido Univ. Ser. VI Zool.* 10:97-105.

Laegdsgaard, P. 1989. *The Reproduction of the Co-occurring Species of the Sea Urchin Heliocidaris in the Sydney Region.* Honors Thesis, University of Sydney, Australia.

Lillie, F.R. 1895. The embryology of the Unionidae. *J. Morphol.* 10:1-100.

Lindahl, P.E. 1932. Zur experimentellen Analyse der Determination der Dorsoventralachse beim Seeigelkeim. I. Versuch mit gestreckten Eiern. *Wilhelm Roux's Arch. Dev. Biol.* 127:300-322.

McMillan, W.O., R.A. Raff, and S.R. Palumbi. 1991. Population genetic consequences of reduced dispersal in a direct-developing sea urchin, *Heliocidaris erythrogramma*. *Evolution*, In press.

Martindale, M.Q., C.Q. Doe, and J.B. Morrill. 1985. The role of animal-vegetal interaction with respect to the determination of dorsoventral polarity in the equal-cleaving spiralian, *Lymnaea palustris*. *Wilhelm Roux's Arch. Dev. Biol.* 194:281-295.

Maruyama, Y.K., Y. Nakaseko, and S. Yagi. 1985. Localization of cytoplasmic determinants responsible for primary mesenchyme formation and gastrulation in the unfertilized egg of the sea urchin *Hemicentrotus pulcherrimus*. *J. Exp. Zool.* 236:155-163.

Maynard Smith, J., R. Burian, S. Kauffman, P. Alberch, J. Campbell, B. Goodwin, R. Laude, D. Raup, and L. Wolpert. 1985. Developmental constraints and evolution. *Q. Rev. Biol.* 60:265-287.

Mladenov, P.V. 1979. Unusual lecithotrophic development of the Caribbean brittle star, *Ophiothrix oerstedi*. *Mar. Biol.* 55:55-62.

Morgan, T.H. and G.B. Spooner. 1909. The polarity of the centrifuged egg. *Wilhelm Roux's Arch. Dev. Biol.* 28:104-117.

Nitecki, M.H. (Ed.). 1990. *Evolutionary Innovations*. University of Chicago Press, Chicago.

Okazaki, K. 1975. Normal development to metamorphosis. p.177-232. *In: The Sea Urchin Embryo*. G. Czihak (Ed.). Springer-Verlag, Berlin.

Okazaki, K. and K. Dan. 1954. The metamorphosis of partial larvae of *Peronella japonica* Mortensen, a sand dollar. *Biol. Bull.* 106:83-99.

Pace, N.R., D.K. Smith, G.J. Olse, and B.D. James. 1989. Phylogenetic comparative analysis and the secondary structure of ribonuclease P RNA–a review. *Gene* 82:65-75.

Parks, A.L., B.W. Bisgrove, G.A. Wray, and R.A. Raff. 1989. Direct development in the sea urchin *Phyllacanthus parvispinus* (Cidaroidea): Phylogenetic history and functional modification. *Biol. Bull.* 177:96-109.

Parks, A.L., B.A. Parr, J.-E. Chin, D.S. Leaf, and R.A. Raff. 1988. Molecular analysis of heterochromic changes in the evolution of direct developing sea urchins. *J. Evol. Biol.* 1:27-44.

Parr, B.P., A.L. Parks, and R.A. Raff. 1990. Promoter structure and protein sequence of msp130, a lipid-anchored sea urchin glycoprotein. *J. Biol. Chem.* 265:1408-1413.

Pehrson, J.R. and L.H. Cohen. 1986. The fate of the small micromeres in sea urchin development. *Dev. Biol.* 113:522-526.

Philip, G.M. 1965. The Tertiary echinoids of South-Eastern Australia III Stirodonta, Aulodonta, and Camarodonta (1). *Proc. R. Soc. Victoria* 78:181-196.

Raff, R.A. 1987. Constraint, flexibility, and phylogenetic history in the evolution of direct development in sea urchins. *Dev. Biol.* 119:6-19.

Raff, R.A. (Ed.). 1990. Heterochromic changes in development. *Semin. Dev. Biol.* 1:(4).

Raff, R.A. and T.C. Kaufman. 1983. *Embryos, Genes, and Evolution*. MacMillan, New York.

Raff, R.A. and G.A. Wray. 1989. Heterochrony: Developmental mechanisms and evolutionary results. *J. Evol. Biol.* 2:409-434.

Raff, R.A., G.A. Wray, and J.J. Henry. 1991. Implications of radical evolutionary changes in early development for concepts of developmental constraint. p. 189-207. *In: New Perspectives on Evolution*. L. Warren and H. Kaprowski (Eds.). Alan R. Liss, New York.

Roux, W. 1895. The problems, methods and scope of developmental mechanics. An introduction to the "Archiv für Entwicklungsmechanik der Organismen," translated by W.M. Wheeler. p. 149-190. *In: Biological Lectures of the Marine Biological Laboratory of Woods Hole, Mass*. Ginn and Company, Boston.

Runnström, J. 1917. Analytische Studien Über die Seeigelenwicklung. III. *Wilhelm Roux's Arch. Dev. Biol.* 43:223-328.

Runnström, J. 1975. Integrating factors. p. 646-670. *In: The Sea Urchin Embryo.*, G. Czihak (Ed.). Springer-Verlag, Berlin.

Scott, L.B., W.J. Lennarz, R.A. Raff, and G.A. Wray. 1990. The "lecithotrophic" sea urchin *Heliocidaris erythrogramma* lacks typical yolk platelets and yolk glycoproteins. *Dev. Biol.* 138:188-193.

Schroeder, T.E. 1980. Expression of the preformation polar axis in sea urchin eggs. *Dev. Biol.* 79:428-443.

Smith, A.B. 1988a. Fossil evidence for the relationships of extant echinoderm classes and their times of divergence. p. 85-97. *In: Echinoderm Phylogeny and Evolutionary Biology*. C.R.C. Paul and A.B. Smith (Eds.). Clarendon Press, Oxford.

Smith, A.B. 1988b. Phylogenetic relationship, divergence times, and rates of molecular evolution for camarodont sea urchins. *Mol. Biol. Evol.* 5:345-365.

Smith, M.J., J.D. G. Boom, and R.A. Raff. 1990. Single copy DNA distance between two congeneric sea urchin species exhibiting radically different modes of development. *Mol. Biol. Evol.* 7:315-326.

Stern, C. 1990. The evolution of segmental patterns. *Semin. Dev. Biol.* 1:75-145.

Sternberg, P.W. and H.R. Horvitz. 1984. The genetic control of cell lineage during nematode development. *Annu. Rev. Genet.* 18:489-524.

Sternberg, P.W. and H.R. Horvitz. 1981. Gonadal cell lineages of the nematode *Panagrellus redivivus* and implications for evolution by modification of cell lineage. *Dev. Biol.* 88:147-166.

Sternberg, P.W. and H.R. Horvitz. 1982. Postembryonic nongonadal cell lineages of the nematode *Panagrellus redivivus*: Description and comparison with those of *Caenorhabditis elegans*. *Dev. Biol.* 93:181-205.

Strathmann, M.F. 1987. *Reproduction and Development of Marine Invertebrates of the Northern Pacific Coast.* University Washington Press, Seattle.

Strathmann, R.R. 1978. The evolution and loss of larval feeding stages of marine invertebrates. *Evolution*. 32:894-906.

Sulston, J.E., E. Schierenberg, J.G. White, and J.N. Thomason. 1983. The embryonic cell lineage of the nematode *Caenorhabditis elegans*. *Dev. Biol.* 100:64-119.

van den Biggelaar, J.A.M. and P. Guerrier. 1979. Dorsoventral polarity and mesentoblast determination as concomitant results of cellular interactions in the mollusc *Patella vulgata*. *Dev. Biol.* 68:462-471.

Waugh, D.S., C.J. Green, and N.R. Pace. 1989. The design and catalytic properties of a simplified ribonuclease P RNA. *Science* 244:1569-1571.

Williams, D.H.C. and D.T. Anderson. 1975. The reproductive system, embryonic development, larval development, and metamorphosis of the sea urchin

Heliocidaris erythrogramma (Val.) (Echinoidea: Echinometridae). *Aust. J. Zool.* 23:371-403.

Woese, C.R., R. Gutell, R. Gupta, and H.F. Noller. 1983. Detailed analysis of higher-order structure of 16S-like ribosomal ribonucleic acids. *Microbiol. Rev.* 47:621-699.

Wray, G.A. and R.A. Raff. 1989. Evolutionary modification of cell lineage in the direct-developing sea urchin *Heliocidaris erythrogramma. Dev. Biol.* 132:458-470.

Wray, G.A. and R.A. Raff. 1990. Novel origins of lineage founder cells in the direct-developing sea urchin *Heliocidaris erythrogramma. Dev. Biol.* 141:41-54.

Wray, G.A. and R.A. Raff. 1991a. The evolution of developmental strategy in marine invertebrates. *Trends Ecol. Evol.* 6:45-50.

Wray, G.A. and R.A. Raff. 1991b. Rapid evolution of gastrulation mechanisms in a direct-developing sea urchin. *Evolution*, In press.

Yang, Q., L.M. Angerer, and R.C. Angerer. 1989. Unusual pattern of accumulation of mRNA encoding EGF-related protein in sea urchin embryos. *Science* 246:806-808.

SEA URCHIN MICROMERES, MESENCHYME, AND MORPHOGENESIS

Fred H. Wilt, Nikolaos C. George, and Oded Khaner

Department of Molecular and Cell Biology
371 LSA, University California, Berkeley
Berkeley, CA 94720

INTRODUCTION

The sea urchin embryo has been a favorite material for the study of gastrulation and morphogenesis. Among the first attempts to understand morphogenesis in terms of cell behavior is found in the pioneering work of Gustafson and Wolpert (1967), which served as a powerful stimulus for more recent studies of morphogenesis. We wish to present here some ideas on what we do not know about sea urchin gastrulation, and to indicate different approaches being used in our laboratory to address these questions about the cellular and molecular basis of morphogenesis in sea urchin embryos.

THE POTENTIAL TO FORM A GUT

The fates of the different cells that constitute the early sea urchin embryo are well known, having been studied by conventional dye marking studies in the 30's (reviewed by Horstadius 1973), as well as more recent work using microinjected lineage tracers (Cameron *et al.* 1987). The entire gut of euechinoid sea urchins that form a feeding pluteus larva is derived from the cells of the vegetal plate of the hatched blastula, which in turn has formed from the so-called veg_2 layer of the 64-cell stage. This octet of cells from the veg_2 layer gives rise to approximately 60 vegetal plate cells that surround the centrally located 40 descendants of the large and small micromeres. Since

Gastrulation, Edited by R. Keller *et al.*
Plenum Press, New York, 1991

Moore and Burt showed many years ago (1939) that isolated vegetal halves of the blastula carry out gastrulation movements, a result confirmed by Ettensohn (1984), it follows that the vegetal plate is probably the source of the gut, and also the source of the morphogenetic forces that build the archenteron.

But many other territories of the early embryo may give rise to guts or gut-like structures, a result obtained hundreds of times by many investigators spanning experiments carried out over many decades. Isolated animal halves may give rise to guts if exposed to sea water containing Li ions (von Ubisch 1929; Horstadius 1973; Livingston and Wilt 1990). Mesomeres from the animal hemisphere may also give rise to guts if subequatorial cleavages have occurred (Driesch 1894; Henry *et al.* 1989), or if subjected to long term culture (Khaner and Wilt 1990). The territory between the animal hemisphere and the veg_2 layer, termed veg_1 by Horstadius, can also give rise to portions of the gut if isolated vegetal halves are cultured (Horstadius 1973) or if lineage tracer marked veg_1 cells are co-cultured with mesomeres (Khaner and Wilt 1991). Hence, almost the entire epithelial surface of the blastula is derived from cells that have the potentiality to form gut, a property we associate with the essence of the gastrulation process. The only cells that have never been observed to form gut-like structures, under any circumstances, are the derivatives of the micromeres [see the review of the older literature by Horstadius (1973) and recent experiments by Livingston and Wilt 1990; Khaner and Wilt 1990, and Bernacki and McClay 1989]. It is rather remarkable that the cells derived from the extreme vegetal pole, the region associated with gastrulation and archenteron formation, are themselves apparently determined at the 4th cell division and are incapable of participating in formation of the archenteron. It is possible the egg domain giving rise to the vegetal most micromeres is "determined" before fertilization, perhaps by specialization of the plasmalemma or cortex.

It is not clear whether formation of a gut in these various experimental situations always occurs by the same morphogenetic maneuvers. Likewise it is not clear whether there is a necessary order of changes in cellular behavior that bring about the formation of an archenteron. For example, in development that occurs from isolated early blastomeres, which may form a gut autonomously in response to Li ion or in response to "induction" from micromeres, gastrulation movements have not been directly observed. Some authors have proposed that development that occurs from dissociation and reaggregation of early blastomeres may involve a delamination or polyingression of surface cells that later undergo an "epithelialization" to form a gut like structure (Bernacki and McClay 1989). Giudice (1973) and his colleagues, who pioneered the use of dissociation and reaggregation of hatched blastulae, reported gut formation occurred by construction of epithelial vesicles and tubes *in situ* from the cells present in the middle of the reaggregate. In this latter instance, cells isolated from mesenchyme blastulae stages were often used, while other studies like those of Bernacki and McClay (1989), and Spiegel and Spiegel (1975), used blastomeres prepared from dissociated 16-cell stage embryos. It is possible gut formation proceeds by different routes in reaggregates formed from blastomeres of different developmental stages. To date, however, documentation of an invagination to form gut like structures from dissociated blastomeres has not been reported. We have observed hollow embryoids formed from reaggregates of various blastomeres of the 16-32-cell stage that possess structures on the epithelial surface that resemble invaginations, but we have never continuously observed these and therefore, do not know if they represent real invaginations leading to gut formation. Events such as these need to be filmed, and we are in the process of doing that now. As it stands, sea urchin embryos and/or the cells prepared from them by dissociation are capable of forming interior epithelia that are gut-like by invagination, by polyingression followed by epithelium formation, or by cavitation of a solid core of interior cells.

SPECIFICATION OF INITIAL GASTRULATION

How is the embryo informed where to begin the formation of an archenteron by gastrulation? In the normal embryo the first phase of gastrulation is the initial inward buckling of the thickened vegetal plate that remains after the ingression of the primary mesenchyme. Neither the animal hemisphere (Moore and Burt 1939) nor the presence of micromeres (von Ubisch 1939; Horstadius 1973) is necessary for this invagination. Micromere formation may be suppressed (Langelan and Whiteley 1985 and gastrulation still occurs at the vegetal pole. It is even possible to remove the primary mesenchyme itself (Fukushi 1962; Ettensohn and McClay 1988) and gastrulation proceeds apace. These studies lead to the conclusion that invagination is a property intrinsic to the cells that actually carry it out, the descendants of the 8 veg_2 cells of the 64-cell stage that give rise to the approximate 64 cells of the vegetal plate.

On the other hand, micromeres can induce invagination of a supernumerary archenteron when transplanted to other sites. For example, Horstadius (reviewed 1973) moved 4 micromeres to a position between the An_2 and Veg_1 layers and obtained a second invagination, which later became amalgamated into the primary archenteron. Movement of 4 micromeres to the animal pole gave rise to a syndrome, which although somewhat variable, included reduction of the apical tuft, formation of supernumerary primary mesenchyme cells in the animal portion of the blastocoel, and an occasional very small invagination. To my knowledge, these difficult transplantations have not been repeated in the past 45 years. In all cases where micromeres were moved, the original vegetal pole of the embryo still carried out gastrulation. It may well be that micromeres may stimulate cells other than the veg_2 layer to organize an invagination leading to archenteron formation, but this needs to be verified with modern methods. The chain of evidence for the ability of micromeres is not completely secure for the following reasons: 1) the actual formation of an invagination is very difficult to document without continuous filmed observation; 2) new lineage tracers are superior to the vital dyes available to workers in the past; and 3) it may be difficult to distinguish the activity of polyingression of primary mesenchyme from limited epithelial invagination. It is very much to be hoped that direct observations of the effects of micromeres on how mesomeres form a gut will soon be carried out.

THE ROLE OF SMALL AND LARGE MICROMERES IN GASTRULATION

Whether micromeres organize an invagination in mesomeres or not, they do truly induce gut formation (reviewed by Horstadius 1973; Wilt 1987). Recently Livingston and Wilt (1990) and Khaner and Wilt (1990) restudied and verified the ability of micromeres to induce guts in animal hemispheres. The structures obtained were positive for the gut specific enzyme, alkaline phosphatase, and also displayed epitopes shown by McClay and his co-workers (1987) to be characteristic of differentiated gut.

We wished to inquire whether the two daughters of the micromeres, which form at the 5th cell division, both had this inducing ability. In normal development, the micromeres divide asymmetrically at the 5th division, giving rise to a very small vegetal most cell, the small micromere, and a sister cell adjacent to the veg_2 layer, the large micromere. It is the large micromere that gives rise to skeletogenic mesenchyme (Okazaki 1975). The small micromeres divide once again, then retire from cell division until the rudiment of the juvenile sea urchin begins to form. This was originally shown by Pehrson and Cohen (1986), and this fate of the small micromeres has been directly confirmed by lineage tracer studies (Cameron, personal communication). Tanaka and Dan (1990) have recently shown the 8 small micromere descendants are retained at the tip of the archenteron, emigrating with the secondary mesenchyme.

Recent experiments (Khaner and Wilt 1991) have compared small and large micromeres when combined with mesomeres. Small and large micromeres were obtained by microdissection of 32 cell stage embryos in calcium free sea water and labeled with the lineage tracer RITC (rhodamine isothiocyante). Various numbers of small and large micromeres were then mixed with 4 unlabeled mesomeres, reaggregated, and cultured for 2 to 3 days. Figure 1 shows a typical result in which 1 large micromere and 4 mesomeres formed an embryoid with spicules derived from the micromere and an induced gut derived from the mesomeres. If two of the small micromeres are used (Figure 2) to stimulate mesomeres, no spicules and few guts are formed. The small micromeres, though somewhat variable in size, may only average about 1/3 of the diameter of the large micromere. We therefore carried out more cultures of 4 mesomeres, but this time with 8 small micromeres. In the cases examined thus far, about 35% formed an internal epithelial structure we classify as gut-like, but these are very small, and are much less well developed than the guts that form with one large micromere (Figure 3). No spicules have been found when small micromeres are combined with mesomeres. Hence, both large and small micromeres, although they have different fates, and undergo different routes of differentiation in reaggregates, can stimulate mesomeres from the animal hemisphere to form gut-like structures. However, the potency of the large micromeres seems very much greater than their smaller and more vegetal sisters. It will be interesting to see if the guts induced by micromeres in these recombinants form by invagination or not.

a

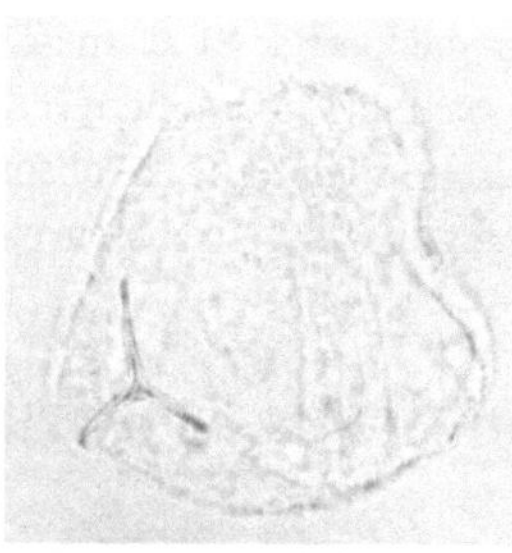

b

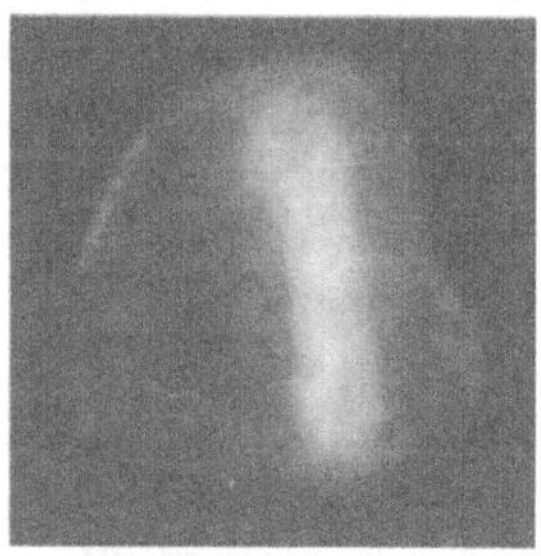

Figure 1. Four mesomeres were combined with one large micromere and cultured for 3 days. The resultant embryoid was photographed in bright field **(a)**, where a prominent spicule that formed from the micromere can be seen. The embryoid was stained with the monoclonal antibody, endo I, specific for gut tissue **(b)**. A prominent tubular gut formed from mesomeres may be observed. 200×.

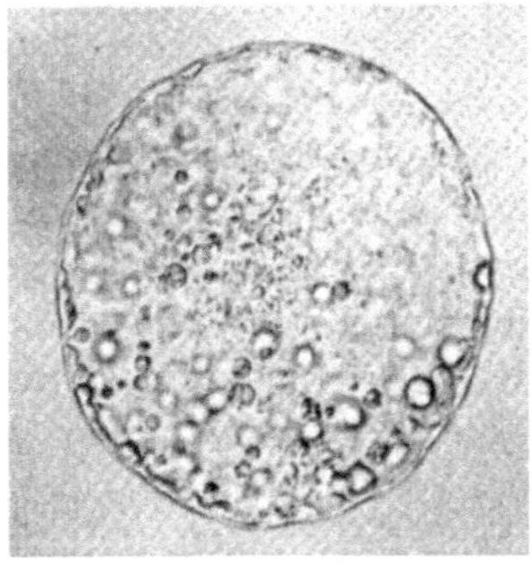

Figure 2. Two small micromeres were combined with 4 mesomeres and cultured for 3 days. The resultant embryoid contains neither spicules nor gut like structures. 200×.

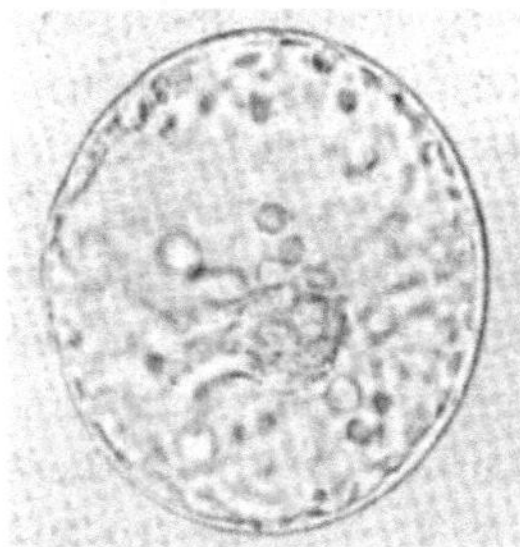

Figure 3. Eight small micromeres were combined with 4 mesomeres and cultured for 3 days. The resultant embryoid contains no spicules, and a very rudimentary gut-like structure may be observed in the center of the embryoid. 200×.

THE ROLE OF CELL INTERACTIONS IN SPICULE MATRIX GENE EXPRESSION

Finally, we turn to a related issue in sea urchin morphogenesis, the terminal differentiation of the micromere derived mesenchyme to form a skeletal spicule. Primary mesenchyme cells enter the interior, populate the blastocoel and move within it, fuse, and secrete a skeleton during archenteron formation. They will also carry out these same events during exogastrulation, so clearly normal invagination, convergence, extension, and towing of the archenteron are not essential for skeletogenesis. Intimate cell-cell interactions leading to syncytium formation are part of skeletal morphogenesis. It is also well known from the study of chimaeras (von Ubisch 1939) and development in isolated bags of basal laminae (Harkey and Whiteley 1980) that mesenchyme interacts with the basal lamina and blastocoelar wall. Inhibitors of normal metabolism of various components of the extracellular matrix have been shown to block morphogenetic movements and/or skeleton formation (See Kabakoff and Lennarz 1990; Blankenship and Benson 1984; Solursh 1986).

We have isolated and characterized two genes that encode distinct matrix proteins that are integral organic components of the spicule. One of these, SM50, encodes a 50 kilodalton matrix protein with an unusual 13 amino acid repeat in the C terminal half of the protein. The gene is transcribed at low levels in the mid-blastula, while intense transcription and protein synthesis occur from the mid-blastula stage until just before the onset of overt spicule construction (Benson *et al.* 1987; Wilt and Benson 1988). The second gene (George and Wilt 1991) dubbed SM30, encodes an unrelated glycoprotein of the spicule matrix. In its glycosylated form it migrates on gels like an approximate 40 kd protein. The SM 30 gene is transcribed a little after SM50 accumulation becomes rapid, and its rapid accumulation coincides well with the course of overt spicule formation. We had previously shown (Stephens, Kitajima and Wilt 1989) that SM50 transcripts accumulate in completely dissociated embryos and concluded that neither cell-cell interactions nor cell-matrix interactions were necessary for regulation of SM50. When this experiment was repeated and SM30 accumulation was examined, its expression seemed nearly completely inhibited by dissociation. Thus, it is likely that either cell-cell interaction or cell-matrix interaction, or both, are important for the regulation of SM30.

Since dissociation involves culture of blastomeres in the absence of calcium, we wished to disrupt cell-cell communication in a different way. This can be accomplished by culturing micromere derivatives in the presence or absence of serum. Micromeres were isolated and cultured as dissociated cells in the absence of calcium until the termination of cleavage. Then, at the equivalent of the early mesenchyme blastula stage, the micromere derivatives were cultured in petri plates in sea water following the general methods of Okazaki (1975). If 3% horse serum is included in the sea water, single cells aggregate, matrix is secreted, and spicules are made. Likewise, both SM50 and SM30 transcripts accumulate. If, however, serum is not included, cells remain dispersed and spicules are not made, though matrix molecules like collagen are still synthesized. Under these circumstances, SM50 is synthesized, SM30 is not. Hence, it is likely that the crucial interaction for regulation of SM30 expression is a cell-matrix interaction, while regulation of SM50 does not require this interaction(s).

This idea was tested by culturing embryos in the lathyrogen, beta amino proprionitrile (BAPN), an inhibitor of lysyl oxidase that prevents collagen cross linking. Several investigators have shown that BAPN inhibits extracellular matrix deposition and interferes with gastrulation and spicule formation (Wessel and McClay 1987; Butler, *et al.* 1987). This inhibitor is not very toxic, and its effects on development are fully reversible. Treatment of embryos with BAPN from fertilization onwards prevents expression of SM30, but not that of SM50. This is exactly what one would expect if interaction of primary mesenchyme cells with intact extracellular matrix was the crucial factor in regulation of SM30 gene expression.

SUMMARY

The potential to form a gut is very widely distributed in the early sea urchin embryo, almost all territories except the extreme vegetal most micromeres being capable under some circumstances of forming a gut. What is not clear is whether the mechanism of initiation of gut formation is always invagination, which is what occurs in the onset of normal gastrulation, or whether polyingression of surface cells or epithelialization of interior cells may also occur.

It is also not clear just what is the relationship between micromeres and gastrulation. Micromeres can stimulate gut formation in cells from the animal hemisphere, but whether invagination and some form of gastrulation occurs is not now known.

Preliminary experiments are discussed in which the ability of the two daughters of the micromere, the small and large micromere, are compared for their ability to induce gut like structures in cells from the animal hemisphere. The ability of small micromeres is very reduced compared to the large micromeres, and the small micromeres do not give rise to spicules.

Finally, the relationships between mesenchyme cells, and between the cells and the extracellular matrix, were explored in experiments in which the transcription of spicule matrix genes can be examined. Disruption of cell-cell interactions does not have a deleterious effect on the transcription of the two spicule matrix genes examined, but disruption of the extra cellular matrix does prevent the onset of transcription of one of the spicule matrix genes (SM30), while having little effect on the other (SM50). We conclude that these different spicule matrix genes are exquisitely and differentially sensitive to their surroundings.

ACKNOWLEDGMENTS

Research for F.H.W. was funded by an NIH grant #HD015043 and NASA. OK was supported by a Fogarty Post-doctoral Fellowship from NIH. The present address of NCG is Department of Biological Sciences, Columbia University, New York, N.Y.

REFERENCES

Benson, S., H. Sucov, L. Stephens, E. Davidson, and F. Wilt. 1987. Lineage specific gene encoding a major matrix protein of the sea urchin embryo spicule. *Dev. Biol.* 120:499-506.

Bernacki, S.H. and D.R. McClay. 1989. Embryonic cellular organization: Differential restriction of fates as revealed by cell aggregates and lineage markers. *J. Exp. Zool.* 251:203-216.

Blankenship, J. and S. Benson. 1984. Collagen metabolism and spicule formation in sea urchin micromeres. *Exp. Cell Res.* 152:98-104.

Butler, E., J. Hardin, and S. Benson. 1987. The role of lysyl oxidase and collagen crosslinking during sea urchin development. *Exp. Cell Res.* 173:174-182.

Cameron, R.A., B.R. Hough-Evans, R.J. Britten, and E.H. Davidson. 1987. Lineage and fate of each blastomere of the eight-cell sea urchin embryo. *Genes Dev.* 1:75-84.

Driesch, H. 1894. *Analytische Theorie der Organischen Entwicklung*. W. Engelmann, Leipzig.

Ettensohn, C.A. 1984. Primary invagination of the vegetal plate during sea urchin gastrulation. *Am. Zool.* 24:571-588.

Ettensohn, C.A. and D.R. McClay. 1988. Cell lineage conversion in the sea urchin embryo. *Dev. Biol.* 125:396-409.

Fukushi, T. 1962. The fates of isolated blastoderm cells of sea urchin blastulae and gastrulae inserted into the blastocoel. *Bull. Mar. Biol. Stn. Asamushi* 11:21-30.

George, N.C., C.E. Killiam, and F.H. Wilt. 1991. Characterization and expression of a gene encoding a 30.6 KD *Strongylocentrotus purpuratus* spicule matrix protein. *Dev. Biol.*, In press.

Giudice, G. 1973. *Developmental Biology of the Sea Urchin*. Academic Press, New York.

Gustafson, T. and L. Wolpert. 1967. Cellular movement and contact in sea urchin morphogenesis. *Biol. Rev.* 42:442-498.

Harkey, M.A. and A.H. Whiteley. 1980. Isolation, culture, and differentiation of echinoid primary mesenchyme cells. *Wilhelm Roux's Arch. Dev. Biol.* 189:111-122.

Henry, J.J., S. Amemyia, G.A. Wray, and R.A. Raff. 1989. Early inductive interactions are involved in restricting cell fates of mesomeres in sea urchin embryos. *Dev. Biol.* 136:140-153.

Horstadius, S. 1973. *Experimental Embryology of Echinoderms.* Clarendon Press, Oxford.

Kabakoff, B. and W.J. Lennarz. 1990. Inhibition of glycoprotein processing blocks assembly of spicules during development of the sea urchin embryo. *J. Cell Biol.* 111:391-400.

Khaner, O. and F.H. Wilt. 1990. The influence of cell interactions and tissue mass on differentiation of sea urchin mesomeres. *Development* 109:625-634.

Khaner, O. and F.H. Wilt. 1991. Interaction of different vegetal cells with mesomers during early stages of sea urchin development. *Development,* In press.

Langelan, R.E. and A.H. Whiteley. 1985. Unequal cleavage and the differentiation of echinoid primary mesenchyme. *Dev. Biol.* 109:464-475.

Livingston, B.T. and F.H. Wilt. 1990. Range and stability of cell fate determination in isolated sea urchin blastomeres. *Development* 108:403-410.

McClay, D.R. and C.A. Ettensohn. 1987. Cell recognition during sea urchin gastrulation. p. 111-128. *In: Genetic Regulation of Development.* W.F. Loomis (Ed.). Alan R. Liss, New York.

Moore, R.R. and A.S. Burt. 1939. On the locus and nature of the forces causing gastrulation in the embryos of *Dendraster excentricus. J. Exp. Zool.* 82:159-171.

Okazaki, K. 1975. Spicule formation by isolated micromeres of the sea urchin embryo. *Am. Zool.* 15:567-427.

Pehrson, J.R. and L.H. Cohen. 1986. The fate of the small micromeres in sea urchin development. *Dev. Biol.* 113:522-526.

Solursh, M. 1986. Migration of sea urchin primary mesenchyme cells. p. 391-431. *In: Developmental Biology: A Comprehensive Synthesis, Vol. 2, The Cellular Basis of Morphogenesis.* L.W. Browder (Ed.). Plenum Press, New York.

Spiegel, M. and E.S. Spiegel. 1975. The reaggregation of dissociated sea urchin cells. *Am. Zool.* 15:583-606.

Stephens, L., T. Kitajima, and F.H. Wilt. 1989. Autonomous expression tissue specific genes in dissociated sea urchin embryos. *Development* 107:299-307.

Tanaka, Y. and K. Dan. 1990. Study of the lineage and cell cycle of small micromeres in embryos of the sea urchin, *Hemicentrotus pulcherrimus. Dev. Growth & Differ.* 32:145-156.

von Ubisch, L. 1929. Uber die Determination der larvalen Organe under der Imaginalanlage bei Seeigelen. *Wilhelm Roux' Arch. Entwicklungsmech. Org.* 117:81-122.

von Ubisch, L. 1939. Kleimblattchimarenforschung an Seeigellarven. *Biol. Rev. Camb. Philos. Soc.* 14:88-103.

Wessel, G.M. and D.R. McClay. 1987. Gastrulation in the sea urchin embryo requires the deposition of crosslinked collagen within the extracellular matrix. *Dev. Biol.* 121:149-165.

Wilt, F.H. 1987. Determination and morphogenesis in the sea urchin embryo. *Development* 100:559-575.

Wilt, F.H. and S.C. Benson. 1988. Development of the endoskeletal spicule of the sea urchin embryo. p. 203-228. *In: Self Assembling Architecture.* J.E. Varner (Ed.). Alan R. Liss, New York.

PRIMARY MESENCHYME CELL MIGRATION IN THE SEA URCHIN EMBRYO

Charles A. Ettensohn

Department of Biological Sciences
Carnegie Mellon University
4400 Fifth Avenue
Pittsburgh, PA 15213

INTRODUCTION

Coordinated migrations of embryonic cells are a hallmark of metazoan development. At no time during embryogenesis are such cell movements more dramatic than during gastrulation, when the simply structured blastula is reorganized to produce a complex, multilayered embryo. For over a century, the gastrulation movements of the sea urchin embryo have both fascinated and challenged developmental biologists. The external development and optical transparency of these embryos mean that the cellular rearrangements of gastrulation are directly accessible to the experimenter. In addition, methods for raising large numbers of synchronously developing embryos and for isolating specific cell types and extracellular matrices facilitate many cellular and molecular approaches with this experimental system.

Gastrulation in the sea urchin embryo begins with the movements of the primary mesenchyme cells (PMCs), the cells that will construct the larval skeleton. (It has been conventional to mark the beginning of gastrulation with the invagination of the prospective gut. Thus the stage of development immediately following PMC ingression is the *mesenchyme blastula* stage, while *early*, *mid-*, and *late gastrula* stages are defined by the increasing depth of the invaginating archenteron. This terminology is so deeply embedded in the literature that it has been used here, although the early movements of the PMCs should be considered part of gastrulation.) The PMCs are perhaps the best characterized population of cells in the sea urchin embryo. The lineage and morphogenesis of these cells have been well described, and there is a growing body of

Gastrulation, Edited by R. Keller *et al.*
Plenum Press, New York, 1991

information concerning their program of gene expression during development. Methods have been developed for isolating PMCs and their embryonic precursors and culturing the cells *in vitro* under conditions that allow them to undergo a normal program of gene expression and morphogenesis. Many of these aspects of PMC development have been reviewed previously (Gustafson and Wolpert 1967; Okazaki 1975a,b; Harkey 1983; Solursh 1986; Wilt and Benson 1988; Decker and Lennarz 1988; Ettensohn and Ingersoll 1991). This paper will focus on the motile behavior of the PMCs and their directed migration in the embryo.

The PMCs are the sole progeny of the four large daughter cells of the micromeres (the VO*M*k, VA*M*k, and two VL*M*k blastomeres after the nomenclature of Cameron *et al.* 1987). The descendants of these cells reside temporarily within the epithelial wall of the embryo at the vegetal pole. At the late blastula stage, however, they undergo a transformation from an epithelial to a mesenchymal phenotype, penetrating the basal lamina that lines the blastocoel and moving into the interior of the embryo by a process known as ingression (see Solursh 1986; Ettensohn and Ingersoll 1991; Solursh, this volume). There are an average of 32 or 64 PMCs in the embryos of a given species, depending upon whether the large micromere daughter cells undergo three or four rounds of cell division. In addition, there is considerable variation in the numbers of PMCs among embryos within a single species (Ettensohn 1990b). There is no further division or death of PMCs after ingression, and therefore throughout the remainder of their morphogenesis the number of these cells remains constant. At the time of ingression, PMCs first begin to express lineage-specific gene products, including a unique family of antigenically-related cell surface glycoproteins (McClay *et al.* 1983; Carson *et al.* 1985; Wessel and McClay 1985; Anstrom *et al.* 1987; Leaf *et al.* 1987; Parr *et al.* 1990; Fuhrman *et al.* 1991), proteins that will form the organic matrix of the skeleton (Benson *et al.* 1987; Killian and Wilt 1989; George *et al.* 1990), and other PMC-specific gene products (Angerer *et al.* 1988).

Ingression is followed by a brief period of quiescence during which the PMCs remain at the vegetal pole in a non-motile state. This transient phase may indicate that changes in the adhesive properties of the PMCs (Fink and McClay 1985) and/or a reorganization of their cytoskeletal machinery are required for protrusive activity and motility (Gustafson and Kinnander 1956; Kinnander and Gustafson 1960). After 1-2 hours, (times are for *Lytechinus variegatus* at 23° C), the cells begin a phase of almost frantic movement. They migrate away from the vegetal pole along the inner surface of the blastocoel wall, dispersing throughout the vegetal region of the blastocoel. The cells translocate by the continuous extension and retraction of slender filopodial cell processes (see below). Gradually the PMCs accumulate at specific target sites near the equator, forming a characteristic ring-like pattern known as the subequatorial PMC ring. As the subequatorial ring forms, filopodial processes of the PMCs fuse, forming an extensive network of syncytial cables. Within these filopodial cables the PMCs secrete the larval skeleton, an array of crystalline rods (spicules) composed of $CaCO_3$ and $MgCO_3$ embedded in a glycoprotein matrix (Decker and Lennarz 1988; Wilt and Benson 1988). As development proceeds, the spicules lengthen and branch, resulting in an elaborate skeletal framework that gives shape and rigidity to the larval body. At the same time that the PMCs undergo their morphogenetic program and construct the larval skeleton, they engage in important cell-cell interactions that regulate the expression of cell fates by other mesenchymal cells in the blastocoel (Ettensohn and McClay 1988; Ettensohn 1990a; Ettensohn 1991).

PMCs move by means of slender filopodial protrusions. These spike-like processes are less than 0.5 μm in thickness (somewhat thicker near the cell body) and often extremely long, extending in some cases over 40 μm, a distance sufficient to span half the diameter of the blastocoel. Studies of PMC motility *in vivo* (Gustafson and Wolpert 1961; Wolpert and Gustafson 1961; Katow and Solursh 1981) and *in vitro* (Karp and Solursh 1985a,b) demonstrate that the filopodia are highly dynamic, and undergo a continuous process of extension and retraction. A single PMC often extends several filopodia in different directions simultaneously. Time-lapse cinemicroscopic studies indicate that the filopodia make stable attachments only to the inner surface of the blastocoel wall; they do not appear to terminate in the fluid-filled space of the blastocoel, and cell bodies of the PMCs are usually found along the blastocoel wall. There is considerable evidence that cell translocation is brought about by contractile forces generated by the filopodia (Gustafson and Wolpert 1967; Tilney and Gibbins 1969; Izzard 1974; Letourneau 1985), although it has also been suggested that these cell processes may act as "sensory" structures that respond to local stimuli and promote directional cell locomotion by mechanisms other than contractility (Solursh and Lane 1988).

The PMCs migrate on the basal surfaces of the ectodermal cells that form the outer wall of the embryo. These cells are organized as a monolayered cell sheet which by the start of gastrulation has differentiated into a polarized epithelium. The basal surface of the epithelium is covered by a thin basal lamina, a continuous layer of extracellular matrix (ECM), probably synthesized and secreted by the ectoderm cells, that lines the blastocoel cavity. It is this basal lamina that appears to serve as the substratum for PMC migration (Katow and Solursh 1981; Galileo and Morrill 1985; Amemiya 1990). At the ultrastructural level the basal lamina is composed of a dense meshwork of fibers and granules, and is continuous with the loose, fibrous ECM known as the blastocoel matrix that fills the interior of the embryo (Katow and Solursh 1979; Galileo and Morrill 1985; Amemiya 1990). Biochemical and immunological studies have demonstrated the presence of proteoglycans, glycoproteins, laminin, and possibly fibronectin in the basal lamina at the time of PMC migration (reviewed by Ettensohn and Ingersoll 1991).

METHODS

We have used fluorescent marking methods and cell transplantation techniques to analyze the regulation of PMC migration *in vivo* (Ettensohn and McClay 1986; Ettensohn and McClay 1988; Ettensohn 1990b). Covalently labeled donor cells are prepared by incubating embryos in rhodamine B isothiocyanate [RITC] (Bonhoeffer and Huf 1980). PMCs are then removed from these embryos and microinjected into the blastocoels of unlabeled, recipient embryos. The developmental stages of donor and recipient embryos can be varied, although it is easiest to remove PMCs from donor embryos soon after ingression, before the cells adhere firmly to the wall of the blastocoel and begin to migrate. Using these methods, single cells can be introduced into recipients or the total numbers of PMCs can be altered more drastically by adding or removing larger numbers of cells. The behavior of the PMCs following such manipulations is then monitored in two ways; 1) The optical clarity of the embryo makes it possible to follow the behavior of the transplanted cells directly by light microscopy (with transmitted light or epifluorescence), either at successive time intervals or continuously by means of time-lapse videomicroscopy. 2) The final distribution of PMCs in fixed samples can be assessed by immunostaining embryo whole mounts with any of several PMC-specific monoclonal antibodies (MAbs). This method has the advantage of being simple and rapid. In the studies described below we have made extensive use of MAbs 6a9 and 6e10, which recognize a family of PMC-specific, cell surface glycoproteins (Ettensohn and McClay 1988; Ettensohn 1990b; Fuhrman *et al.* 1991).

RESULTS

1. PMCs are guided to target sites by directional cues

In an undisturbed embryo, PMCs originate at the vegetal pole and migrate in a predominantly animal direction to the position of the subequatorial ring. The time-lapse cinemicrographic studies of Gustafson and coworkers (Gustafson and Wolpert 1961; Wolpert and Gustafson 1961) revealed the dynamic nature of this migration and showed that the filopodia of the PMCs actively explore the surface of the blastocoel wall. This behavior suggests that the cells are sensing directional cues in their local environment as they migrate. Okazaki *et al.* (1962) observed that PMCs displaced to the animal pole of the embryo by centrifugation would return to the correct position. Likewise, when PMCs are microinjected into the blastocoel near the animal pole, they migrate in a predominantly vegetal direction to the correct target sites and form a typical subequatorial ring (Figure 1, [2]). Examples of cell trajectories following such experiments are shown in Figure 2. Some cells move directly to the target site while others follow a more circuitous route, but all the transplanted cells eventually find their way to the correct position on the blastocoel wall. The PMCs are therefore not "programmed" with a specific pathway of migration but respond to external cues that direct them to the correct target sites. The initial distribution of PMCs in the blastocoel and their specific routes of migration can be highly variable, yet the cells still find their way to the proper target sites and a correct cellular pattern is produced.

2. PMC guidance cues are expressed in a temporal sequence

Transplantations of PMCs into recipient embryos of different developmental stages reveal that the directional cues governing PMC migration are expressed in a specific temporal sequence (Ettensohn and McClay 1986). When fluorescently labeled PMCs are transplanted from mesenchyme blastula stage embryos into the blastocoels of early (pre-hatching) blastulae, they migrate directionally, accumulating at the vegetal pole within 3-4 hours (Figure 1, [1]). The donor cells do not at first arrange themselves in a ring pattern, however. Instead, they remain near the vegetal plate until the PMCs of the host embryo ingress. Then the donor cells mingle with those of the host and migrate to the position of the subequatorial ring, contributing to all parts of the pattern. These observations show that a limited set of directional signals- those sufficient to guide the PMCs to the vicinity of the vegetal plate- are present in the embryo even before ingression. Subsequent changes in the guidance cues, however, are

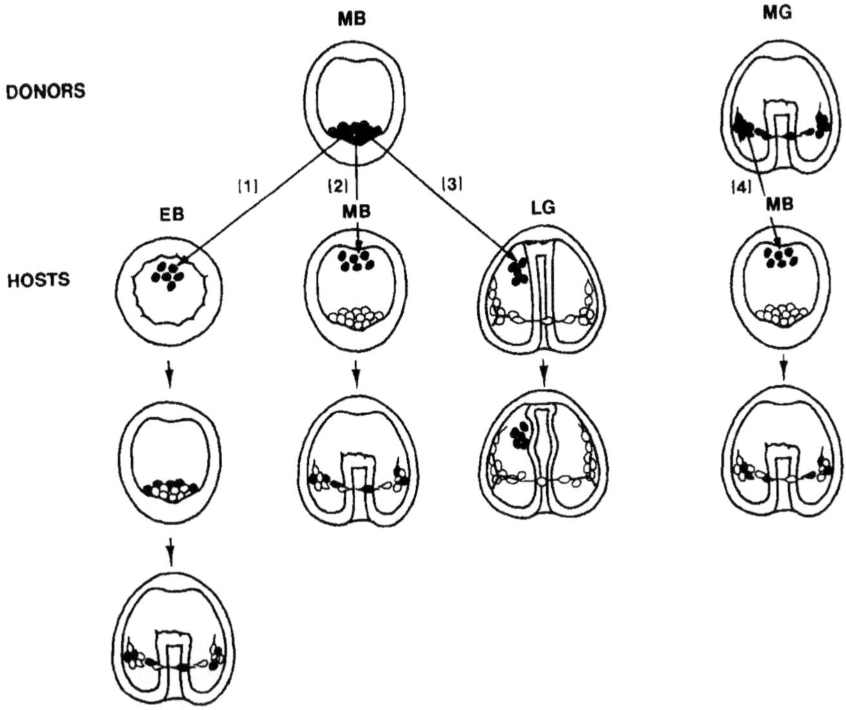

Figure 1. PMC transplantation experiments (*L. variegatus*). EB = early blastula, MB = mesenchyme blastula, MG = mid-gastrula, LG = late gastrula. Black cells represent rhodamine-labeled donor PMCs. PMCs from MB stage donor embryos when microinjected into EB hosts move to the vegetal plate and later join with host PMCs in the formation of the subequatorial ring **[1]**. Following transplantation into MB hosts, PMCs move directly to the subequatorial ring target sites **[2]**. When transplanted into older (LG) hosts, the cells do not translocate **[3]**. Donor PMCs removed from MG stage embryos (after the formation of the ring pattern) and introduced into younger (MB) hosts repeat their patterning process and give rise once again to a subequatorial ring **[4]** (see text for details).

required for the elaboration of a complete ring pattern. It is not necessarily the case that qualitatively different signals are used at different developmental stages; instead, the same signal might be used but the position of its source might change.

Heterochromic transplants of PMCs from mesenchyme blastula stage embryos into older (late gastrula or early prism stage) recipients show that in such an environment, PMCs have a reduced ability to migrate and most donor cells remain near the site of injection (Figure 1, [3]). Measurements of the average velocities of cells from time-lapse videorecordings indicate that the while PMCs microinjected into mesenchyme blastula stage embryos move at an average rate of almost 50 μm/hour, their velocity in older embryos is only 2-3 μm/hour. Thus there is a window of developmental time lasting some 8 hours during which the blastocoel is capable of supporting PMC migration. This does not mean that the motile activity of the PMCs ceases after ring formation. It is more limited than at earlier stages, however, and is restricted to the extension of filopodial processes along the blastocoel wall at the sites of spicule elongation and branching (Gustafson and Wolpert 1961; Wolpert and Gustafson 1961).

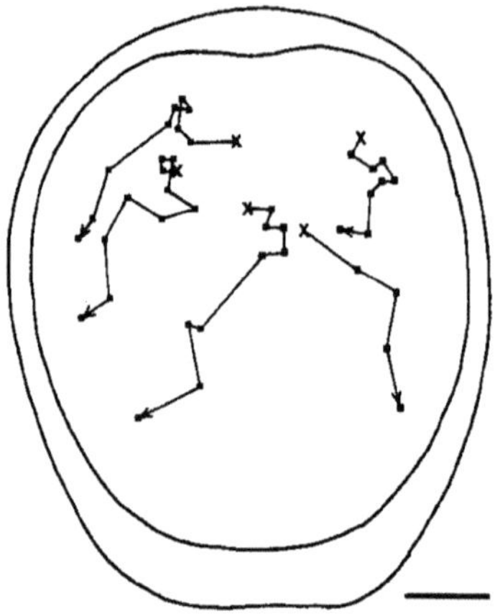

Figure 2. PMC trajectories following transplantation near the animal pole of mesenchyme blastula stage recipients (Experiment [2] in Figure 1). The trajectories of 5 donor PMCs as determined from time-lapse videorecordings are shown. The initial position of each cell is marked by an X, and squares represent the position of the center of each cell body at successive 10 minute intervals. Scale bar = 25 μm.

These studies reveal a remarkable degree of flexibility in the morphogenetic behavior of the PMCs. As illustrated in Figure 1 [1], after transplantation into early embryos PMCs are capable of *delaying* their normal patterning process by several hours. The flexibility of the cells is also shown by experiments in which PMCs are removed from midgastrula stage embryos (after the formation of the subequatorial ring) and microinjected into the blastocoel of younger recipients (Figure 1, [4]). Under these conditions, the PMCs *repeat* their patterning process, responding once again to the directional cues in the blastocoel and giving rise for a second time to a subequatorial ring pattern. These cell movements take place even when the endogenous complement of PMCs is first removed from the recipient embryo, demonstrating that the patterning of the cells is the result of their active directional migration and not a passive adherence to PMCs of the recipient embryo. The plasticity of the PMCs with respect to their morphogenetic behavior is in apparent contrast to the rigidly determined program of gene expression exhibited by this cell lineage (Harkey 1983; Davidson 1986; Ettensohn and Ingersoll 1991).

3. Response to the guidance cues is cell-type specific

To determine whether the PMCs are unique in their ability to respond to guidance cues in the blastocoel, the behaviors of other cell types and inert latex beads were examined after microinjection into the blastocoel. Two cell types were tested, ectoderm cells (a non-motile cell type) removed from the outer wall of the embryo, and secondary mesenchyme cells (SMCs), isolated from the tip of the archenteron at the late gastrula stage. SMCs, like PMCs, migrate actively within the blastocoel by filopodial extension and contraction, but because most of these cells do not ingress into the blastocoel until the late gastrula stage they are not normally exposed to the same set of directional cues that confront the PMCs. As a control, latex beads with a diameter of 10 μm (the approximate size of a PMC) were also microinjected into the blastocoel. As shown in Figure 3, when exposed to the same directional cues, only the PMCs migrated to the vegetal region of the embryo and subsequently to the position of the subequatorial ring. Ectoderm cells and latex beads remained near the site of injection and did not translocate, while SMCs migrated actively but instead of moving to the PMC target sites, remained scattered in the blastocoel or moved into the epithelial wall of the embryo, behaviors that are normally exhibited by SMCs after their ingression at the late gastrula stage.

Nature has provided another example of the remarkable specificity of the interaction between the PMCs and their migratory environment. During normal development a small number of SMCs leave the tip of the archenteron early in gastrulation, at the same time that the PMCs are undergoing their patterning process (Gibson and Burke 1985; Ettensohn and McClay 1988). Although these cells migrate within the same limited region of the blastocoel during the early gastrula stage, they show a very different pattern of movement. The early-ingressing SMCs migrate within the blastocoel for 1-4 hours before penetrating the basal lamina and intercalating into the ectodermal layer, where they differentiate as pigment cells. Thus, even though both the PMCs and pigment-forming SMCs are exposed to the same ensemble of guidance cues, only the PMCs move to target sites within the subequatorial ring.

4. PMC patterning involves site-selection by equivalent cells, but not competition for target sites

As described above, the migratory phase of the PMCs culminates in their arrangement in a ring-like configuration near the equator of the embryo. The patterning of the PMCs is a critical morphogenetic process since the arrangement of these cells serves as a direct template for the formation of the larval skeleton. The ring pattern has a characteristic substructure; two aggregates of PMCs (the ventrolateral clusters) form along the ventral aspect of the ring, while sparse chains of cells join these clusters ventrally and dorsally (the ventral and dorsal chains, respectively). In *L. variegatus*, each ventrolateral cluster consists of approximately 20 PMCs, while 5-10 cells are found in the ventral chain and 15-20 cells complete the ring dorsally. Skeletogenesis begins in the ventrolateral clusters with the formation of a single triradiate spicule rudiment in each. As development proceeds, the spicule rods lengthen and branch, always following the filopodial cables of the PMCs.

At least at a first level of analysis, the movements of PMCs to specific target sites within the subequatorial ring might be envisioned to take place by either of two basic mechanisms. First, the PMCs might consist of a homogeneous population of equivalent cells, and each PMC might choose a target site from multiple options. Alternatively, the primary mesenchyme might consist of distinct subpopulations of cells, each predetermined to migrate to a specific target site within the ring. According to this model, PMC patterning would take place by a process of cell sorting. To distinguish between these possibilities, PMCs were removed from specific sites within the ring pattern of mid-gastrula stage embryos and reintroduced into younger (mesenchyme blastula stage) recipients (Ettensohn 1990b). These experiments were designed to test whether PMCs from a specific site in the ring would always return to that same site. This was not the case; instead, PMCs isolated from either the ventrolateral clusters or the dorsal chain contributed to other parts of the pattern when microinjected into younger recipient embryos. The clearest demonstration of this came from the observation that PMCs isolated specifically from the ventrolateral clusters reformed a complete ring pattern when microinjected into recipient embryos from which the endogenous complement of PMCs had been removed. These experiments indicate that the PMCs consist of a homogeneous population of cells that are equivalent with respect to their patterning behavior.

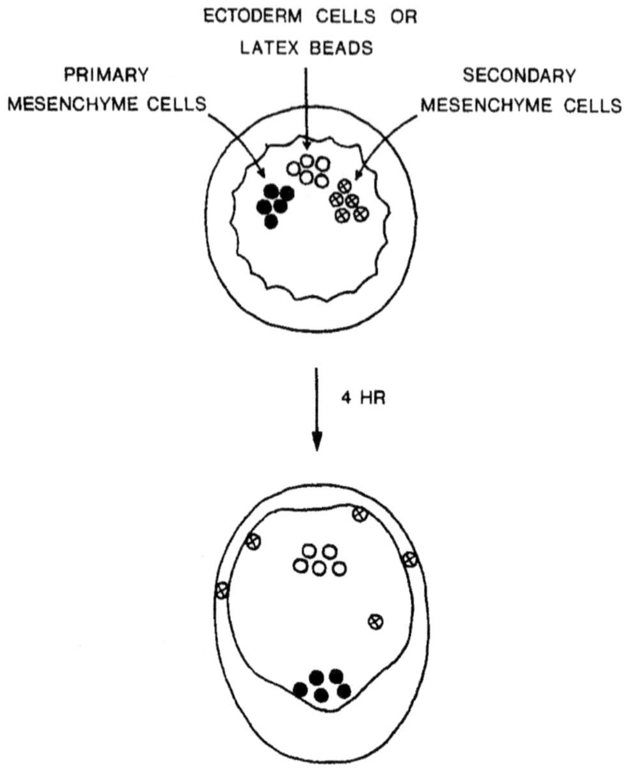

Figure 3. Transplantations of different cell types and latex beads into the blastocoel of early blastula stage embryos. PMCs (●) show directed movements to the vegetal plate and later to the subequatorial ring. Secondary mesenchyme cells (⊗) migrate to sites in the blastocoel and within the epithelial wall of the embryo but not to the PMC target sites, while ectoderm cells and latex beads (○) do not translocate. Modified from McClay and Ettensohn (1987).

If the PMCs select target sites from several possible options, then how is the final distribution of cells within the pattern achieved? One possible mechanism of PMC patterning is that as the cells migrate they compete with one another for preferred sites. The ventrolateral clusters, for example, where most of the PMCs become localized, are good candidates for such sites. A necessary correlate of such a view is that different sites within the ring pattern have a limited capacity for cells (that is, sites can "fill up"), and that PMCs that cannot be accommodated at preferred sites instead move to other positions within the ring. To test whether PMC patterning involves a competition of cells for a limited number of preferred target sites, the distribution of PMCs was examined in embryos with increased or decreased complements of these cells (Ettensohn 1990b). The relative distribution of PMCs in different parts of the ring pattern was found to be the same in embryos with half the usual complement of PMCs and in control embryos, a result inconsistent with a competition model (Figure 4). Moreover, the distribution of PMCs in PMC-supplemented embryos showed that all sites in the ring pattern have the capacity to accept more cells than they normally do. The PMCs therefore do not compete with one another for preferred sites. There may still be preferences for certain sites that are expressed only under unusual

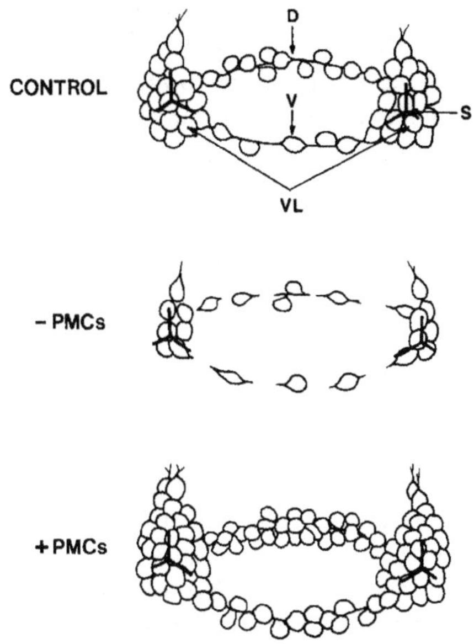

Figure 4. PMC patterning in control, PMC-depleted and PMC-supplemented embryos. These diagrams are based upon MAb 6a9-stained whole mounts of mid-gastrula stage embryos with varying numbers of PMCs. The distribution of the cells within the ring pattern of PMC-depleted embryos is that of a reduced ring pattern in which the proportion of the cells in different regions of the ring is the same as in control embryos. In PMC-supplemented embryos, increased numbers of PMCs are added to all parts of the subequatorial ring pattern. Regardless of the numbers of PMCs, triradiate spicule rudiments form at the positions of the ventrolateral clusters. S = spicule rudiment, VL = ventrolateral cluster, V = ventral chain, D = dorsal chain.

circumstances (for example, if a cell finds itself equidistant between the position of the ventrolateral cluster and the dorsal chain it may tend to move to the former) but such behavior must play a very limited role in determining the overall spatial distribution. Either all sites are equally attractive to the PMCs or few cells actually explore all possible options before joining the pattern.

5. Sites of spicule elongation and the pattern of the larval skeleton are independent of PMC number

In undisturbed embryos, skeleton formation begins with the appearance of one triradiate spicule rudiment in each ventrolateral cluster. The spicule rudiments are initially oriented randomly within the ventrolateral clusters, but undergo a rapid rotation that orients their radii properly (Wolpert and Gustafson 1961). The three arms of each rudiment subsequently elongate, giving rise to a ventral-transverse rod, dorso-ventral connecting rod, and a process that will branch to form the anal and body rods (Wolpert and Gustafson 1961; Okazaki 1975a). Spicule rudiments form at the correct locations and at the normal time whether the ventrolateral clusters contain as few as 4-5 cells or as many as 40 (Ettensohn 1990b). In addition, in PMC-supplemented embryos, the ventral and dorsal chains of the subequatorial ring accumulate much larger numbers of PMCs than normal, yet triradiate spicule rudiments nevertheless form only at the sites of the ventrolateral clusters. These observations demonstrate that the formation of the skeletal rudiments at the positions of the ventrolateral clusters is independent of the numbers of PMCs and that an interaction with the overlying basal lamina/ectodermal cells at those sites promotes spiculogenesis. Several investigators have noted that small calcareous granules often arise in other locations along the PMC cables but that these almost always fail to elongate. The influence of the ectoderm/basal lamina at the sites of the ventrolateral clusters is therefore to enhance spicule elongation. This effect may be related to the serum-dependence of spiculogenesis shown by PMCs *in vitro* (Okazaki 1975b).

The overall size and shape of the skeleton is the same in embryos with twice or even three times the usual complement of PMCs (Ettensohn 1990b). Fluorescent labeling experiments show that in such embryos, all PMCs take part in skeletogenesis. The size and shape of the skeleton are therefore controlled by other cell populations in the embryo that modulate the skeletogenic properties of the PMCs. The effect of reducing the number of PMCs is to delay the elaboration of the skeleton, but this effect is temporary as SMCs are eventually recruited to participate in skeletogenesis (Ettensohn and McClay 1988).

6. Interspecific transplants show that mechanisms of PMC guidance are shared among closely related species.

To test whether the guidance cues that direct PMC migration are shared by different species of sea urchins, we examined the movements of PMCs following interspecific transplantations. In these studies, donor RITC-labeled PMCs were removed from mesenchyme blastula stage *L. variegatus* embryos and microinjected into embryos of either a congeneric species, *L. pictus*, or a more distantly related sea urchin, *Eucidaris tribuloides*. *E. tribuloides* is a cidaroid urchin; species of this subclass resemble the ancestral stock from which most present day urchins probably arose. *E. tribuloides* and the two *Lytechinus* species are separated from one another by approximately 200 million years (see Raff 1987).

As shown in Figure 5 (A-C), microinjection of *L. variegatus* PMCs into the blastocoel of an *L. pictus* embryo results in the directional movement of the donor cells to the position of the subequatorial ring. *L. variegatus* cells mingle with the PMCs of the recipient embryo and contribute to all parts of the subequatorial ring pattern. The directional signals governing PMC patterning therefore appear to be held in common by these two species. The subsequent morphogenesis of the larval skeleton has not been followed in such embryos, although the blastomere recombination experiments of v. Übisch (1939) suggest that in interspecific chimeras the basic morphology of the skeletal rods (*i.e.*, simple *vs.* fenestrated) is determined by the PMCs, while the arrangement of the rods is at least partly controlled by the overlying ectodermal cells.

The behavior of *L. variegatus* PMCs in a *E. tribuloides* recipient is very different. In this environment, donor PMCs remain near the site of injection and fail to migrate, even when monitored for many hours (Figure 5D-F). Eventually they appear to fuse, as the borders of individual cells become less distinct, but no skeletal elements form. At the same time, skeletogenic mesenchyme cells of the host embryo migrate actively and localize at the correct target sites, forming spicule rudiments. These experiments demonstrate that there are profound differences in the blastocoel environment in these two species. It is not necessarily the case that the guidance cues that direct PMC migration are fundamentally different; rather, it may be that the composition of the extracellular matrix of the *E. tribuloides* blastocoel is simply not suitable to support *L. variegatus* motility. Directional cues might be overlaid on a general requirement for a blastocoel environment that can support cell migration.

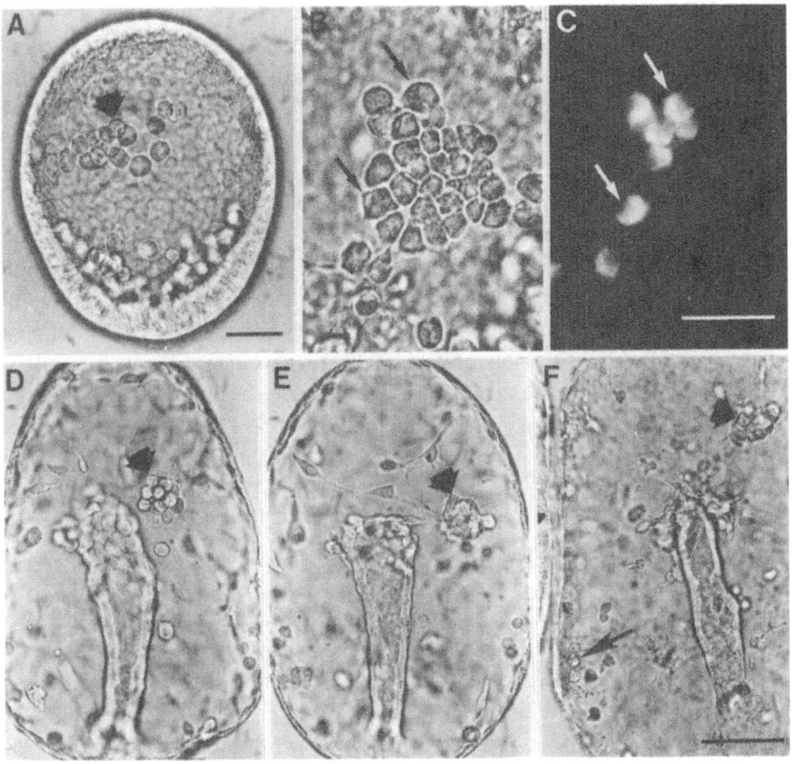

Figure 5. Interspecific transplantations. Rhodamine-labeled PMCs were removed from mesenchyme blastula stage *L. variegatus* embryos and microinjected into the blastocoel of *L. pictus* (**A-C**) or *E. tribuloides* (**D-F**) embryos. (**A**) Recipient *L. pictus* mesenchyme blastula immediately after injection with 17 donor *L. variegatus* PMCs (arrow). Scale bar = 25 μm. (**B,C**) Brightfield and epifluorescence micrographs, respectively, of one ventrolateral cluster of such an embryo 6 hours after the operation. The host embryo has reached the mid-gastrula stage, and fluorescently-labeled *L. variegatus* PMCs have joined with host PMCs in the formation of the ventrolateral cluster. Arrows indicate the corresponding positions of two cells. Scale bar = 25 μm. (**D-F**) Injection of *L. variegatus* PMCs into a *E. tribuloides* mid-gastrula stage (41 h) host, shown 0 h (**D**), 4 h (**E**), and 16 h (**F**) after the operation. At this stage, the host embryo is releasing mesenchyme cells into the blastocoel but the endogenous PMCs have not migrated to their target sites. The transplanted *L. variegatus* PMCs (short arrows) do not migrate or synthesize spicules, but appear to fuse into a compact mass. At the same time, the PMCs of the recipient embryo migrate to the positions of the ventrolateral clusters and synthesize spicule rudiments (long arrow). Note that there are relatively few PMCs in *E. tribuloides* and the ventrolateral clusters are correspondingly small. Transplantations were also carried out into recipient embryos of earlier developmental stages (mesenchyme blastula and early gastrula) with the same result. Scale bar = 50 μm.

DISCUSSION

The development of the PMCs provides an example of the interplay of morphogenetic processes in the early embryo. The early determination of the PMCs appears to be based upon the inheritance of maternal determinants partitioned to the micromere-PMC lineage (Davidson 1986). These early determinative events and the subsequent program of gene expression they evoke are responsible for early changes in the adhesive properties, cell surface composition, and motile behavior of the PMCs (Okazaki 1975b; Fink and McClay 1985; DeSimone and Spiegel 1986b), probably including those required for ingression and migration. The pattern of PMC migration, however, is regulated with exquisite precision by directional cues derived from the embryonic ectoderm and/or the basal lamina that it produces (see below). These interactions determine the arrangement of the PMCs in the blastocoel and the elaborate branching pattern of the larval skeleton. In addition, there is evidence that the PMCs regulate their own morphogenesis by secreting ECM components that condition the local microenvironment (Blankenship and Benson 1984; Angerer *et al.* 1988; D'Alessio *et al.* 1989; Benson *et al.* 1990; reviewed by Ettensohn and Ingersoll 1991). The further development of the skeleton is important in the morphogenesis of the ectoderm and in sculpting the angular form of the larva. As the PMCs play out their morphogenetic program, they also participate in interactions that specify the fates of other cell populations (Ettensohn and McClay 1988; Ettensohn 1990a). The migration and patterning of the PMCs is therefore part of an interrelated sequence of events that regulate the formation of the larval skeleton and the shaping of the larval body.

In one sense, the elaboration of the larval skeleton is a testament to the precision with which developmental processes are controlled. However, a detailed analysis of the behavior of the PMCs demonstrates a surprising element of flexibility in this system. As noted by Wolpert and Gustafson (1961), while the end product of morphogenesis is relatively constant, the events leading to it show considerable variation. For example, the pathways that individual cells take to reach the target sites can vary tremendously, as illustrated by the cell trajectories shown in Figure 2. In addition, the PMCs remain capable of responding to directional signals for many hours, as shown by transplantations of these cells into early embryos and from late stage donors into early recipients (Figure 1). In addition, there is a surprising flexibility in the ability of the skeletal pattern to regulate in response to widely varying numbers of skeletogenic cells. Together these characteristics confer a high degree of responsiveness on this morphogenetic system. Variability in numbers of cells and the timing of various processes (ingression, the onset of migration, the expression of directional cues, etc.) can be accommodated and still successfully produce the final state.

Probably the key piece still missing in the puzzle of PMC migration is the nature of the guidance cues that direct the movements of these cells. This is a central problem in development, since directional cell movements are an integral feature not only of gastrulation but of many later morphogenetic processes, including the formation of the nervous system (Trinkaus 1984). One possible mechanism of PMC guidance is that of selective "trapping" based on the differential affinity of the PMCs to sites in the vicinity of the ring (Gustafson and Wolpert 1967). This mechanism is consistent with the dynamic, apparently exploratory activity of the PMC filopodia. Such adhesive sites might be localized in the basal lamina, which appears to serve as the substratum for PMC migration. There is now evidence that the molecular composition of the basal lamina varies in different regions of the embryo (Katow and Solursh 1982; DeSimone and Spiegel 1986a; Ingersoll and Ettensohn, unpublished observations), and therefore there may be specific adhesive components associated with the target sites, although as yet good candidates for PMC patterning molecules have not been identified.

Ultrastructural studies have thus far not revealed any distinguishing characteristics of the basal lamina in the vicinity of the PMC ring.

In several species, the position of the PMC ring is predicted by an annular pattern of thickened ectoderm in the wall of the blastula (Gustafson and Wolpert 1961; Okazaki *et al.* 1962; Galileo and Morrill 1985). The pattern is the result of a fan-like, overlapping arrangement of ectoderm cells in this region of the embryo. Chemical treatments that result in a displacement of the PMC ring also cause a displacement of the annular pattern of ectoderm (Gustafson and Wolpert 1961; Okazaki *et al.* 1962). In *L. variegatus*, the annular pattern of ectodermal cells forms prior to PMC ingression, a time when directional cues are also expressed (Galileo and Morrill 1985; Ettensohn and McClay 1986). Whether (or how) the PMCs might recognize this distinctive region of the epithelial wall has not been established, although the temporal and spatial correlations between the ectoderm and PMC patterns are provocative. One indication that the ectoderm is involved in signalling comes from the observations of Harkey and Whiteley (1980), who noted that PMCs isolated in intact basal lamina "bags" migrated but did not accumulate in a ring pattern unless the basal laminae were allowed to reassociate with epithelial cells. These observations suggest that the basis of PMC patterning lies in the specific molecular interactions between the PMC filopodia and the epithelial cells in the region of the prospective subsequatorial ring.

ACKNOWLEDGEMENTS

The author gratefully acknowledges the support of D.R. McClay, in whose laboratory this work was initiated. The author's research is currently supported by a March of Dimes Basil O'Connor Starter Scholar Award, a N.S.F. Presidential Young Investigator Award, and N.I.H. Grant HD24690.

REFERENCES

Amemiya, S. 1990. Development of the basal lamina and its role in migration and pattern formation of primary mesenchyme cells in sea urchin embryos. *Dev. Growth & Differ.* 31:131-145.

Angerer, L.M., S.A. Chambers, Q. Yang, M. Venkatesan, R.C. Angerer, and R.T. Simpson. 1988. Expression of a collagen gene in mesenchyme lineages of the *Strongylocentrotus* embryo. *Genes Dev.* 2:239-246.

Anstrom, J.A., J.E. Chin, D.S. Leaf, A.L. Parks, and R.A. Raff. 1987. Localization and expression of msp 130, a primary mesenchyme lineage-specific cell surface protein of the sea urchin embryo. *Development* 101:255-265.

Benson, S., L. Smith, F. Wilt, and R. Shaw. 1990. The synthesis and secretion of collagen by cultured sea urchin micromeres. *Exp. Cell Res.* 188:141-146.

Benson, S.C., H.M. Sucov, L. Stephens, E.H. Davidson, and F.H. Wilt. 1987. A lineage specific gene encoding a major matrix protein of the sea urchin embryo spicule. I. Authentication of the cloned gene and its developmental expression. *Dev. Biol.* 120:499-506.

Blankenship, J. and S. Benson. 1984. Collagen metabolism and spicule formation in sea urchin micromeres. *Exp. Cell Res.* 152:98-104.

Bonhoeffer, F. and J. Huf. 1980. Recognition of cell types by axonal growth cones *in vitro*. *Nature* 288:162-164.

Cameron, R.A., B.R. Hough-Evans, R.J. Britten, and E.H. Davidson. 1987. Lineage and fate of each blastomere of the eight-cell sea urchin embryo. *Genes Dev.* 1:75-84.

Carson, D.D., M.C. Farach, D.S. Earles, G.L. Decker, and W.J. Lennarz. 1985. A monoclonal antibody inhibits calcium accumulation and skeleton formation in cultured embryonic cells of the sea urchin. *Cell* 41:639-648.

D'Alessio, M., F. Ramirez, H.R. Suzuki, M. Solursh, and R. Gambino. 1989. Structure and developmental expression of a sea urchin fibrillar collagen gene. *Proc. Natl. Acad. Sci. USA* 86:9303-9307.

Davidson, E.H. 1986. *Gene Activity in Early Development*, 3rd ed. p. 213-246, 493-504. Academic Press, Orlando.

Decker G.L. and W.J. Lennarz. 1988. Skeletogenesis in the sea urchin embryo. *Development* 103:231-247.

DeSimone, D.W. and M. Spiegel. 1986a. Concanavalin A and wheat germ agglutinin binding to sea urchin embryo basal laminae. *Wilhelm Roux's Arch. Dev. Biol.* 195:433-444.

DeSimone, D.W. and M. Spiegel. 1986b. Wheat germ agglutinin binding to the micromeres and primary mesenchyme cells of sea urchin embryos. *Dev. Biol.* 114:336-346.

Ettensohn, C.A. 1990a. Cell interactions in the sea urchin embryo studied by fluorescence photoablation. *Science* 248:1115-1118.

Ettensohn, C.A. 1990b. The regulation of primary mesenchyme cell patterning. *Dev. Biol.* 140:261-271.

Ettensohn, C.A. 1991. Mesenchyme cell interactions in the sea urchin embryo. *In: Cell Interactions in Early Development.* J. Gerhart, (Ed.). 49th Symp. Soc. Dev. Biol. Alan R. Liss, New York.

Ettensohn, C.A. and E.P. Ingersoll. 1991. Morphogenesis of the sea urchin embryo. *In: Morphogenesis: Analysis of the Development of Biological Structures.* E.F. Rossomando and S. Alexander (Eds.). Marcel Dekker, New York.

Ettensohn, C.A. and D.R. McClay. 1986. The regulation of primary mesenchyme cell migration in the sea urchin embryo: Transplantations of cells and latex beads. *Dev. Biol.* 117:380-391.

Ettensohn, C.A. and D.R. McClay. 1988. Cell lineage conversion in the sea urchin embryo. *Dev. Biol.* 125:396-409.

Fink, R.D. and D.R. McClay. 1985. Three cell recognition changes accompany the ingression of sea urchin primary mesenchyme cells. *Dev. Biol.* 107:66-74.

Fuhrman, M.H., A. Knecht, and C.A. Ettensohn. 1991. A family of cell surface glycoproteins specifically expressed by primary mesenchyme cells of the sea urchin embryo. *Dev. Biol.* In press.

Galileo, D.S. and J.B. Morrill. 1985. Patterns of cells and extracellular matrix material of the sea urchin *Lytechinus variegatus* (Echinodermata: Echinoidea) embryo, from hatched blastula to late gastrula. *J. Morphol.* 185:387-402.

George, N.C., C.E. Killian, and F.H. Wilt. 1990. Differential regulation of two sea urchin spicule matrix genes. *J. Cell Biol.* 111:484a.

Gibson, A.W. and R.D. Burke. 1985. The origin of pigment cells in the sea urchin *Strongylocentrotus purpuratus*. *Dev. Biol.* 107:414-419.

Gustafson, T. and M. Kinnander. 1956. Microaquaria for time-lapse cinematographic studies of morphogenesis in swimming larvae and observations of gastrulation. *Exp. Cell Res.* 11:36-51.

Gustafson, T. and L. Wolpert. 1961. Studies on the cellular basis of morphogenesis in the sea urchin embryo. Directed movements of primary mesenchyme calls in normal and vegetalized larvae. *Exp. Cell Res.* 24:64-79.

Gustafson, T. and L. Wolpert. 1967. Cellular movement and contact in sea urchin morphogenesis. *Biol. Rev. Camb. Philos. Soc.* 42:441-498.

Harkey, M.A. 1983. Determination and differentiation of micromeres in the sea urchin embryo. p. 131-155. *In: Time, Space, and Pattern in Embryonic Development.* W.R. Jeffery and R.A. Raff (Eds.). Alan R. Liss, New York.

Harkey M.A. and A.H. Whiteley. 1980. Isolation, culture, and differentiation of echinoid primary mesenchyme cells. *Wilhelm Roux's Arch. Dev. Biol.* 189:111-122.

Izzard, C.S. 1974. Contractile filopodia and *in vivo* cell movements in the tunic of the ascidian *Botryllus schlosseri. J. Cell Sci.* 15:513-535.

Karp, G.C. and M. Solursh. 1985a. Dynamic activity of the filopodia of sea urchin embryonic cells and their role in directed migration of the primary mesenchyme *in vitro. Dev. Biol.* 112:276-283.

Karp, G.C. and M. Solursh. 1985b. *In vitro* fusion and separation of sea urchin primary mesenchyme cells. *Exp. Cell Res.* 158:554-557.

Katow, H. and M. Solursh. 1979. Ultrastructure of blastocoel material in blastulae and gastrulae of the sea urchin, *Lytechinus pictus. J. Exp. Zool.* 210:561-567.

Katow, H. and M. Solursh. 1981. Ultrastructural and time-lapse studies of primary mesenchyme cell behavior in normal and sulfate deprived sea urchin embryos. *Exp. Cell Res.* 136:233-245.

Katow, H. and M. Solursh. 1982. *In situ* distribution of Con A binding sites in mesenchyme blastulae and gastrulae of the sea urchin *Lytechinus pictus. Exp. Cell Res.* 139:171-180

Killian, C.E. and F.H. Wilt. 1989. The accumulation and translation of a spicule matrix protein mRNA during sea urchin embryo development. *Dev. Biol.* 133:148-156.

Kinnander, H. and T. Gustafson. 1960. Further studies on the cellular basis of gastrulation in the sea urchin larva. *Exp. Cell Res.* 19:278-290.

Leaf, D.S., J.A. Anstrom, J.E. Chin, M.A. Harkey, R.M. Showman, and R.A. Raff. 1987. Antibodies to a fusion protein identify a cDNA clone encoding msp 130, a primary mesenchyme-specific cell surface protein of the sea urchin embryo. *Dev. Biol.* 121:29-40.

Letourneau, P.C. 1985. Axonal growth and guidance. p. 269-293. *In: Molecular Bases of Neural Development.* G.M. Edelman, W.E. Gall, and W.M. Cowan (Eds.). John Wiley, New York.

McClay, D.R., G.W. Cannon, G.M. Wessel, R.D. Fink, and R.B. Marchase. 1983. Patterns of antigenic expression in early sea urchin development. p. 157-169. *In: Time, Space, and Pattern in Embryonic Development.* W.R. Jeffery and R.A. Raff (Eds.). Alan R. Liss, New York.

McClay, D.R. and C.A. Ettensohn. 1987. Cell recognition during sea urchin gastrulation. p 111-128. *In: Genetic Regulation of Development.* W. Loomis (Ed.). 45th Symp. Soc. Dev. Biol. Alan R. Liss, New York.

Okazaki, K. 1975a. Normal development to metamorphosis. p. 177-232. *In: The Sea Urchin Embryo: Biochemistry and Morphogenesis*. G. Czihak (Ed.). Springer-Verlag, New York.

Okazaki, K. 1975b. Spicule formation by isolated micromeres of the sea urchin embryo. *Am. Zool.* 15:567-581.

Okazaki, K., T. Fukushi, and K. Dan. 1962. Cyto-embryological studies of sea urchins. IV. Correlation between the shape of the ectodermal cells and the arrangement of the primary mesenchyme cells in sea urchin larvae. *Acta Embryol. Morphol. Exp.* 5:17-31.

Parr, B.A., A.L. Parks, and R.A. Raff. 1990. Promoter structure and protein sequence of msp130, a lipid-anchored sea urchin glycoprotein. *J. Biol. Chem.* 265:1408-1413.

Raff, R.A. 1987. Constraint, flexibility, and phylogenetic history in the evolution of direct development in sea urchins. *Dev. Biol.* 119:6-19.

Solursh, M. 1986. Migration of sea urchin primary mesenchyme cells. p. 391-431. *In: Developmental Biology: A Comprehensive Synthesis, vol. 2. The Cellular Basis of Morphogenesis*. L.W. Browder (Ed.). Plenum Press, New York.

Solursh, M. and M.C. Lane. 1988. Extracellular matrix triggers a directed cell migratory response in sea urchin primary mesenchyme cells. *Dev. Biol.* 130:397-401.

Tilney, L.G. and J.R. Gibbins. 1969. Microtubules and filaments in the filopodia of the secondary mesenchyme cells of *Arbacia punctulata* and *Echinarachnius parma*. *J. Cell Sci.* 5:195-210.

Trinkaus, J.P. 1984. *Cells Into Organs*. 2nd edition. Prentice Hall, Englewood Cliffs, New Jersey.

von Übisch, L. 1939. Keimblattchimarenforschung an Seeigellarven. *Biol. Rev. Camb. Philos. Soc.* 14:88-103.

Wessell, G.M. and D.R. McClay. 1985. Sequential expression of germ layer specific molecules in the sea urchin embryo. *Dev. Biol.* 111:451-463.

Wilt, F.H. and S.C. Benson. 1988. Development of the endoskeletal spicule of the sea urchin embryo. p. 203-227. *In: Self-Assembling Architecture*. J.E. Varner (Ed.). Alan R. Liss, New York

Wolpert, L. and T. Gustafson. 1961. Studies on the cellular basis of morphogenesis of the sea urchin embryo. Development of the skeletal pattern. *Exp. Cell Res.* 25:311-325.

CELL-EXTRACELLULAR MATRIX INTERACTIONS DURING PRIMARY MESENCHYME FORMATION IN THE SEA URCHIN EMBRYO

Michael Solursh and Mary Constance Lane*

Department of Biology
University of Iowa
Iowa City, Iowa 52242

*Dr. Lane's current address:
Department of Molecular and Cell Biology
University of California
Berkeley, California 94720

INTRODUCTION

Gastrulation in different organisms can involve distinct morphogenetic mechanisms, at least at the cellular level. In the sea urchin, two distinct periods of gastrulation are observed (Solursh 1986). One involves the formation of the primary mesenchyme cells, which give rise to the larval skeleton, and the other involves the formation of the archenteron. Each of these illustrate different sorts of cellular activities. The formation of the primary mesenchyme involves an epithelial-mesenchymal transition followed by the apparently random migration of individual mesenchymal cells in the blastocoel until they form two ventral-lateral clumps connected by a ring of cells. The formation of the archenteron involves the infolding of cells at the vegetal plate followed by cellular rearrangements as the archenteron elongates (see chapter by McClay).

Primary mesenchymal cells arise *via* the conversion of a group of epithelial cells in the vegetal plate region to mesenchymal cells, which subsequently migrate on the basal lamina lining the blastocoel. This epithelial-mesenchymal transition is a fundamental, widespread morphogenetic process. This transition occurs for example, in the primitive streak of avian and mammalian embryos as cells move into the interior of the embryo (Solursh and Revel 1978).

Gastrulation, Edited by R. Keller *et al.*
Plenum Press, New York, 1991

PRIMARY MESENCHYME CELL FORMATION

The sequence of events surrounding primary mesenchyme cell formation has been described in the sea urchin embryo in detail at the ultrastructural level (Gibbons *et al.* 1969; Katow and Solursh 1980) and in living embryos (Gustafson and Kinnander 1956; Gustafson and Wolpert 1963). The basal lamina subjacent to the presumptive primary mesenchyme cells is disrupted and the cells exhibit blebbing activity at their basal surfaces (Figure 1). The presumptive primary mesenchyme cells detach from the outer hyaline layer and lose apical cell junctions as they elongate into the blastocoel (Katow and Solursh 1980). This behavior is due in part to a decrease in affinity of the mesenchymal cells for hyaline, measured by a cell attachment assay (Fink and McClay 1985). Adjacent cells, which will not become mesenchyme, actually exhibit increased numbers of hyaline associated microvilli (Amemiya 1989). The formation of the primary mesenchyme cells does not depend on microtubules (Anstrom 1989), as was previously believed, and the role of other specific cytoskeletal components still requires analysis. Like other cells undergoing epithelial-mesenchymal transitions, the Golgi apparatus remains at the trailing end of the cells (Anstrom and Raff 1988). Once the cytoplasmic mass has shifted into the blastocoel, with the narrowing of the apical end, detachment from the apical side occurs rapidly, followed by retraction of the apical portion of the cell. This process results in the transposition of the primary mesenchymal cell into the interior as a round cell resting on the vegetal plate. The adjacent epithelial cells quickly fill in the spaces left by the ingressing primary mesenchymal cells (Katow and Solursh 1980). The factors which regulate the breakdown of the basal lamina, the onset of blebbing behavior and cell elongation, changes in cell-cell and cell-extracellular matrix affinity and the stability of cell junctions are all likely to be involved in this morphogenetic process.

MIGRATION OF PRIMARY MESENCHYME CELLS

After ingression the primary mesenchyme cells migrate extensively. Migratory behavior begins after a distinct lag period during which the ingressed cells remain at the vegetal pole. Once migration begins, the cells move on the basal lamina that lines the wall of the blastocoel. Each cell moves in a distinct, different pattern, suggesting that the movement is random (Gustafson and Kinnander 1956). Eventually, however, most of the primary mesenchyme cells arrange themselves in a ring-like configuration just below the equator of the embryo where the larval skeleton is elaborated within two ventrolateral clumps of cells.

Primary cell migration *in vivo* can be inhibited by several treatments. These include sulfate deprivation (Herbst 1904; Karp and Solursh 1974) and exposure to β-D-xyloside (Akasaka *et al.* 1980; Solursh *et al.* 1986). Ingression occurs normally but the primary mesenchyme cells fail to move off of the vegetal plate (Katow and Solursh 1981). These two treatments are likely to interfere with the synthesis of sulfated glycoproteins and proteoglycans. In other systems, xyloside treatment results in the synthesis of free glycosaminoglycan chains and glycosaminoglycan-depleted proteoglycan core protein (Galligani *et al.* 1975; Fukunaga *et al.* 1975). With the addition of sulfate (Karp and Solursh 1974) or the removal of xyloside (Solursh *et al.* 1986), normal migration ensues.

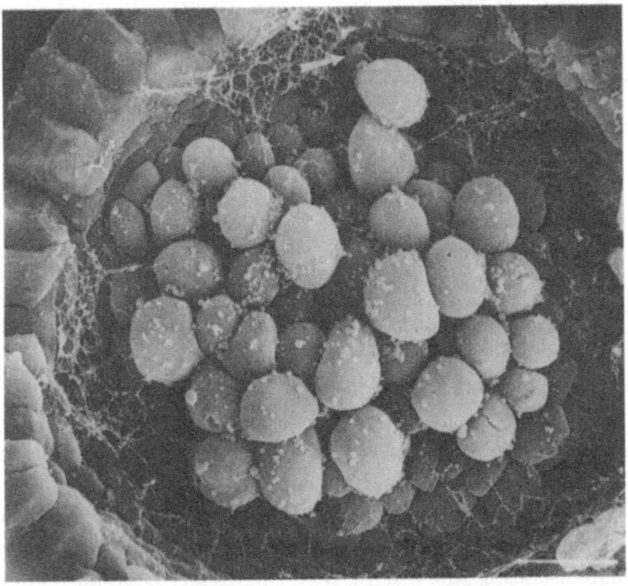

Figure 1. Scanning electron micrograph of an early mesenchyme blastula stage *Lytechinus pictus* illustrating the ingressing primary mesenchyme cells from their surface facing the blastocoel. One primary mesenchyme cell is apparently beginning to migrate from the vegetal plate, as suggested by the presence of a lamellipodium (arrow) (bar - 10 μ). (Micrograph taken by H. Katow.)

In order to approach the basis for the inhibition of the initiation of migration by sulfate deprivation or xyloside treatment, an *in vitro* migration assay was developed (Venkatasubramanian and Solursh 1984). Primary mesenchyme cells can be readily isolated by their differential ability to attach to tissue culture plastic in the presence of horse serum. The isolated cells from normal embryos will migrate extensively on a variety of substrata, including tissue culture plastic coated with human plasma fibronectin or blastocoelic extracellular matrix isolated from dissociated sea urchin embryos. Under these *in vitro* test conditions, cells isolated from sulfate-deprived embryos or xyloside-treated embryos attach to the substratum but fail to migrate. Most interesting is the observation that migration can be initiated *in vitro* by the addition of sulfate to primary mesenchyme cells isolated from sulfate-deprived embryos (Venkatasubramanian and Solursh 1984) or the removal of xyloside from treated primary mesenchyme cells (Lane and Solursh 1988). Recovery takes about two hours. These results support the hypothesis that the primary mesenchyme cells themselves synthesize a sulfated proteoglycan that is required to initiate cell migration.

CELL SURFACE SULFATED PROTEOGLYCAN

As an initial approach to the characterization of the sulfate/xyloside-sensitive components that are synthesized by primary mesenchyme cells and required for migratory behavior, components released by treatment of normal mesenchyme cells with 1 M urea in sea water were tested for their ability to correct the migratory behavior of cells isolated from sulfate-deprived or xyloside-treated embryos. This extraction procedure was utilized since it is known to solubilize peripheral membrane components and extracellular matrix components such as fibronectin (Yamada and Weston 1975). It was found that treatment *in vitro* of mesenchyme cells from sulfate-deprived embryos or xyloside-treated embryos with extract from normal primary

Table 1. *In vitro* PMC Migration Recovery. Effects of Fractionated PMC-urea Extract Separated on DEAE-Sephacel.

DEAE-Sephacel fraction	% migrating[a]	N[b]
0.25 M NaCl	0	27
0.55 M NaCl	36	67
1.00 M NaCl	0	37

[a] Migration is defined as movement greater than one cell diameter.
[b] Number of cells observed.
(From Lane and Solursh 1991)

mesenchyme cells corrects their migratory deficency *in vitro* (Lane and Solursh 1988). Similar extracts from sulfate-deprived or xyloside-treated primary mesenchyme cells were not effective in the same migration assay. In addition, extract from epithelial cells was also ineffective in correcting migratory deficency of mesenchyme cells, demonstrating the specificity of the extract from normal mesenchymal cells.

The content of the 1 M urea extract was explored further, as described by Lane and Solursh (1991). The extract was applied to a DEAE-Sephacel column and eluted by a NaCl step gradient. As can be seen in Table 1, the migration promoting activity was recovered in the fraction eluting in 0.55 M NaCl, where proteoglycans are typically eluted. Additional analysis of the extract demonstrated that incorporated sulfate was completely sensitive to treatment with chondroitinase ABC, but only partially sensitive to chondroitinase AC, suggesting that a cell surface chondroitin sulfate/dermatan sulfate proteoglycan is the active component (Figure 2).

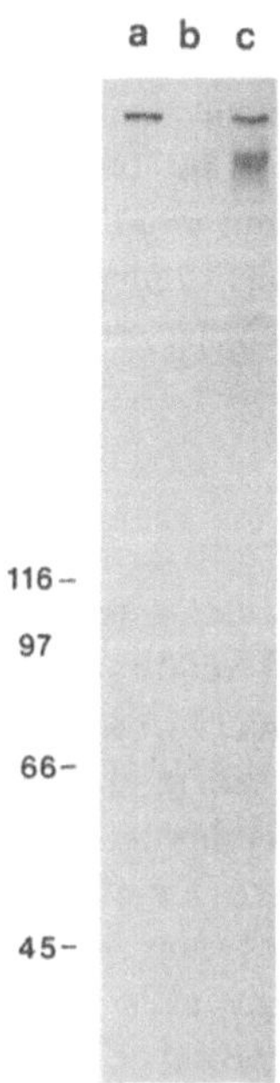

Figure 2. SDS-PAGE fluorogram of $^{35}S0_4$-labeled PMC urea extract, digested by chondroitinases. The urea extract contains a very large, sulfated species (Lane a), that is completely digested by chondroitinase ABC (Lane b), and partially digested by chondroitinase AC (lane c) pretreatments. These results indicate that the PMC extract contains a chondroitin sulfate/dermatan sulfate proteoglycan. (From Lane and Solursh 1991.)

Table 2. Effects of Chondroitinase ABC on *In vitro* PMC Migration

Treatment	% migrating[a]	N[b]
controls	41	125
+ chond'ase ABC (2 hrs)	3	110
+chond'ase ABC (4 hrs)	2	61
remove chond'ase ABC, recover (2 hrs)	41	51

[a] Migration is defined as movement greater than one cell diameter.

[b] Number of cells observed.

[c] Primary mesenchyme cells on a human plasma fibronectin substratum were cultured in either sea water or sea water containing 1 U/ml chondroitinase ABC and migration monitored by video microscopy (from Lane and Solursh 1991).

As an independent test for a role of a chondroitin sulfate/dermatan sulfate proteoglycan in cell migration, the effect on migration of treatment of living cells with chondroitinase ABC was examined (Lane and Solursh 1991). Cells in enzyme continued to migrate for about 1.5 hr after the addition of enzyme and then became immobile (Table 2). If the enzyme was removed, migratory behavior was again detected within an additional hour. These results again suggest that glycosaminoglycan chains such as chondroitin sulfate and dermatan sulfate might be required at the surface of migrating mesenchyme cells. Their covalent attachment to a proteoglycan core protein may serve to anchor these highly polyanionic moieties to the cell surface.

FUTURE STUDIES

The major immediate questions concern the structure and function of the mesenchymal cell surface proteoglycan. Structural analysis of the molecule is likely to provide insight into how it interacts with components in or on the surface of primary mesenchyme cells and with components in the extracellular matrix during the process of cell migration.

The current working hypothesis is that the mesenchymal cell surface chondroitin sulfate/dermatan sulfate proteoglycan may be required to reduce cell surface-extracellular matrix interactions to the extent that detachment and movement is possible. Migratory behavior requires a delicate balance between cell-substratum adhesion and cell detachment. While such a balance could readily play an important role in several morphogenetic processes observed during gastrulation in many systems, the primary mesenchyme cells of the sea urchin embryo provide a useful system for detailed analysis of the mechanisms.

REFERENCES

Akasaka, K., S. Amemiya, and H. Terayama. 1980. Scanning electron microscopical study of the inside of sea urchin embryos (*Pseudocentrotus depressus*). Effects of aryl β-xyloside, tunicamycin and deprivation of sulfate ions. *Exp. Cell Res.* 129:1-13.

Amemiya, S. 1989. Electron microscopic studies on primary mesenchyme cell ingression and gastrulation in relation to vegetal pole cell behavior in sea urchin embryos. *Exp. Cell Res.* 183:453-462.

Anstrom, J.A. 1989. Sea urchin primary mesenchyme cells: Ingression occurs independent of microtubules. *Dev. Biol.* 131:269-275.

Anstrom, J.A. and R.A. Raff. 1988. Sea urchin primary mesenchyme cells: Relation of cell polarity to the epithelial-mesenchymal transformation. *Dev. Biol.* 130:57-66.

Fink, R.D. and D.R. McClay. 1985. Three cell recognition changes accompany the ingression of sea urchin primary mesenchyme cells. *Dev. Biol.* 107:66-74.

Funkunaga, Y., M. Sobue, N. Suzuka, H. Kushida, and S. Suzuki. 1975. Synthesis of a fluorogenic mucopolysaccharide by chondrocytes in cell culture with 4-methylumbelliferyl β-D-xyloside. *Biochim. Biophys. Acta* 381:443-447.

Galligani, L., J. Hopwood, N.B. Schwartz, and A. Dorfman. 1975. Stimulation of synthesis of free chondroitin sulfate chains by β-D-xyloside in cultured cells. *J. Biol. Chem.* 250:5400-5406.

Gibbons, J.R., L.G. Tilney, and K.R. Porter. 1969. Microtubules in the formation and development of the primary mesenchyme in *Arbacia punctulata*. I. The distribution of microtubules. *J. Cell Biol.* 41:201-226.

Gustafson, T. and H. Kinnander. 1956. Microaquaria for time-lapse cinematographic studies of morphogenesis in swimming larvae and observations on sea urchin gastrulation. *Exp. Cell Res.* 11:36-51.

Gustafson, T. and L. Wolpert. 1963. The cellular basis of morphogenesis and sea urchin development. *Int. Rev. Cytol.* 15:139-214.

Herbst, C. 1904. Über die zur Entwicklung der Seeigelarven nothwendigen anorganischen Stoffe, ihre Rolle und Vertretbarkeit. III. Theil. Die Rolle der nothwendigen anorganischen Stoffe. *Wilhelm Roux's Arch. Dev. Biol.* 17:306-520.

Karp, G.C. and M. Solursh. 1974. Acid mucopolysaccharide metabolism, the cell surface, and primary mesenchyme cell activity in the sea urchin embryo. *Dev. Biol.* 41:110-123.

Katow, H. and M. Solursh. 1980. Ultrastructure of primary mesenchyme cell ingression in the sea urchin, *Lytechnius pictus*. *J. Exp. Zool.* 213:231-246.

Katow, H. and M. Solursh. 1981. Ultrastructural and time-lapse studies of primary mesenchyme cell behavior in normal and sulfate-deprived sea urchin embryos. *Exp. Cell Res.* 136:233-245.

Lane M.C. and M. Solursh. 1988. Dependence of sea urchin primary mesenchyme cell migration on xyloside- and sulfate-sensitive cell surface-associated components. *Dev. Biol.* 127:78-87.

Lane, M.C. and M. Solursh. 1991. Primary mesenchyme cell migration requires a chondroitin sulfate/dermatan sulfate proteoglycan. *Dev. Biol.* 143:389-397.

Solursh, M. 1986. Migration of sea urchin primary mesenchyme cells. p. 391-431. *In: Developmental Biology: A Comprehensive Synthesis, vol. 2, The Cellular Basis of Morphogenesis*. L.W. Browder (Ed.). Plenum Press, New York.

Solursh, M., S.L. Mitchell, and H. Katow 1986. Inhibition of cell migration in sea urchin embryos by β-D-xyloside. *Dev. Biol.* 118:325-332.

Solursh, M. and J.P. Revel. 1978. A scanning electron microscope study of cell shape and cell appendages in the primitive streak region of the rat and chick embryo. *Differentiation* 11:185-190.

Venkatasubramanian, K. and M. Solursh. 1984. Adhesive and migratory behavior of normal and sulfate-deficient sea urchin cells in vitro. *Exp. Cell Res.* 154:421-431.

Yamada, K.M. and J.A. Weston. 1975. The synthesis, turnover, and artificial restoration of a major cell surface glycoprotein. *Cell* 5:75-81.

THE ROLE OF CELL ADHESION DURING GASTRULATION IN THE SEA URCHIN

David R. McClay

Department of Zoology
Duke University
Durham, NC 27706

INTRODUCTION

For those who are interested in morphogenetic movements, the gastrula has long been a favorite embryonic stage for study. Gastrulation occurs early in morphogenesis so that the movements are rather simple to describe, though even then the descriptions can be long and detailed [Gustafson and Wolpert 1967; Gerhart and Keller 1986]. The general wisdom has been that since gastrulation is a time when there are relatively few cell types in the embryo, one ought to be able to explain how various morphogenetic movements occur in the emergence of pattern. Often there is an attempt to simplify, though in actuality, simplicity need not be a component of any movement. In fact, if one begins to study a single morphogenetic movement in some detail, that movement begins to appear more and more complicated, especially when one considers gene expression, changing adhesions, shifting cell positions, cell motility, net directionality of cell movements, etc. The morphogenetic event soon becomes bewilderingly complicated. And if one then projects to the complex anatomy that eventually carries the organism through its somatic lifetime, the complexity seems even more impressive. Yet at the same time we are persuaded that there must be some simplicity in the rules that govern pattern because, to use the same argument that led immunologists to look for simplicity in the immunoglobulin gene, there is not enough genomic space for each cell to be given its own complete set of instructions that are qualitatively different from those of any other cell.

In spite of the hope that there are unifying principles, several different fields of research each claim experts on the study of morphogenetic movements. It is likely that almost every cellular system is involved in some aspect of the changes that lead to the

Gastrulation, Edited by R. Keller *et al.*
Plenum Press, New York, 1991

simplest of morphogenetic movements. With such a sweeping utilization of cellular and molecular properties it is evident that there must be coordinate control at several levels. These statements are rather daunting when one thinks of how morphogenesis might work. The only way to make sense of the process, therefore, is to concentrate on small portions of model systems in order to learn, mechanistically, how a developing organism actually pulls off its magic. In that spirit this review will focus on cell interactions that participate in the movements of ingression and invagination that characterize gastrulation in the sea urchin embryo. The transparency of this embryo allows one to film the process by time-lapse, and the embryo is rather easy to manipulate experimentally to test hypotheses based on the observations. What follows is a current review of information on cell interactions during ingression and invagination, the two major morphogenetic movements at gastrulation in the sea urchin.

The term 'cell interactions' as given above, has been used to describe several properties that may or may not relate to one another. At one end of the spectrum, cell interaction refers to adhesions between cells or between a cell and its substrate. 'Cell interactions' also refers to an interaction in which information transfer occurs between two cells. In this kind of cell interaction the cells must come into close contact for there to be a transfer between the two cells. Recent examples of such interactions under current scrutiny are the lin 12 and the Glp-1 genes in *Caenorhabditis elegans* [Austin and Kimble 1989], the notch locus in *Drosophila* [Xu *et al.* 1990; Fehon *et al.* 1990, and recombination experiments in amphibian embryos in which growth factors induce presumptive ectoderm to become mesoderm as a result of an interaction with information transfer from the underlying yolky endoderm [Smith *et al.* 1989; Cooke 1989].

An interaction may include both an adhesion <u>and</u> information transfer, although it is often difficult to decide whether the adhesion and the transfer are independent events. In an effort to begin to clarify this area in the sea urchin embryo we have been examining a number of interactions, some of which may be adhesive alone, some may involve information transfer alone, and others may perform both functions simultaneously. An effort has been made to quantify the adhesive interactions and to design approaches that experimentally address the problem of information transfer. The goal is to find objective criteria that distinguish between these phenomena and then to fully describe the molecular basis of the interaction. This chapter reports on the approaches and the progress of that effort.

CELL ADHESION

Adhesion has been quantified in a number of ways, beginning with simple observation of cell aggregate formation [Wilson 1907; Moscona and Moscona 1952]. Using cell aggregation, early studies examined organisms, tissues, and embryos to ask questions about the capacity of cells to reassociate with one another. At first it was important simply to document the ability of cells to sort out from one another in rather specific ways [Holtfreter 1943; Moscona 1956; Steinberg 1970]. As such, aggregation was used as a model for tissue associations in morphogenesis beginning with the classic studies of Holtfreter as the model example [Holtfreter 1943; Townes and Holtfreter 1955]. As this process of aggregation and sorting out has been reexamined in recent years, the original assumption, that aggregation was measuring cell adhesion, has had to be reconsidered. Aggregation lasts a long time. Often the assays are performed for hours or even days. In that time much more than adhesion takes place. For example, it is known that aggregate stability involves junction formation [Sheffield and Moscona 1989]. Cell motility is involved in packing the cells in an aggregate, and in the sorting out process [Trinkaus 1984]. A number of signals have been reported to be passed between cells or between cells and the substrate [Spray *et al.* 1987], and in a number of cases, the interactions of cells in an aggregate are necessary for the expression of certain genes which, in turn, could affect other properties of the aggregates [Rutishauser 1989]. Thus the picture about what aggregates can reveal becomes confusing, and it becomes difficult to learn which event is directly responsible for a consequential event. The embryo presents a similar problem. Lots of events are occurring simultaneously so that if one wants to learn about a complicated morphogenetic process it is necessary to understand the component parts. It therefore becomes necessary to simplify experimental approaches to adhesion, and at the same time in ways that reveal properties of morphogenetic processes.

How might one measure cell adhesion in isolation from other processes? One goal might be to isolate the molecules that mediate adhesions. Using antibodies as blocking reagents of cell associations a number of cell adhesion molecules have been isolated [Edelman 1985; Takeichi 1988; Hynes 1987; Buck and Horowitz 1987]. At the molecular level many of these molecules are associated in 'homotypic' interactions (a molecule on one cell interacts with an identical molecule on the adhering cell), or a 'heterotypic' interaction (the interacting molecules are different). In the case of a number of adhesion molecules it is known that carbohydrates are involved in the adhesion [Jessell 1990]. Also, a number of receptor-ligand complexes have been described at the molecular level, especially in the interaction of integrins and their variety of substrates [Yamada and Akiyama 1984]. In most cases the adhesion molecule is defined experimentally as a molecule that is present on the cell surface that, if blocked by an antibody specific to that molecule, results in a reduced level of adhesion. For some molecules there is the additional evidence that if the molecule is expressed in a cell that normally does not adhere, and the transfected cells become adhesive, then the molecule in question truly is an adhesion molecule [Nose *et al.* 1988].

For the antibody approach to identify adhesion molecules an important assumption had to be correct. It was necessary for the antibodies to block the first step in the process. This may sound like a sure bet but it is not a trivial assumption. Consider, for example the case of lymphocyte interactions. The T-cell receptor remained elusive for a number of years because it was assumed that it was involved in the first interaction that took place between two cells. As it turned out, an adhesive interaction had to occur first before the T-cell receptor could interact effectively with its ligand on the opposite cell [Springer *et al.* 1987]. In similar fashion, an assumption is made that if an antibody can block a cell interaction the antibody must recognize an adhesion molecule. In most cases this assumption is probably correct but one must verify the interaction by independent means (such as by the transfection experiments mentioned above), and further, one must learn how the particular antigen fits into the overall adhesive process that functions *in vivo*. A case in point is the 22s particle that was isolated from sea urchin cell membranes [Noll 1979; Noll *et al.* 1985]. Originally, an antibody to that preparation blocked adhesion [Noll *et al.* 1979], thereby satisfying the criterion of antibody blocking. Later, it was found that the major component of the 22s particle was antibody that partially blocked adhesion of developing cells. Using the antibody, a cDNA for the molecule was identified. From that point all information that accumulated was based on homology. Yet, beyond the antibody data no information was with radioactive iodine on cell surfaces, even though, if isolated, the 22s particle was labeled easily [Lee *et al.* 1989]. Since the molecule could not be demonstrated ever to be on the surface of cells in an embryo, it had to be concluded that the adhesive effect of the 22s particle was artifactual. Thus, although a number of putative adhesion molecules may have been isolated there is no single way of knowing whether the antigen actually functions in adhesion, nor is it known whether the antigen is the only adhesion molecule on the cell in question. Again, using the lymphocyte model, there are several adhesion molecules involved in addition to the T-cell receptor, each of which has a specific adhesive function [Springer *et al.* 1987].

If one measures inhibition of adhesion with antibodies the only function observed is partial or complete blocking of the whole process. Many cell types have been found that bear multiple adhesion molecules [Bixby *et al.* 1987]. Is the function of each molecule the same? One hypothesis is that adhesion of cells is redundant so that a number of molecules participates in each interaction [Bixby *et al.* 1987]. The idea is that each of several molecules (in the example used by Bixby *et al.*, [Bixby *et al.* 1987]) can promote an adhesion to allow growth cone migrations. This conclusion was reached when it was observed that growth cone migration continued when antibodies to any one adhesion molecule was present, and only when antibodies to three were present was growth cone migration blocked. A second hypothesis might be that the actual function of each of the several adhesion molecules is different. Unfortunately, in order to test such a hypothesis antibody blocking usually is insufficient since the antibody blocking looks only at the initial 'stickiness' of cells and the term adhesion is usually asserted to mean more than that stickiness.

How then can one measure the properties of adhesion and then factor in the different antigens and their potentially different functions? One approach is to quantify the distinct events of an adhesion and then ask how each of the known molecules might participate in the several parameters. That approach is illustrated below.

A series of assays has enabled us to divide adhesions into an initial binding phase in which the molecular interactions of adhesion molecules binds cells together. An example of this approach is adhesion of CHO cells. Using variants bearing different amounts of the a5ß1 integrin (fibronectin receptor), initial binding strengths (at 4°) were compared on a fibronectin substrate. The cells bearing the most receptor stuck more tightly than those bearing 20% or 2% the number of a5ß1 receptors. When the cells were incubated at 37° while in contact with the substrate the interaction strengthened enormously, although the cells bearing fewer fibronectin receptors strengthened more slowly relative to those with the full complement. A number of studies have shown that it is the cytoplasmic portion of the molecules that is important for the strengthening [Burridge *et al.* 1987, 1988; Lotz *et al.* 1989]. In yet other studies one can follow motility, speed of strengthening, rate of movement of an adhesion molecule in the plane of the membrane, and relative contribution of several adhesion molecules to an adhesion in question [Lotz *et al.* 1989].

How does such an approach enter into the problems of morphogenesis? By quantifying the interactions and by learning the parameters provided by each participating molecule, one can have a better picture of how each molecule might be contributing to the process. Consider the following contrast. Suppose one found an yolk [Lee 1989 *et al.*; Cervello and Matranga 1989]. This raised the question as to whether the molecule was actually an adhesion molecule or a protein that artifactually stuck cells together. It was finally learned that the 22s particle never could be labeled known about the putative function of the molecule aside from inferences based on sequence homology. Ultimately it is necessary to return to an examination of the partial inhibition of adhesion, of the function of other antigens, and of other parameters in the interaction to gain an assessment of the actual function (which might not be adhesive after all). Below, the sea urchin embryo will be used as an example to illustrate this principle in our attempt to reduce morphogenetic behavior to a series of molecular interactions.

ADHESIVE PROPERTIES IN GASTRULATION

Although a similar history could be constructed for several embryos, this review will focus on the sea urchin embryo. Early in this century it was shown that the sea urchin embryo could be dissociated if calcium were removed from the medium [Herbst 1900]. This began the thinking that calcium was somehow important in the cell adhesion process. Models of how calcium might work have included divalent cation bridges between adjacent cells, or other charge-related activities. As it turns out, many of the calcium dependent adhesive events in vertebrate cells have an explanation that has more to do with trypsin sensitivity of the adhesion molecule than direct participation of calcium in the adhesion event itself (some molecules are sensitive to trypsin in the presence of calcium, others are protected from trypsin degradation in the presence of calcium, [Takeichi *et al.* 1981]). In the sea urchin embryo, removal of calcium is all that is necessary to effect dissociation [McClay and Matranga 1986], though the story is not simple. The hyaline layer is sensitive to calcium [Kane and Stephens 1969; Stephens and Kane 1970] such that in the absence of calcium the protein hyalin loses an ability to associate with other hyalin molecules, and the extraembryonic matrix separates from the embryo. At the same time cells can be observed to round up and it takes little effort to then dissociate the embryo. Cells in suspension appear to require calcium for cell-cell adhesion [McClay and Matranga 1986]. If calcium is added to the medium where cells are suspended, the cells immediately become adhesive, without the addition of hyalin, or any other extracellular matrix protein, and without the opportunity to make any changes metabolically. Thus, for cells that have been isolated in the absence of protease treatment, addition of calcium seems to have a direct participation in the adhesive process. The details of that participation are unknown.

The cells of the early embryo interact with the extracellular matrix and that interaction appears to have an important role in several aspects of development. Dan [Dan and Ono 1952] suggested that if embryos were placed into hypotonic sea water, the cells tended to separate from one another but they continued to adhere to the hyaline layer. Dan proposed that perhaps the cell-substrate interaction was stronger at this stage than the cell-cell interactions. Whether this interpretation is correct or not it showed that the cells of the embryo do have an active interaction with the extracellular matrix.

Several molecules in the extracellular matrix have been found to serve as adhesive substrates for sea urchin cells. Hyalin and echinonectin in the extraembryonic matrix

[McClay and Fink 1982; Alliegro *et al.* 1988], and fibronectin and laminin in the basal lamina [Fink and McClay 1985; McCarthy and Burger 1987] have been shown experimentally to support adhesion. Vertebrate laminin and fibronectin molecules were used in the adhesion demonstrations so it remains an open question as to whether the equivalent molecules in the sea urchin support adhesion. There is evidence for the existence of those molecules natively in the sea urchin [McCarthy *et al.* 1987; Spiegel *et al.* 1980; Spiegel *et al.* 1983; Wessel *et al.* 1984].

HYALIN IN CELL INTERACTIONS

Using the cell adhesion assay it was possible to ask which extracellular matrix molecules might serve as substrates for cells. Focus was on the mesenchyme blastula stage where primary mesenchyme cells were known to ingress at a precise time in morphogenesis. We were able to isolate a number of substrates and learned that the ingression movements involved at least four simultaneous adhesion changes. The primary mesenchyme cells lost an adhesion for hyalin [McClay and Fink 1982], echinonectin [Alliegro *et al.* 1988], for other cells of the embryo [Fink and McClay 1985], and they gained an adhesion for basal lamina, especially for fibronectin [Fink and McClay 1985]. We looked at other extracellular matrix proteins such as collagen, chondroitin sulfate and several other proteoglycans and found that only the substrates given above supported significant adhesion, although even in these cases there were differences. For example, on an equimolar basis, cells adhere 2 to 3 times better to hyalin than to echinonectin [Budsal *et al.* 1991] (although this changes with cell type during development–see below).

To further ask how the adhesive molecules might participate in the morphogenetic properties we have isolated hyalin and echinonectin and characterized them further. Hyalin was isolated using the methods of Kane [Kane 1973]. The protein normally is in a matrix that has about 12-15 proteins in it [McCarthy and Spiegel 1983; Hall and Vacquier 1982]. Hyalin was purified on the basis of its ability to polymerize in the presence of calcium and to depolymerize in EDTA. Hyalin, thus purified continued to serve as a support for adhesion. Monoclonal antibodies to the hyalin were prepared and one antibody was found to inhibit adhesion to a hyalin substrate [Adelson and Humphreys 1988]. That antibody was used to screen a lambda GT-11 library and identify a cDNA that, by a number of criteria, was the hyalin gene. The gene has been sequenced (about 80% complete) and the portion recognized by the antibody (the same antibody that inhibited adhesion) was expressed in a bacterial expression system. The expressed protein was an adhesive substrate and the adhesive activity was blocked by the monoclonal antibody [McClay *et al.*, in preparation]. In other studies hyalin was viewed ultrastructurally after rotary shadowing purified molecules. The protein is a club headed filament about 75 nm long. The antibody binds to a region near the head (when antibodies were added to the purified molecular preparation). Multiple antibodies bind to that region in agreement with the sequencing data which shows that the cell binding region contains a series of amino acid repeats [Adelson *et al.*, in preparation]. We have learned little about how the molecule associates with itself in the presence of calcium. The ultrastructural data indicates that the molecule can associate head to head in multiple complexes in the presence of a larger protein that joins the cluster of hyalin heads, and the protein also seems to associate tail to tail. Whether either or both of those associations requires calcium is as yet unknown.

The cell adhesive properties of hyalin have been documented above as an example of how one can quantify cell adhesion and learn about the molecular characteristics of that association. If one examines the hyalin association in other ways the molecule takes on other dimensions in this exercise toward understanding cell interactions. Those data are included below.

Does hyalin have a distinct adhesive role in development? Going back to the mesenchyme blastula stage it was shown that primary mesenchyme cells lost an adhesiveness to hyalin as they ingressed. In contrast, the presumptive ectoderm and endoderm continued to adhere to hyalin at a constant strength all through development. Thus, only the primary mesenchyme cells lost the capacity to adhere to this protein. The reason for that loss is not known but it must reside in the membranes of the primary mesenchyme cells because the experimental conditions were such that the hyalin substrate remained unchanged at all times through the experiments. The simple explanation, conceptually, is that a hyalin receptor was somehow disengaged or lost from the primary mesenchyme cells at the time of ingression. Whatever the explanation, it is clear that primary mesenchyme cells undergo a number of changes, simultaneously, in order to convert from an epithelial cell in the blastula, to a mesenchyme cell following the invasive movements of ingression.

Another kind of cell interaction can also be detected in experiments with hyalin. It was noticed that when the adhesive-inhibiting antibody was added directly to cultures of embryos the blastomeres pulled away from the hyaline layer and the embryo failed to gastrulate [Adelson and Humphreys 1988]. If added at fertilization the embryo developed until the mesenchyme blastula stage (with a smaller blastocoel) but it was arrested at that stage. If the antibody were withdrawn, even after the embryo had been arrested for a number of hours, the embryo then continued to develop and gastrulate normally [Adelson and Humphreys 1988]. These observations raise the question as to whether the cell-hyalin interaction has some sort of transducing function. The hyalin effect is reminiscent of systems where there is a genetic lesion and the embryo develops to a critical stage and then an aspect of development is arrested. In the present case hyalin is known as an adhesion molecule. Does this mean that the hyalin receptor is a transducing molecule that somehow governs a critical step leading to gastrulation? Perhaps, but a number of other interpretations have equal, or more compelling merit. These will be reviewed below. Suppose, however, we were faced with the question of an interaction that was known only from the genetic lesion. How could we decide which of the possible interactions is the correct function for the molecule in question? Could adhesion be involved, or might adhesion only very indirectly contribute to the lesion? The following paragraphs attempt to offer guidelines that can be used to unscramble this dilemma.

First, suppose we did not know that the molecule was an adhesive substrate, but that adhesion was one of the possibilities for its function. If we had an antibody we could be lucky and learn that the antibody partially blocks adhesion. In the present case that enables us to conclude, by the antibody blocking model, that hyalin is an adhesion molecule. But, let us further suppose that if we were able to measure other adhesion molecules in the system we could actually determine that the molecule in question contributed a certain percentage of the total adhesiveness to the native substrate. This would lend some confidence not only in the adhesive function but also in the relative contribution of the specific molecule to the total interaction between a cell and its substrate. Quantitative measurements are now possible [McClay *et al.* 1981;

Lotz *et al.* 1989] and can contribute to this kind of an interpretation of adhesion as a potential function for a molecule. When such measurements are made, hyalin offers an adhesive strength that is about 2/3 that measured when cells are placed on complete hyaline layers with all 15 proteins. Since echinonectin is about 1/3 to 1/2 as strong an adhesive substrate as hyalin when the two are compared on an equimolar basis [Burdsal *et al.* 1991], the data suggest that hyalin is the major adhesive protein of the extraembryonic matrix during early development and echinonectin is the perhaps the only other major adhesive protein in that matrix. This lends confidence to the notion that hyalin is an adhesion substrate, but what of the transducing activity?

Is there a similar quantitative approach to determining whether a molecule has a transducing function? In the case of hyalin only very indirect clues exist. First, other inhibitions of extracellular matrix have a similar effect to that of hyalin [Wessel and McClay 1987]. This forces one to the conclusion that it is the extracellular matrix and not necessarily hyalin that is somehow involved in the signalling that is necessary for the embryo to begin gastrulation. Even if hyalin is only permissive for the expression and action of some other signal, it is necessary experimentally to reach that conclusion. In similar fashion genes such as lin 12 and glp 1 in *C. elegans*, [Austin and Kimble 1989], or notch in *Drosophila* [Fehon *et al.* 1990] are thought to be involved in cell interactions that have developmental consequences, and in these cases it is difficult to conclude, as it is with hyalin, that a molecule is purely adhesive, or is involved directly in the transfer of information.

ECHINONECTIN IN CELL INTERACTIONS

Echinonectin is the second molecule of the hyaline layer that is known experimentally to have adhesion function [Allegro *et al.* 1988]. Echinonectin is released several minutes after hyalin in response to the fertilization reaction. The protein is deposited in the innermost area of the hyaline extracellular matrix [McClay *et al.* 1990a]. The protein is multifunctional in that it has lectin function as well as an apparent adhesion function [Alliegro *et al.* 1988]. Is echinonectin important for development? Hyalin is more adhesive than echinonectin and therefore, it could be argued, more important. Perhaps, however, echinonectin has two properties that suggest a role in morphogenesis. If one measures the adhesion to echinonectin quantitatively, each of the three germ layers respond differently to it. Primary mesenchyme cells lose an affinity for echinonectin as they ingress [Burdsal *et al.* 1991]. Later, during gastrulation, ectoderm cells acquire an increased affinity for echinonectin [Burdsall *et al.* 1991]. Meanwhile, endoderm retains an affinity for echinonectin of approximately the same strength as pre-gastrular cells. Without a complete knowledge of affinities of all the potential cellular substrates it is difficult to assess the relative merit of these observations. They do suggest, however, a role for echinonectin in morphogenesis.

ADHESION DURING INVAGINATION OF THE ARCHENTERON

Elsewhere we have dealt with the known morphogenetic movements of the archenteron [McClay *et al.* 1990b]. Briefly, during the first two-thirds of invagination the primary morphogenetic activity is a cell rearrangement by the endoderm cells [Ettensohn 1985; Hardin 1989]. During the last third of the invagination, filopodia extended by secondary mesenchyme cells at the tip of the archenteron play a major role in the completion of archenteron extension, and in locating the anatomical target for positioning the archenteron near the future site for the stomodaeum [Hardin and McClay 1990]. In all of these interactions cell adhesion occurs, and in striking the target it appears that some kind of transduction occurs. Again, what are the molecular explanations for these morphogenetic phenomena?

Efforts toward finding cell-cell adhesion molecules that govern invagination have been very preliminary. Aggregation studies have shown that cells have the ability to recognize germ layer differences such that cells sort out appropriately [Giudice and Mutolo 1970; Spiegel and Spiegel 1975; Bernacki and McClay 1989]. Immunological approaches have shown that antibodies can be made that preferentially inhibit the adhesion of one germ layer or another [McClay *et al.* 1977; McClay and Chambers 1978]. Studies with hybrid embryos have shown that expression of paternal components affect cell interactions [McClay and Hausman 1975; McClay *et al.* 1977]. Also inhibitors of various sorts, such as those that inhibit glysosylation events, affect cell-cell adhesion and inhibit invagination. Nevertheless, the actual molecules that participate in these adhesions are not known. Efforts have been made to look for crossreactive molecules to vertebrate counterparts such as neural-cell adhesive molecule (N-CAM) or cadherins, but as of this writing none of these efforts has been successful in uncovering a related sea urchin molecule.

Several adhesive properties of sea urchin endoderm and secondary mesenchyme cells are worth mentioning. First, endoderm cells can form archenterons by mechanisms that do not involve invagination [Spiegel and Spiegel 1975; Bernacki and McClay 1989]. It is known, for example, that in aggregates embryoids form that are very close to being normal embryos. In these embryoids the archenteron forms by first sorting out to the inside of the aggregate and then forming a gut tube by the process of cavitation. We have shown that it is the original endoderm cells that are able to do this if mixtures of cells from all three germ layers are added together [Bernacki and McClay 1989], (this eliminates the possibility that the embryo actually dedifferentiates and reorganizes itself in the aggregate). We have also learned that in the aggregate one of the first properties to be exercised by cells after initially adhering to one another is to reestablish cell polarity [Nelson and McClay 1988]. Actually it appears that the cells retain polarity and simply rearrange in an aggregate so that all cells reconform to boundaries with the same original polarity [Nelson and McClay 1988]. In fact, it could be that the mechanism by which cavitation works is that endoderm cells that end up in the interior of an aggregate and associate with one another become repolarized by recognizing with the basal ends of their cells the extracellular matrix. This would result in the apical ends of the cells aligning in parallel, aiding in the formation of a lumen. Whether recognition of the basal end of the cell is the mechanism by which this rearrangement phenomenon works is not known. It is clear, however, that there is a matrix recognition property that can be demonstrated experimentally, and there is a cell-cell association that preferentially allows endoderm cells to associate with one another.

The biggest mystery in the invagination of the archenteron is the question of direction. How do cells know how to rearrange such that the tube becomes elongated? It is hard to imagine how, in the microenvironment of the cell, that cells can undergo convergent-extension in the proper direction. The only experiments that have been performed are those that tend to rule out hypothetical mechanisms. For example, it was proposed that secondary mesenchyme cells, in their attachment to the wall of the blastocoel might pull the archenteron [Guftafson and Wolpert 1967]. That pulling force might help elongate the archenteron as well as provide a direction for the rearrangements [Guftafson and Wolpert 1967]. Such a model is almost certainly wrong. First, archenterons elongate in absence of filopodial extension [Hardin 1988], and in the absence of traction or pulling. Thus, external pulling forces can be ruled out. Diffusible signals from the target region also can be ruled out, since a coordinated rearrangement occurs even when archenterons evaginate in exogastrulation [McClay *et al.* 1990b]. Exogastrulae also rule out apical-basal polarity as being especially important for directionality since the tube is inside out in exogastrulae. At the vegetal plate, the presumptive endoderm cells are part of the dorso-ventral axis. Is it possible that the cells can somehow read dorsal or ventral locally and use that to rearrange? Apparently not since invagination continues normally in embryos in which the dorso-ventral axis has been abolished [Hardin, *et al.*, in preparation]. It would appear from these experiments that a global signal is not the explanation for directionality of invagination. If not, what is the signal?

The cells of the endoderm are located in a microenvironment in which they are surrounded by cells to all sides. They border on the basal lamina and on the lumen of the archenteron on the apical side. The rearrangement is such that the cells move not toward the vegetal pole but they converge along concentric rings surrounding the vegetal pole. Experiments have shown that just at about the time of invagination the endoderm cells, even those in isolation, begin to be motile [Hardin 1989]. The motility can provide a jostling kind of force and all that is necessary for a cell rearrangement is for the cells to intermingle in a net direction. What, along the concentric rings surrounding the vegetal pole, provides the directional information? One could model that the cells increase adhesions to one class of molecules and decrease an adhesion to another. This would imply that there is an animal-vegetal distribution of certain adhesion molecules. No data exists for such molecules but the possibility has not been ruled out either. Even if such molecules were to exist, there would have to be an ability for the cells to detect and respond to very subtle differences in adhesion molecule distribution. This too is possible though very little work has been done to examine the ability of cells to respond to adhesion gradients of adhesion molecules (in artificial gradients cells are known to respond rather impressively to charge or to other artificial circumstances [Harris 1973]). Until more is known about the normal adhesive properties of the endoderm cells the question of direction of cell rearrangement will remain open.

The second kind of adhesion that occurs during invagination is that of the secondary mesenchyme cells. These cells remain bound to the archenteron at its tip throughout much of gastrulation (although not in all species, [Armstrong, Hardin and McClay, unpublished observations]). The behavior of the secondary mesenchyme cells has been studied by time lapse videomicroscopy, after which a number of parameters were quantified [Hardin and McClay 1990]. Early in gastrulation the secondary mesenchyme cells begin to send out long filopodia. The filopodia extend up to about 35 μm. If the filopodia make contact with the blastocoel within that distance, they attach for from 2-10 min and then are withdrawn. If no contact is made within 35 μm the filopodia are withdrawn immediately. When contact is finally made with the target region the duration of contact becomes indefinite and the secondary mesenchyme cells undergo a change in morphology and begin to behave like fibroblasts.

The filopodia demonstrate two kinds of adhesion. The contacts on the lateral sides of the blastocoel are short-lived but appear to be strong enough to support some contractile tension. The contacts that form in the target region appear to be far more stable that those that form on other regions of the blastocoel wall suggesting perhaps a qualitative difference in the substrate [Hardin and McClay 1990]. Experimental evidence strongly supports this hypothesis in that precocious contact with the target region brings about a precocious cell behavior change, and removal of the target region from the reach of secondary mesenchyme cells extends the filopodial extension behavior for several hours beyond the time at which it normally would end (as a result of reaching the target) [Hardin and McClay 1990] . There is no evidence to support notion that filopodia adhere at the site of junction between two cells (as had been suggested by Gustafson and Wolpert, [Gustafson and Wolpert 1967]. Collagen, fibronectin, and laminin all are present as potential substrates for the secondary mesenchyme cells though there may be a special component at the target that somehow promotes the behavior change. At present there are no candidate molecules that participate in that behavior change.

Although we have identified several adhesion molecules in the morphogenetic sequence of gastrulation in the sea urchin, many behaviors remain to be characterized at the molecular level. The lesson in this model organism is that even for morphogenetic movements that are fairly simple such as ingression and invagination, there are multiple adhesion molecules, several receptors that must be present to explain the observed behaviors, several different substrates, and an important coordination between motility and adhesion for the cells to rearrange in the correct pattern. In addition a number of other cellular properties also contribute to the patterns that emerge. Cell polarity and the primary axes of symmetry are involved. Substrates on both sides of the cells are involved. And, as discussed above, several transducing activities may be involved in the coordination of the morphogenetic movements. Thus, we have come full circle. Its complicated. However there are signs of hope that significant understanding of morphogenesis is possible. There are not too many adhesion molecules. The cell behaviors are relatively few in number. There may be a fairly small number of transductions, and molecular biologists are approaching a significant understanding of coordinate gene regulation.

Support for this work was provided by NIH grants HD14483 and HD24199.

REFERENCES

Adelson, D.L. and T. Humphreys. 1988. Sea urchin morphogenesis and cell-hyalin adhesion are perturbed by a monoclonal antibody specific for hyalin. *Development* 104:391-402.

Alliegro, M.C., C.A. Burdsal, and D.R. McClay. 1990. *In vitro* biological activities of echinonectin. *Biochemistry* 29:2135-2141.

Alliegro, M.C., C.A. Ettensohn, C.A. Burdsal, H.P. Erickson, and D.R. McClay. 1988. Echinonectin: A new embryonic substrate adhesion protein. *J. Cell Biol.* 107:2319-2327.

Austin, J. and J. Kimble. 1989. Transcript analysis of glp-1 and lin-12, homologous genes required for cell interactions during development of *C. elegans*. *Cell* 58:565-571.

Bernacki, S.H. and D.R. McClay. 1989. Embryonic cellular organization: Differential restrictions of fates as revealed by cell aggregates and lineage markers. *J. Exp. Zool.* 251:203-216.

Bixby, J.L., R.S. Pratt, J. Lilien, and L.F. Reichardt. 1987. Neurite outgrowth on muscle cell surfaces involves extracellular matrix receptors as well as $Ca2^{+}$-dependent and -independent cell adhesion molecules. *Proc. Natl. Acad. Sci. USA* 84:2555-2559.

Buck, C.A. and A.F. Horowitz. 1987. Cell surface receptors for extracellular matrix molecules. *Annu. Rev. Cell Biol.* 3:179-205.

Burdsal, C.A., M.C. Alliegro, and D.R. McClay. 1991. Tissue-specific, temporal changes in cell adhesion to echinonectin within the mesoderm and ectoderm lineages of the sea urchin embryo. *Dev. Biol.* 144:327-334.

Burridge, K., K. Fath, T. Kelly, G. Nuckolls, and C. Turner. 1988. Focal adhesions: Transmembrane junctions between the extracellular matrix and the cytoskeleton. *Annu. Rev. Cell Biol.* 4:487-526.

Burridge, K., L. Molony, and T. Kelly. 1987. Adhesion plaques: Sites of transmembrane interaction between the extracellular matrix and the actin cytoskeleton. *J. Cell Sci. Suppl.* 8:211-229.

Cervello, M. and V. Matranga. 1989. Evidence of a precursor-product relationship between vitellogenin and toposome, a glycoprotein complex mediating cell adhesion. *Cell Differ. Dev.* 26:67-76.

Cooke, J. 1989. Mesoderm-inducing factors and Spemann's organiser phenomenon in amphibian development. *Development.* 107:229-241.

Dan, K. and T. Ono. 1952. Cyto-embryological studies of sea urchins. I. The means of fixation of the mutual position among the blastomeres of sea urchin larvae. *Biol. Bull.* 102:58-73

Edelman, G.M. 1985. Cell adhesion and the molecular process of morphogenesis. *Annu. Rev. Biochem.* 54:135-169.

Ettensohn, C.A. 1985. Gastrulation in the sea urchin is accompanied by the rearrangement of invaginating epithelial cells. *Dev. Biol.* 112:383-390.

Fehon, R.G., P.J. Kooh, I. Rebay, C.L. Regan, T. Xu, A.T. Muskavitch, and S. Artavanis-Tsakonas. 1990. Molecular interactions between the protein products of the neurogenic loci notch and delta, two EGF-Homologous genes in *Drosophila*. *Cell* 61:523-534.

Fink, R.D. and D.R. McClay. 1985 Three cell recognition changes accompany the ingression of sea urchin primary mesenchyme cells. *Dev. Biol.* 107:66-74.

Gerhart, J. and R. Keller. 1986. Region-specific cell activities in amphibian gastrulation. *Annu. Rev. Cell Biol.* 2:201-229.

Giudice, G. and V. Mutolo. 1970. Reaggregation of dissociated cells of sea urchin embryos. *Adv. Morphog.* 10:115-158.

Gustafson, T. and L. Wolpert. 1967. Cellular movement and contact in sea urchin morphogenesis. *Biol. Rev. Camb. Philos. Soc.* 42:442-498.

Hall, G. and V. Vacquier. 1982. The apical lamina of the sea urchin embryo: Major glycoproteins associated with the hyaline layer. *Dev. Biol.* 89:160-178.

Hardin, J. 1988. The role of secondary mesenchyme cells during sea urchin gastrulation studied by laser ablation. *Development* 103:317-324.

Hardin, J. 1989. Local shifts in position and polarized motility drive cell rearrangement during sea urchin gastrulation. *Dev. Biol.* 136:430-445.

Hardin, J. and D.R. McClay. 1990. Target recognition by the archenteron during sea urchin gastrulation. *Dev. Biol.* 142:86-102.

Harris, A. 1973. Behavior of cultured cells on substrata of variable adhesiveness. *Exp. Cell Res.* 77:285-296.

Herbst, C. 1900. Über das Auseinandergehen von Furchungszellen und Gewebezellen in kalkfreiem Medium. *Wilhelm Roux' Arch. Entwicklungsmech Org.* 9:424-463.

Holtfreter, J. 1943. A study of the mechanics of gastrulation. Part I. *J. Exp. Zool.* 94:261-318.

Hynes, R.O. 1987. Integrins: A family of cell surface receptors. *Cell* 48:549-554.

Jessell, T.M., M.A. Hynes and J. Dodd. 1990. Carbohydrates and carbohydrate-binding proteins in the nervous system. *Annu. Rev. Neurosci.* 13:227-256.

Kane, R.E. 1973. Hyalin release during normal sea urchin development and its replacement after removal at fertilization. *Exp. Cell Res.* 81:301-311.

Kane, R.E. and R.E. Stephens. 1969. A comparative study of the isolation of the cortex and the role of the calcium-insoluble protein in several species of sea urchin egg. *J. Cell Biol.* 41:133-144.

Lee, G.F., E.W. Fanning, M.P. Small, and M.B. Hille. 1989. Developmentally regulated proteolytic processing of a yolk glycoprotein complex in embryos of the sea urchin, *Strongylocentrotus purpuratus*. *Cell Differ.* 26:5-18.

Lotz, M.M., C.A. Burdsal, H.P. Erickson, and D.R. McClay. 1989. Cell adhesion to fibronectin and tenascin: Quantitative measurements of initial binding and subsequent strengthening response. *J. Cell Biol.* 109:1795-1805.

McCarthy, R.A., K. Beck, and M.M. Burger. 1987 Laminin is structurally conserved in the sea urchin basal lamina. *EMBO J.* 6:1587-1593.

McCarthy, R.A. and M.M. Burger. 1987 *In vivo* embryonic expression of laminin and its involvement in cell shape change in the sea urchin *Sphaerechinus granularis*. *Development* 101:659-671.

McCarthy, R.A. and M. Spiegel. 1983. Protein composition of the hyaline layer of sea urchin embryos and reaggregating cells. *Cell Differ.* 13:93-102.

McClay, D.R., M.C. Alliegro, and S.D. Black. 1990a. The ontogenetic appearance of extracellular matrix during sea urchin development. p. 1-15. *In: Organization and Assembly of Plant and Animal Extracellular Matrix*. S. Adair and R. Mecham (Eds.). Academic Press, New York.

McClay, D.R., M.C. Alliegro, and J.D. Hardin. 1990b. Cell interactions as epigenetic signals in morphogenesis of the sea urchin embryo. p. 70-87. *In: The Cellular and Molecular Basis of Pattern Formation*. D. Stocum and T. Karr (Eds.). Oxford University Press, Oxford.

McClay, D.R. and A.F. Chambers. 1978. Identification of four classes of cell surface antigens appearing at gastrulation in sea urchin embryos. *Dev. Biol.* 63:179-186.

McClay, D.R., A.F. Chambers, and R.G. Warren. 1977. Specificity of cell-cell interactions in sea urchin embryos. Appearance of new cell-surface determinants at gastrulation. *Dev. Biol.* 56:343-355.

McClay, D.R., J.C. Coffman, and J.D. Hardin. 1990c. Epigenetic signals at gastrulation in the sea urchin. *UCLA Symp. Mol. Cell. Biol.* 125:251-256.

McClay, D.R. and R.D. Fink. 1982. Sea urchin hyalin: Appearance and function in development. *Dev. Biol.* 92:285-293.

McClay, D.R. and R.E. Hausman. 1975. Specificity of cell adhesion: Differences between normal and hybrid sea urchin cells. *Dev. Biol.* 47:454-460.

McClay, D.R. and V. Matranga. 1986. On the role of calcium in the adhesion of embryonic sea urchin cells. *Exp. Cell Res.* 165:152-164.

McClay, D.R., G.M. Wessel and R.B. Marchase. 1981. Intercellular recognition: Quantitation of initial binding events. *Proc. Natl. Acad. Sci. USA* 78:4975-4979.

Moscona, A. 1956. Development of heterotypic combinations of dissociated embryonic chick cells. *Proc. Soc. Exp. Biol. Med.* 92:410-419.

Moscona, A. and H. Moscona. 1952. Dissociation and aggregation of cells from organ rudiments of the early chick embryos. *J. Anat.* 86:287-301.

Nelson, S.H. and D.R. McClay. 1988. Cell polarity in sea urchin embryos: Reorientation of cells occurs quickly in aggregates. *Dev. Biol.* 127:235-247.

Noll, H., M. Cervelo, T. Humphreys, B. Kuwasaki, and D. Adelson. 1985. Characterization of toposomes from sea urchin blastula cells: A cell organelle mediating cell adhesion and expressing positional information. *Proc. Natl. Acad. Sci. USA* 82:8062-8066.

Noll, H., V. Matranga, D. Cascino, and L. Vittorelli. 1979. Reconstitution of membranes and embryonic development in dissociated blastula cells of the sea urchin by reinsertion of aggregation-promoting membrane proteins extracted with butanol. *Proc. Natl. Acad. Sci. USA* 72:288-292.

Nose, A., A. Nagafuchi, and M. Takeichi. 1988. Expressed recombinant cadherins mediate cell sorting in model systems. *Cell* 54:993-1001.

Rutishauser, U. 1989. Membrane apposition as a regulating parameter in cell-cell interactions. p. 137-150. *In: The Assembly of the Nervous System.* L. Landmesser (Ed.). Alan R. Liss, New York.

Sheffield, J.B. and A.A. Moscona. 1989. Electron microscopic analysis of aggregation of embryonic cells: The structure and differentiation of aggregates of neural retina cells. *Dev. Biol.* 23:36-61.

Smith, J.C., J. Cooke, J.B.A. Green, G. Howes, and K. Symes. 1989. Inducing factors and the control of mesodermal pattern in *Xenopus laevis*. *Development Suppl.* 8:149-159.

Spiegel, E., M. Burger, and M. Spiegel. 1980. Fibronectin in the developing sea urchin embryo. *J. Cell Biol.* 87:309-313.

Spiegel, E., M.M. Burger, and M. Spiegel. 1983. Fibronectin and laminin in the extracellular matrix and basement membrane of sea urchin embryos. *Exp. Cell Res.* 144:47-55.

Spiegel, M. and E. Spiegel. 1975. The reaggregation of dissociated embryonic sea urchin cells. *Am. Zool.* 15:583-606.

Spray, D.C., M. Fujita, J.C. Saez, H. Choi, T. Watanabe, E. Hertzberg, L.C. Rosenberg, L.M. Reid. 1987. Proteoglycans and glycosaminoglycans induce gap junction synthesis and function in primary liver cultures. *J. Cell Biol.* 105:541-551.

Springer, T.A., M.L. Dustin, T.K. Kishimoto, and S.D. Marlin. 1987. LFA-1, CD2, and LFA-3 molecules: Cell adhesion receptors of the immune system. *Annu. Rev. Immunol.* 5:223-252.

Steinberg, M. 1970. Does differential adhesion govern self-assembly processes in histogenesis? Equilibrium configurations and the emergence of a hierarchy among populations of embryonic cells. *J. Exp. Zool.* 173:395-434.

Stephens, R.E. and R.E. Kane. 1970. Some properties of hyalin. *J. Cell Biol.* 44:611-617.

Takeichi, M. 1988. The cadherins: Cell-cell adhesion molecules controlling animal morphogenesis. *Development* 102:639-655.

Takeichi, M., T. Atsumi, C. Yoshida, K. Uno, and T.S. Okada. 1981. Selective adhesion of embryonal carcinoma cells and differentiated cells by Ca2+- dependent sites. *Dev. Biol.* 87:340-350.

Townes, P.L. and J. Holtfreter. 1955. Directed movements and selective adhesion of embryonic amphibian cells. *J. Exp. Zool.* 128:53-120.

Trinkaus, J.P. 1984. *Cells into Organs.* Prentice-Hall, Englewood Cliffs, New Jersey.

Wessel, G.M., R.B. Marchase, and D.R. McClay. 1984. Ontogeny of the basal lamina in the sea urchin embryo. *Dev. Biol.* 103:235-245.

Wessel, G.M. and D.R. McClay. 1987. Gastrulation in the sea urchin embryo requires the deposition of crosslinked collagen within the extracellular matrix. *Dev. Biol.* 121:149-165.

Xu, T., I. Rebay, R.J. Fleming, T.N. Scottgale, and S. Artavanis-Tsakonas. 1990. The Notch locus and the genetic circuitry involved in early *Drosophila* neurogenesis. *Genes Dev.* 4:464-475.

Yamada, K.M. and S.K. Akiyama. 1984. The interactions of cells with extracellular matrix components. p. 77-148. *In: Cell Membranes, vol. 2.* E. Elson, W. Frazier, and L.Glaser (Eds.). Plenum Press, New York.

INDEX

GPSR Compliance

The European Union's (EU) General Product Safety Regulation (GPSR) is a set of rules that requires consumer products to be safe and our obligations to ensure this.

If you have any concerns about our products, you can contact us on ProductSafety@springernature.com

In case Publisher is established outside the EU, the EU authorized representative is:

Springer Nature Customer Service Center GmbH
Europaplatz 3
69115 Heidelberg, Germany

Batch number: 10692380

Printed by Printforce, the Netherlands